Enteric Infections and Immunity

INFECTIOUS AGENTS AND PATHOGENESIS

Series Editors: Mauro Bendinelli, *University of Pisa*
Herman Friedman, *University of South Florida*

COXSACKIEVIRUSES
A General Update
Edited by Mauro Bendinelli and Herman Friedman

DNA TUMOR VIRUSES
Oncogenic Mechanisms
Edited by Giuseppe Barbanti-Brodano, Mauro Bendinelli, and Herman Friedman

ENTERIC INFECTIONS AND IMMUNITY
Edited by Lois J. Paradise, Mauro Bendinelli, and Herman Friedman

FUNGAL INFECTIONS AND IMMUNE RESPONSES
Edited by Juneann W. Murphy, Herman Friedman, and Mauro Bendinelli

MYCOBACTERIUM TUBERCULOSIS
Interactions with the Immune System
Edited by Mauro Bendinelli and Herman Friedman

NEUROPATHOGENIC VIRUSES AND IMMUNITY
Edited by Steven Specter, Mauro Bendinelli, and Herman Friedman

PSEUDOMONAS AERUGINOSA AS AN OPPORTUNISTIC PATHOGEN
Edited by Mario Campa, Mauro Bendinelli, and Herman Friedman

PULMONARY INFECTIONS AND IMMUNITY
Edited by Herman Chmel, Mauro Bendinelli, and Herman Friedman

VIRUS-INDUCED IMMUNOSUPPRESSION
Edited by Steven Specter, Mauro Bendinelli, and Herman Friedman

---

Enteric Infections and Immunity

Edited by

Lois J. Paradise
University of South Florida College of Medicine
Tampa, Florida

Mauro Bendinelli
University of Pisa
Pisa, Italy

and

Herman Friedman
University of South Florida College of Medicine
Tampa, Florida

Plenum Press • New York and London

Library of Congress Cataloging in Publication Data

On file

ISBN 0-306-45242-1

A Division of Plenum Publishing Corporation
233 Spring Street, New York, N.Y. 10013

10 9 8 7 6 5 4 3 2 1

Printed in the United States of America

Contributors

DAVID W. K. ACHESON • Division of Geographic Medicine and Infectious Diseases, Tupper Research Institute, New England Medical Center, Boston, Massachusetts 02111

SERGIO ARIAS-NEGRETE • Institute of Parasitology, McGill University, Ste-Anne de Bellevue, Quebec H9X 3V9, Canada

MERLIN S. BERGDOLL • Food Research Institute, University of Wisconsin, Madison, Wisconsin 53706

KRIS CHADEE • Institute of Parasitology, McGill University, Ste-Anne de Bellevue, Quebec H9X 3V9, Canada

SHUNZO CHIBA • Department of Pediatrics, Sapporo Medical University, School of Medicine, Sapporo, Hokkaido 060, Japan

WITOLD CIEPLAK, JR. • Laboratory of Intracellular Parasites, Rocky Mountain Laboratories, National Institute of Allergy and Infectious Diseases, National Institutes of Health, Hamilton, Montana 59840

MICHAEL S. DONNENBERG • Division of Infectious Diseases, University of Maryland School of Medicine, R. A. Cowley Shock Trauma Center, Baltimore, Maryland 21201

TOBY K. EISENSTEIN • Department of Microbiology and Immunology, Temple University, School of Medicine, Philadelphia, Pennsylvania 19140

ROBERT M. GENTA • Departments of Pathology, Medicine, and Microbiology and Immunology, Baylor College of Medicine, and Center for Infectious Diseases, University of Texas School of Public Health, Houston, Texas 77030

MARTIN F. HEYWORTH • Department of Veterans Affairs Medical and Regional Office Center, Fargo, North Dakota 58102; and Department of Internal Medicine, University of North Dakota, Grand Forks, North Dakota 58203

DUAN HUANG • Department of Microbiology and Immunology, Temple University, School of Medicine, Philadelphia, Pennsylvania 19140

GERALD T. KEUSCH • Division of Geographic Medicine and Infectious Diseases, Tupper Research Institute, New England Medical Center, Boston, Massachusetts 02111

HIROSHI KIYONO • The Immunobiology Vaccine Center, Departments of Oral Biology and Microbiology, University of Alabama at Birmingham, Birmingham, Alabama 35294-2170; and Department of Mucosal Immunology, Research Institute for Microbial Disease, Osaka University, Osaka 565, Japan

MICHAEL E. KONKEL • Department of Microbiology, Washington State University, Pullman, Washington 99164-4233

DAVID M. LYERLY • TechLab, Inc., Corporate Research Center, Blacksburg, Virginia 24060

JERRY R. McGHEE • The Immunobiology Vaccine Center, Departments of Oral Biology and Microbiology, University of Alabama at Birmingham, Birmingham, Alabama 35294-2170; and Department of Mucosal Immunology, Research Institute for Microbial Disease, Osaka University, Osaka 565, Japan

JOHN J. MEKALANOS • Department of Microbiology and Molecular Genetics, Harvard Medical School, Boston, Massachusetts 02115

LOIS J. PARADISE • Department of Medical Microbiology and Immunology, University of South Florida College of Medicine, Tampa, Florida 33612

DAVID W. PASCUAL • Veterinary Molecular Biology, Montana State University, Bozeman, Montana 59717-0360

RONALD P. RABINOWITZ • Division of Infectious Diseases, University of Maryland School of Medicine, R. A. Cowley Shock Trauma Center, Baltimore, Maryland 21201

MARTIN G. SCHWACHA • Department of Microbiology and Immunology, Temple University, School of Medicine, Philadelphia, Pennsylvania 19140

ROGER L. VAN TASSELL • Virginia Polytechnic Institute and State University, Blacksburg, Virginia 24061

MATTHEW K. WALDOR • Department of Microbiology and Molecular Genetics, Harvard Medical School, Boston, Massachusetts 02115

TRACY D. WILKINS • Fralin Biotechnology Center, Virginia Polytechnic Institute and State University, Blacksburg, Virginia 24061

KENNETH H. WILSON • Infectious Diseases Section, VA Medical Center, and Duke University, Durham, North Carolina 27705

Preface

As we approach the end of this millennium, enteric diseases remain important public health problems. In many parts of the world, sanitary measures have advanced little over the last century, although some of the governments in those areas are striving to improve facilities for sanitation and to educate their people in proper handling of food, water, sewage, and other modes of transmission of pathogenic microbes. Even in highly developed countries, outbreaks of diarrheal diseases occur today. Globally, the annual morbidity from enteric infections is estimated at several billion and deaths at several million per year. In this volume, descriptions of some of these diseases, of immunity that results from them, of clinical studies that promote understanding of individual and community immunity, of molecular factors of pathogenesis, and/or of advances in vaccine development have been provided by leading researchers. At present, the application of molecular methods is enhancing the identification of protective antigens of many microorganisms. In addition, new methods for design and delivery of vaccines are being devised. Perhaps then more effective tools for reducing at least some of these diseases will be available within the next decade.

Lois J. Paradise
Herman Friedman
Mauro Bendinelli

Contents

Introduction

LOIS J. PARADISE

Gastrointestinal diseases are both a bane of the developing areas of the world and a disturbing problem in industrialized regions, despite greater emphasis on sanitary measures to promote safe food-handling and water distribution. Diarrheal disease is the most common acute infectious disease in the world and is responsible for the considerable mortality in developing countries.[1] In these countries, continuing sanitation problems and failure to prevent exposure of babies to enteric pathogens result in acute, life-threatening and/or fatal diseases. If the individual survives acute diarrheal disease, chronic disease may persist for the remainder of life. In the aged population, acute infectious enteritis is again a major cause of death. In the United States and other developed countries, many niches for traditional enteric organisms remain, although sanitation generally is better and inspection of food and water resources aimed at reducing the risk of exposure to infection is more common. In addition, outbreaks occur with newly recognized enteric pathogens. Complacency, undoubtedly, is another factor. Because many people believe that food and water supplies are entirely safe today, they may be careless about precautions such as cooking food sufficiently to destroy pathogens. In the industrialized world, both the incidence and mortality are higher among infants and young children and the very elderly.[2]

Microorganisms associated with human diarrhea include various species of bacteria, fungi, viruses, and protozoa. Some helminths also cause diarrheal disease. The acquisition of diarrheal diseases depends on both broad and specific aspects of the characteristics of individuals and on their socioeconomic environment and activities.[3] Age is a determining factor for susceptibility to infection by some organisms. With some enteric pathogens,

---

LOIS J. PARADISE • Department of Medical Microbiology and Immunology, University of South Florida College of Medicine, Tampa, Florida 33612.

Enteric Infections and Immunity, edited by Lois J. Paradise *et al.* Plenum Press, New York, 1996.

very young individuals are particularly at risk; for other pathogens, the elderly are most at risk; and with still other pathogens, both age groups are readily affected. Regardless of age, severely immunocompromised patients, such as those with acquired immunodeficiency syndrome (AIDS), are particularly susceptible to secondary infectious diseases, including diarrheal diseases. Immunocompromised patients not only are more susceptible to some microbial species that are indigenous organisms for immunocompetent populations, but also are at risk for infections caused by microorganisms once considered environmental microbes unable to initiate human disease. A number of these agents cause enteric diseases. In addition to host-specific factors (*who*), *when* and *where* exposure takes place are epidemiologically important.[3] In tropical and temperate zones, seasons of higher incidence of enteric diseases differ, occurring more in summer in the former and in winter in the latter areas. However, this may be an oversimplification since other climatic conditions, such as rainfall, may introduce variations into this pattern. The incidence of diarrheal disease also differs with the particular species of microorganisms present in the population and environment of a given location.

As might be expected, mechanisms that affect the intestine and result in diarrhea vary with the organism or group of organisms. Diarrhea is increased frequency and/or fluidity of bowel movements.[4] Discharges may be watery and contain no evidence of an inflammatory response to the infection. This picture is found with organisms that produce enterotoxins or other chemicals capable of affecting intestinal epithelial cells directly and thereby alter their secretory functions. Watery, noninflammatory diarrheas occur with cholera, rotaviral disease, and *Clostridium difficile* and enterotoxigenic *Escherichia coli* infections, among others. With microorganisms such as shigellae, *Salmonella enteritidis*, enteroinvasive *E. coli*, and *Entamoeba histolytica*, invasion of epithelial cells and/or the effects of cytotoxins produced by the microorganisms on the epithelium result in inflammatory diarrheal disease. Some gastrointestinal pathogens penetrate more deeply into tissues than others. *Salmonella typhi* produces a systemic febrile disease and is a member of this group. With the last two groups of agents, inflammatory diarrhea is the norm. Leukocytes appear in the stool, indicating that polymorphonuclear leukocytes have been recruited to the area and have escaped into the intestinal lumen. This does not occur with watery diarrhea. With helminthic infections, mast cells and their mediators may play roles in diarrhea production. Studies of various bacterial, viral, protozoan, and helminthic organisms have demonstrated that they also interfere with normal intestinal motility and the resulting dysfunction promotes diarrhea.[4]

Understanding of the pathogenesis of some enteric diseases has advanced markedly at the cellular and molecular levels within the last couple of decades. For enteric microorganisms more recently recognized as being pathogenic, knowledge has not developed to an equivalent state, but technical advances of the last decade should allow faster progress toward understanding than was possible in the past. In the chapters that follow, the state of knowledge concerning mechanisms of action of toxins, virulence factors such as adhesions, nutritional requirements, mechanisms of invasion, effects on eukaryotic cell signaling, and stimulation of the inflammatory response by induction and the actions of cytokines is discussed for each group of organisms. The host response to cholera and cholera toxin (CT) was the first to be described and provided a model for studying similar aspects of some other enteric diseases. In Chapter 3, the molecular characteristics of CT and virulence factors of *V. cholerae* that enable colonization of the intestines and molecular aspects of genetic control of these factors are delineated. For *E. coli* strains associated with diarrheal diseases, more is known about enterotoxigenic (ETEC) and enteropathogenic (EPEC) strains than for more recently described virulent strains (Chapter 6), but progress is escalating. ETEC strains cause a diarrheal disease similar to, but less severe than, cholera. CT enhances intracellular adenylate cyclase activity; ETEC enterotoxin increases guanylate cyclase activity. The net result in each case is increased secretion of water and electrolytes, but the two differ in electrolytes affected and the extent of the responses induced. In EPEC diarrhea, toxins produced are cytotoxins that directly affect intestinal epithelial cells. The result is an inflammatory diarrhea rather than a watery, noninflammatory one. In its pathogenesis, the disease resembles shigellosis rather than cholera. Other *E. coli* strains produce hemorrhagic diarrheas. It is obvious that within this one species, various mechanisms of action on the human gastrointestinal tract have developed. It is important to note, however, that most strains of *E. coli* are not pathogenic and many are members of the human indigenous, intestinal microflora and as such undoubtedly participate in protecting the host from infection with pathogens (Chapter 1).

Genetic control of the various virulence factors in strains of one species or among groups of pathogenic species may be plasmid or chromosomal in origin or both. Identification of genetic bases for formation and/or release of components related to virulence will enable performance of experiments, e.g., using deletion mutants or transfectants, to discover whether they also are protective antigens. When immune responses that protect against infectious challenge have been identified, vaccines can be designed from the identified components. The vaccines can be tested in model systems not just for protection against challenge inoculations of the microorganisms, but also for molecular analysis of the protection afforded. These results can be compared to findings in some humans challenged after vaccination and others immunized by naturally acquired disease. Successful and useful vaccines are available for few of the enteric pathogens. Using molecular techniques to study these aspects of active immunization, it should be possible to prepare vaccines that are highly effective and yet have few side effects.

Both clinical symptoms and immune responses are results of the molecular and cellular interactions that pathogenesis comprises. Studies that determine the interrelationships of these aspects eventually will point out likely routes for design and development of optimally safe and protective vaccines for these diseases. The fact that understanding of the pathogenesis and immune responses of the enteric diseases is incomplete has hindered vaccine development, as implied above. For many years humoral immunity was the focus for evaluation of vaccines. In time, cellular immune functions were observed to be involved in some cases. Still more recently, mucosal immunity has been in the forefront. Although knowledge of the existence, functions, and capabilities of the mucosal immune system has advanced markedly during the past few decades (Chapter 2), the relative importance of humoral and mucosal antibody responses as well as cellular immune mechanisms in protecting against primary infection or reinfection with enteric agents is only partly understood. This is described for several diseases in the chapters of this volume. For example, the pathogenic mechanisms of rotavirus and the Norwalk group of viral diarrheas (Chapter 10) have been considerably less well clarified than for many bacterial diseases. There are several reasons for this, but one is the lack of certainty that animals serve as adequate models for human studies. In addition, it has not always been certain that clinical studies have been dealing with primary infection rather than reinfection. For other diseases, controversies exist because of differing observations that cannot be resolved within the prevailing theories. Increasing recognition that mucosal and humoral antibodies may have synergistic effects on intestinal immunity and that cellular immunity may further enhance protection may lead to investigations that will provide solution.[5]

In summary, the potential for effective vaccine development has improved greatly. Conditions that continue to inhibit immediate development include: (1) identification of protective antigen(s) for many organisms is still in progress; (2) understanding of those immune functions that are active in protection is incomplete; and (3) some effective, molecularly derived vaccines require modification to eliminate deleterious side effects. Progress is being made in these same areas, e.g., (1) specific epitopes of protective antigens have been isolated for some organisms and are being tested for efficacy; (2) the proposition that interactions among mucosal, humoral, and cellular immune functions provide the gastrointestinal protective barrier, not just one or the other, is being examined; and (3) investigators are examining purified antigenic molecules to modify their toxic properties.

REFERENCES

1. Walsh, J. S., and Warren, K. S., 1979, Selective primary health care, *N. Engl. J. Med.* **301:** 967–974.
2. Bennett, R. G., and Greenough, W. B., III, 1993, Approach to acute diarrhea in the elderly, *Gastroenterol. Clin. North Am.* **22:**517–533.
3. Guerrant, R. L., 1990, Principles and syndromes of enteric infection, in: *Principles and Practice of Infectious Diseases*, 3rd ed. (G. L. Mandell, R. G. Douglas, Jr., and J. E. Bennett, eds.), Churchill Livingstone, Edinburgh, pp. 837–851.
4. O'Loughlin, E. V., Scott, R. B., and Gall, D. G., 1991, Pathophysiology of infectious diarrhea: Changes in intestinal structure and function, *J. Pediatr. Gastroenterol. Nutr.* **12:**5–20.
5. Monath, T. P., and Neutra, M. R., 1994, Introduction to Symposium on Mucosal Immunity: Protection Against Pathogens, *Am. J. Trop. Hyg.* **50**(Suppl.):1–2.

1

Indigenous Microorganisms as a Host Defense

KENNETH H. WILSON

1. INTRODUCTION

A wide variety of bacteria are able to infect the human gastrointestinal tract. For many of these pathogens we have detailed knowledge at the molecular level of their pathogenic mechanisms. A common theme has emerged, namely, the process of pathogenesis is not simple. This complexity may seem curious. Why go to the trouble of traveling down chemical gradients through dense mucin to adhere to the small bowel mucosa as *Vibrio cholerae* does (see Chapter 3)? Why take the approach of *Shigella* species and penetrate the colonic mucosa, multiply intracellularly, and spread from cell to cell (see Chapter 5)? Why not simply enter the colonic lumen and multiply in a rich medium at incubator temperature and at a site that is relatively inaccessible to the immune system? In fact, no pathogen is known to perform this feat in the normal host. The presence of a complex biota appears to be such an effective host defense that pathogens have been forced to develop more devious adaptations. Diarrhea enhances the ability of an organism to spread from host to host; thus, causing diarrhea may make an organism more fit. To some extent, disease probably also occurs serendipitously when organisms are faced with an overwhelming force in the obvious place to colonize and develop elaborate ways to elude it. This chapter will attempt to clarify why pathogens have to develop the complicated pathogenic mechanisms described elsewhere in this volume.

KENNETH H. WILSON • Infectious Diseases Section, VA Medical Center, and Duke University, Durham, North Carolina 27705.

Enteric Infections and Immunity, edited by Lois J. Paradise *et al.* Plenum Press, New York, 1996.

2. COMPOSITION OF INTESTINAL BIOTA

The vast majority of bacteria in the GI tract reside in the colon. The stomach is not known to have an indigenous population of bacteria aside from *Helicobacter* species. These urease-producing organisms circumvent the stomach's acid pH and short retention time by residing below the thick layer of mucin.[(1)] They presumably neutralize the pH by producing ammonia from urea. Because peristalsis propels contents rapidly through the small bowel, the upper small bowel is relatively devoid of bacteria. Viable counts are usually less than 10^5 colony forming units (CFU)/ml. Neither helicobacters nor the low populations of bacteria within the small bowel are known to perform a protective role for the host.

Passage through the colon is generally slow enough for appropriately adapted bacterial species theoretically to reach maximal population densities. However, most species never actually reach maximal density in the normal individual because they are limited by competition with other community members of the gut ecosystem. The total bacterial population size is reliably around 10^{11} CFU/ml throughout the colon. The redox potential of the colonic contents is less than -200 mV, and over 99% of the bacteria present are strictly anaerobic. Table I shows the breakdown of bacterial species found in unbiased sampling of the culturable organisms found in fecal samples from the Apollo astronauts.[(2)] These organisms have been found to be typical of humans from around the world. Although there is some overlap in species names between the biota of humans and those of various animals, overall the human biota is unique to humans.[(3,4)] Over 500 species have been isolated from predominant human biota; a given person probably harbors around 200 species (W. E. C. Moore and L. V. H. Moore, personal communication). No one species dominates the ecosystem; the most numerous species usually constitutes only 15–20% of bacterial cells. Though there is considerable variation from person to person, the makeup of an individual's biota remains stable over long periods of time.

Although great strides have been made in describing the community members of this complex ecosystem, much work remains. Out of the 100 most numerous isolates found by W. E. C. Moore and L. V. H. Moore (unpublished data), around one-fourth have not yet been officially named. Among less numerous organisms, unnamed species are even more frequent. Taxonomy of some named organisms has recently come into question as ribosomal RNA-based phylogenies sometimes are at odds with established taxonomic schemes. Some of the most obvious examples are that the genera *Clostridium, Ruminococcus, Peptostreptococcus,* and *Eubacterium* are polyphyletic.[(5,6)] It is not certain what proportion of bacterial cells found in the colon can be cultivated, but the best results to date cite 70–90% recovery.[(2,7)]

However, because these high rates are based on counts of Gram-stained organisms, and because it is possible that organisms wash off a slide at the time of Gram stain, the real rate could be lower. It is not known to what extent unculturable taxa are a significant part of the biota.

3. MOLECULAR APPROACHES TO DETERMINE COMPOSITION OF THE BIOTA

Several new approaches have been developed in recent years in which ribosomal RNA sequences are used to determine the composition of ecosystems. The strategy grew from the field of molecular evolution when it became clear that rRNA sequences were helpful in determining the phylogeny of all life forms.(8,10) Small subunit (SSU) rRNA sequences vary enough that usually a species can be identified by its rRNA sequence. Thus, it is possible to blot rRNA from an ecosystem and probe with taxon-specific oligonucleotide probes to estimate the relative population size of given taxa.(11) It is also possible to label an oligonucleotide with a fluorophore and perform *in situ* hybridization to visualize individual bacterial cells belonging to a taxon(12) or to use flow cytometry.(13)

Another approach has been to study the relative abundance of ribosomal DNA genes. This can be done by cloning chromosomal DNA fragments and looking for those that contain rDNA.(14,15) The advent of polymerase chain reaction (PCR) has given this strategy a boost. Thus, it is possible to specifically amplify bacterial SSU rDNA, clone the amplicon, and then determine relative abundance by sequencing the clones. This technique has been applied to sea water,(16,17) soil,(18) and ocean floor sediment(19) as well as numerous individual bacteria that cannot be cultured.(5,20–24) To date, all studies involving ecosystems have shown that previous culture-based studies failed to detect the vast majority of species present. Despite the very extensive past studies of human-associated bacteria, initial application of rDNA analysis to a human fecal sample also indicates that major components of the colonic biota can be underrepresented by traditional culture methods (K. H. Wilson and R. B. Blitchington, unpublished data).

4. ROLE OF THE HOST IN DETERMINING THE COMPOSITION OF THE BIOTA

As mentioned above, each individual has a characteristic biota that changes little over long periods of time. Though the biota varies among

TABLE I

Relative Frequency of Bacterial Species of the Fecal Flora of Three North American Men, Each Sampled 8 or 9 Times during a 5-Month Period, on Normal Diet, Astronaut Diet, and in Land-Based Skylab Chamber (1442 Isolates)[a,b]

Rank	No. isol.	Count[c]	% of flora[d]	Organism(s)
		(× 10^{10})		
1	171	3.04 (0.23)	11.8 (0.90)	*Bacteroides fragilis* ss. *vulgatus*
2	144	2.56 (0.21)	9.9 (0.83)	*Eubacterium aerofaciens*
3	129	2.29 (0.20)	8.9 (0.78)	*Bacteroides fragilis* ss. *thetaiotaomicron*
4	96	1.70 (0.17)	6.6 (0.68)	*Peptostreptococcus productus* II
5	87	1.54 (0.17)	6.0 (0.64)	*Bacteroides fragilis* ss. *distasonis*
6	64	1.13 (0.14)	4.4 (0.55)	*Fusobacterium prausnitzii*
		(× 10^{9})		
7	51	8.95 (1.3)	3.5 (0.49)	*Coprococcus eutactus*
8	44	7.70 (1.2)	3.0 (0.45)	*Eubacterium aerofaciens* III
9	41	7.17 (1.1)	2.8 (0.44)	*Peptostreptococcus productus* I
10	40	6.99 (1.1)	2.7 (0.43)	*Ruminococcus bromii*
11	39	6.81 (1.1)	2.6 (0.43)	*Bifidobacterium adolescentis*
12–13[e]	33	5.74 (1.0)	2.2 (0.39)	*Gemmiger formicilis, Bifidobacterium longum*
14	31	5.38 (1.0)	2.1 (0.38)	*Eubacterium siraeum*
15	27	4.66 (0.09)	1.8 (0.35)	*Ruminococcus torques*
16	25	4.31 (0.09)	1.7 (0.34)	*Eubacterium rectale* III-H
17–18	24	4.13 (0.09)	1.6 (0.33)	*Eubacterium rectale* IV, *E. eligens*
19	22	3.77 (0.08)	1.5 (0.32)	*Bacteroides eggerthii*
20	21	3.59 (0.08)	1.4 (0.31)	*Clostridium leptum*
21	19	3.24 (0.08)	1.3 (0.29)	*Bacteroides fragilis* ss. *a*
22	18	3.06 (0.07)	1.2 (0.29)	*Eubacterium biforme*
23	14	2.35 (0.06)	0.91 (0.25)	*Bifidobacterium infantis*
24	13	2.17 (0.06)	0.84 (0.24)	*Eubacterium rectale* III-F
25–26	9	1.46 (0.05)	0.57 (0.20)	*Coprococcus comes, Bacteroides capillosus*

27–31	8	1.29 (0.05)	0.50 (0.18)	*Ruminococcus albus, Eubacterium formicigenerans, E. hallii, E. ventriosum* I, *Fusobacterium russii*
32–35	7	1.11 (0.04) ($\times 10^8$)	0.43 (0.17)	*Ruminococcus obeum, Eubacterium rectale* II, *Clostridium ramosum* I, *Lactobacillus leichmannii*
36–37	6	9.39 (0.40)	0.36 (0.16)	*Ruminococcus callidus, Butyrivibrio crossotus*
38–41	5	7.67 (0.36)	0.30 (0.14)	*Acidaminococcus fermentans, Eubacterium ventriosum, Bacteroides fragilis* ss. *fragilis, Bacteroides* AR
42–47	4	5.96 (0.32)	0.23 (0.12)	*Coprococcus catus, Eubacterium hadrum, E. cylindroides, E. ruminantium, Eubacterium* CH-1, *Staphylococcus epidermidis*
48–57	3	4.29 (0.27)	0.17 (0.10)	*Peptostreptococcus* BL, *Eubacterium limosum, Bacteroides praeacutus, Bacteroides* L, *Fusobacterium mortiferum* I, *F. naviforme, Clostridium innocuum, C. ramosum, Propionibacterium acnes, Ruminococcus flavefaciens*
58–73	2	2.68 (0.21)	0.10 (0.08)	*Ruminococcus* AT, *Peptococcus* AU-1, *Eubacterium* AG, -AK, -AL, -AL-1, -AN, *Bacteroides fragilis* ss. *ovatus*, -ss. *d*, -ss. *f*, *Baceroides* L-1, -L-5, *Fusobacterium nucleatum, F. mortiferum, Escherichia coli, Steptococcus morbillorum*
74–101	1	1.19 (0.13)	0.05 (0.05)	*Peptococcus magnus, Peptococcus* G, -AU-2, *Streptococcus intermedius, Ruminococcus lactaris, Ruminococcus* CO, *Gemmiger* X, *Coprococcus* BH, -CC, *Eubacterium tenue, E. ramulus, Eubacterium* AE, -AG-H, -AG-M, -AJ, -BW-1, *Bacteroides clostridiiformis* ss. *clostridiiformis, B. coagulans, B. oralis, B. ruminicola* ss. *brevis*, -ss. *ruminicola, Bacteroides splanchnicus, Desulfomonas pigra, Bacteroides* L-4, -W-1, *Fusobacterium* H, *Lactobacillus* G, *Succinivibrio* A

[a]Table modified from Ref. 2, Holdeman, L. V., Cato, E. P., and Moore, W. E. C., 1976, Human fecal flora: Variation in bacterial composition within individuals and a possible effect of emotional stress, *Appl. Environ. Microbiol.* **32:**359–375, with permission from the American Society for Microbiology.
[b]Several unnamed isolates in this table were subsequently named; some species have been renamed. For further details, see the following references: (1) Johnson, J. L., 1978, Taxonomy of the bacteroides. I. Deoxyribonucleic acid homologies among *Bacteroides fragilis* and other saccharolytic *Bacteroides* species, *Int. J. Syst. Bacteriol.* **28:**245–256. (2) Johnson, J. L., and Ault, D. A., 1978, Taxonomy of the bacteroides. II. Correlation of phenotypic characteristics with deoxyribonucleic acid homology groupings for *Bacteroides fragilis* and other saccharolytic *Bacteroides* species, *Int. J. Syst. Bacteriol.* **28:**257–268. (3) Moore, W. E. C., Johnson, J. L., and Holdeman, L. V., 1976, Emendation of *Bacteroidaceae* and *Butyrivibrio* and descriptions of *Desulfomonas* gen. nov. and ten new species in the genera *Desulfomonas, Butyrivibrio, Eubacterium, Clostridium,* and *Ruminococcus, Int. J. Syst. Bacteriol.* **26:**238–252.
[c]The estimated count per gram of fecal dry matter (the standard deviation of the estimate is given in parentheses).
[d]The percentage of the fecal population (the standard deviation of the estimate is given in parentheses).
[e]Where two rank numbers are listed, each organism cited was detected with equal frequency.

individuals, the variation between different animal species is even greater. This finding is not solely related to the fact that members of the same animal species tend to associate with one another and thus exchange bacterial strains. Germfree mice can be colonized with human intestinal biota.[25,26] However, the relative populations established are quite different from those in the human donor's colon (in some of the work cited, the data are at odds with the conclusions; we are assuming here that the data are correct). When such mice are exposed to mouse biota, the bacteria indigenous to the mouse displace those from humans.[26]

The host factors that help determine the composition of the biota are not well understood. Certainly, diet has some role in that a few organisms are favored by a given diet.[27,28] However, the effect is minor in humans; the population sizes of most community members do not change significantly even when the diet is changed radically. Flow dynamics of the cecum are an obvious variable. Rodents eat almost continuously and radioactive tracers entering the cecum are eliminated exponentially.[29] The flow of contents is less continuous in the human cecum. Despite this difference, there are no data directly showing that flow dynamics is a determinant of composition of the intestinal biota. The composition of gastrointestinal mucin probably plays a major role. Mucin consists of a protein backbone covered by branching oligosaccharide side chains. The specific oligosaccharide branching patterns are variable even from person to person. In the case of blood group antigens, which are often expressed in gastrointestinal mucin, it has been shown that people are colonized specifically with bacteria that break down these oligosaccharide linkages.[30,31] The presence of the monosaccharide components that comprise mucin is important for the maintenance of organisms in mouse biota that suppress *Clostridium difficile.*[32] Since most gut anaerobes do not appear to penetrate the thick mucous gel to adhere to the mucosal cells,[33] it is likely that they adhere to mucin. A factor determining the fitness of *Escherichia coli* in the GI tract is its ability to adhere to and metabolize mucin.[34] *Helicobacter pylori* appears to colonize preferentially people who express the Lewis^b antigen on gastric epithelial cells.[35] The effect of the secretor state, in which Lewis^b is secreted in mucin, has not been determined. All of these studies indicate that the composition of mucin plays a role in determining the composition of the biota; further work is needed.

5. COLONIC BIOTA AS A HOST DEFENSE

Bacteria found in the colon normally have a low pathogenic potential. Most do not appear to cause disease even if inoculated directly into sterile tissues. The major advantage to the host of having a complex and stable ecosystem associated with mucosal surfaces is that the natural populations exclude other organisms that do have pathogenic potential. It is not hard to see that *Bacteroides uniformis* and *Eubacterium aerofaciens* are much better companions than *Clostridium botulinum* or *C. difficile*, which colonize only those who have not yet acquired a biota[36,37] or have been treated with antibiotics.[37,38] As mentioned above, there are no bacterial pathogens that can establish themselves in the colonic lumen by competing directly with the normal populations of bacteria. The phenomenon of bacterial antagonism was discovered in the mid-1950s independently by Bohnhoff, who studied salmonella,[39] and Freter, who studied cholera.[40] There are many other potentially pathogenic organisms that have been experimentally shown to be excluded by the colonic biota or are known to colonize the colon and are associated strongly with antibiotic use. These include *Candida albicans*,[41] *Pseudomonas aeruginosa*,[42] *Clostridium tetani*,[43] and antibiotic-resistant enteric gram-negative rods.[44] *Enterococcus fecalis*, *Enterococcus fecium*, and *E. coli* are members of the normal colonic community, but are common causes of nosocomial infections. These organisms are normally suppressed by the predominant, anaerobic species, but can reach large populations when the biota is simplified.

Out of the hundreds of species present in the biota, it has been difficult to determine which ones are necessary and sufficient for suppressing a given pathogen. Indeed, it appears that the complexity itself plays a role in suppression. When organisms are picked at random from the predominant biota of mice, the more complex the combination introduced into gnotobiotic mice, the more *E. coli* and *C. difficile* are suppressed.[45,46] On the other hand, not all organisms are equally suppressive. For instance, Freter and Abrams introduced 95 anaerobic isolates into gnotobiotic mice to suppress *Shigella flexneri* to a population size of 10^4 per cecum. The addition of *E. coli* to the collection of anaerobes led to a further drop of shigella to an undetectable level. In a similar experiment, Wilson and Freter inoculated gnotobiotic mice with 167 isolates from mouse biota to suppress *C. difficile* to 0.3% of its maximal population size. Bourlioux was able to attain a greater degree of suppression by inoculating gnotobiotic mice with fecal suspensions that had been treated to eliminate everything but clostridia.[47]

Thus, for both *S. flexneri* and *C. difficile* it appears that organisms similar to the pathogens are particularly suppressive. It is not known how much this conclusion can be generalized. However, the principle has been employed successfully in clinical practice. In the hamster model of *C. difficile* colitis, it is possible to suppress toxigenic *C. difficile* and prevent colitis by intentionally colonizing antibiotic-treated animals with a nontoxigenic strain of the pathogen before they are exposed to a toxin-producing strain.[48] Though tried to only a limited extent, this approach appears to work with humans as well.[49] A similar approach has been used successfully to control colonization with *Staphylococcus aureus.* In several well-documented studies, Shinefield and co-workers were able to arrest epidemics of staphylococcal infections in newborn nurseries by colonizing infants with a relatively nonpathogenic strain of staphylococcus before they were exposed to the epidemic strain.[50,51] It was also possible to prevent the recurrence of furuncles by using this approach.[52,53]

6. CONTROL MECHANISMS

How does the biota suppress all other populations in the colon? For the most part, the mechanisms by which they maintain exclusive colonization are the same processes by which these large and complex populations maintain their own stability. These mechanisms are very redundant.[54]

As in other ecosystems, the ability of indigenous populations to monopolize available nutrients is critical. This concept grew from experiments in which various culture methods were compared for their ability to artificially simulate the interactions between competing strains of bacteria in the gut of the gnotobiotic mouse.[55] It was found that many organisms failing to suppress shigella in the mouse gut could antagonize the pathogen on agar plates or broth cultures. Furthermore, in these systems, organisms that were highly antagonistic to shigella in the mouse gut often failed to suppress it *in vitro.* Interestingly, there was excellent correlation between antagonism in continuous-flow (CF) culture and antagonism *in vivo.* Because of chemostat theory,[56,57] which was well developed by the early 1960s, this result directly suggested the hypothesis that the mechanism of suppression observed in the mouse gut and CF culture involved nutrient competition. When shigella was given a carbon source not available to the suppressive organisms, it was able to colonize CF cultures and mice freely.[58] This experiment established the importance of competition for growth-limiting nutrients as a mechanism of homeostasis for the colonic biota and supported the validity of the CF culture as a model. It also raised the question of how widespread was the phenomenon of nutrient competition in this ecosystem. The latter question has only been partially answered, but it appears that the implications are extensive, for CF cultures can serve as good models of the entire colonic ecosystem.[59–61] For instance, human or rat fecal material inoculated into such a system leads to long-term communities that carry out many of the characteristic metabolic activities of the colonic biota. Detailed bacteriologic studies have shown that the composition of CF cultures harboring mouse cecal bacteria is very similar to that of the original cecal biota of mice. Even months after inoculation, such cultures are able to convert a germfree mouse to the normal state. This result has never been achieved by administering even a complex collection of isolates, thus indicating that a complex ecosystem is maintained in CF culture.

The growth-limiting nutrients for organisms in these model systems have not been determined for most bacteria. However, in the case of *E. coli*, growth is limited by the carbon/energy source.[62,63] Guiot has confirmed that in the human gut, nutrient depletion is a mechanism by which *E. coli* is suppressed.[64] *C. difficile* also is limited by carbon source.[32] CF cultures incorporating in their media the carbohydrates that *C. difficile* can metabolize maintain a biota that is able to suppress the pathogen. When introduced into gnotobiotic mice, the populations in such CF cultures totally suppress the pathogen in the mice as well. The organisms maintained in CF cultures not containing all of the carbohydrates that *C. difficile* utilizes also suppress the pathogen in the culture vessel, but do not antagonize it as effectively when transferred into mice. This result is consistent with chemostat theory. Presumably, in the absence of certain carbohydrates, the organisms that normally scavenge these carbohydrates fail to colonize the culture vessel and are missing from the biota when it is inoculated into the mouse intestinal tract. The carbohydrates are then not depleted *in vivo* and *C. difficile* is able to utilize them.

It is likely from what is known of bacteria growing in CF cultures that all bacteria in the gut are growth limited by some substrate, but whether the substrate is always a carbon source is not known. Indeed, *Bacteroides* species tend to be able to metabolize several complex carbon sources that are present in large concentrations. It is probable that their growth is limited by something other than carbohydrate.

Results in CF culture models suggest that the adhesion of bacteria to the wall of the gut plays a role in determining the composition of the biota. The saturation of adhesion sites is the probable explanation for the experimental observation that, even if two similar organisms seem identically fit, the one to colonize an animal first has a great advantage over a subsequent invader.[48,65,66] Furthermore, there is an interplay between this factor and the competition for substrates. Theoretically, adhesion also adds to the complexity of the biota. An organism that is unable to maintain a growth rate adequate to keep from being washed out of the lumen could still effectively compete at a slower growth rate if it were able to maintain a population on the wall of the colon.[67] Without postulating such a mechanism, a CF culture could theoretically maintain only as many organisms as there were growth-limiting substrates. With adhesion, the complexity could increase severalfold.

The chemical environment produced by the predominant biota adds several layers to interrelated and redundant control mechanisms. For instance, the limit on *E. coli*'s growth rate at low carbohydrate concentrations is much more severe in the presence of H_2S tensions found in the colon than in the absence of H_2S. Because *E. coli*'s maximal growth rate is much faster in the presence of oxygen, the simple fact that the environment is anaerobic puts some organisms at a competitive disadvantage. Short-chain volatile fatty acids (VFAs) have been studied extensively for their ability to suppress growth of bacteria. At concentrations found in the colon, VFAs can profoundly suppress the growth of a variety of organisms *in vitro.*[68,69] However, *in vivo,* particularly in long-term experiments, many of the same organisms have been able to colonize in large numbers despite the presence of high concentrations of VFAs.[70,71] One possible explanation for these overall results is that VFAs extend the period that invading organisms spend in lag phase.[65] As opposed to many other types of bacteria, the anaerobes indigenous to the gut are notable for their ability to grow in the presence of secondary bile acids. Their resistance to these detergent molecules is to be expected since it is the anaerobes of the gut that metabolize primary bile acids to produce the secondary ones. Though it has not been demonstrated experimentally, it is likely that secondary bile acids also antagonize many nonindigenous organisms *in vivo.*[72]

The physiological state of an invading pathogen is an important factor that has been investigated relatively little. Typically, in an experiment where the intestinal tract is inoculated with a new strain of bacterium, that strain is grown to high concentrations in a rich medium. Little attention has been paid to the growth stage of the organism. Yet physiological adaptation is known to be an important determinant of fitness. For instance, if an inoculum of *E. coli* is grown in CF culture, it is a much more fit challenge than its broth-grown counterpart. Indeed, it is clear now that *E. coli* growing on rich media in static cultures in the laboratory is quite different from the organism surviving *in vivo.* The bacterium changes from a rod-shaped to a coccoid form, a phenotype that is stable enough to study further in the laboratory.[73] The coccoid-shaped organism has a much slower maximal growth rate. Other traits that are expressed under typical laboratory conditions are downregulated *in vivo.* Though motile and chemotactic *in vitro,* *E. coli* F-18 does not appear to express these characteristics in the mouse gut.[74] One would expect that if an organism entered the colon in an inappropriate adaptive state, the lag time before onset of cell division would be prolonged. This phenomenon may explain the susceptibility of laboratory-grown strains to VFAs.

7. OVERALL CONCLUSIONS

The colonic biota is a highly effective host defense. This complex ecosystem of relatively nonpathogenic microbes efficiently prevents establishment of more harmful populations. Despite the complexity of this ecosystem, it comprises one of the best characterized natural bacterial communities. New molecular approaches are raising new taxonomic questions in the process of establishing a consistent taxonomic order in this field. However, the final resolution promises to offer more certainty and ease in identifying community members. Many of the control mechanisms that underlie the stability of this ecosystem are understood. The challenge currently facing this field is how to approach this very important natural bacterial community with important new questions and novel approaches.

REFERENCES

1. Marshall, B. J., and Warren, J. R., 1984, Unidentified curved bacilli in the stomach of patients with gastritis and peptic ulceration, *Lancet* **1:**1311–1314.
2. Holdeman, L. V., Cato, E. P., and Moore, W. E. C., 1976, Human fecal flora: Variation in bacterial composition within individuals and a possible effect of emotional stress. *Appl. Environ. Microbiol.* **32:**359–375.
3. Moore, W. E. C., 1977, Anaerobes as normal flora: Gastrointestinal tract, in: *Metronidazole: Proceedings of the International Conference* (S. M. Finegold, ed.), Excerpta Medica, Amsterdam, pp. 222–228.
4. Moore, W. E. C., Moore, L. V. H., and Cato, E. P., 1988, You and your flora, *The U.S. Federation for Culture Collections Newsletter* **18:**7–22.
5. Amann, R., Springer, N., Ludwig, W., Gortx, H.-D., and Schleifer, K.-H., 1991, Identification *in situ* and phylogeny of uncultured bacterial endosymbionts, *Nature* **351:**161–164.
6. Rainey, F. A., and Stackebrandt, E., 1993, 16S rDNA analysis reveals phylogenetic diversity among the polysaccharolytic clostridia, *FEMS Microbiol. Lett.* **113:**125–128.
7. Moore, W. E. C., and Holdeman, L. V., 1974, Human fecal flora: The normal flora of 21 Japanese-Hawaiians, *Appl. Microbiol.* **27:**961–979.
8. Fox, G. E., Peckman, K., and Woese, C. R., 1977, Comparative cataloguing of 16S ribosomal ribonucleic acid: Molecular approach to prokaryotic systematics, *Int. J. Syst. Bacteriol.* **27:** 44–57.
9. Woese, C. R., 1987, Bacterial evolution, *Microbiol. Rev.* **51:**221–271.
10. Woese, C. R., and Olsen, G. J., 1993, Ribosomal RNA: A key to phylogeny, *FASEB J.* **7:**113–123.
11. Stahl, D. A., Flesher, B., Mansfield, H. R., and Montgomery, L., 1988, Use of phylogenetically based hybridization probes for studies of ruminal ecology, *Appl. Environ. Microbiol.* **54:**1079–1084.
12. DeLong, E. F., Wickham, G. S., and Pace, N. R., 1989, Phylogenetic strains: Ribosomal RNA-based probes for the identification of single cells, *Science* **243:**1360–1363.
13. Amann, R. I., Binder, B. J., Olson, R. J., Chisholm, S. W., Devereux, R., and Stahl, D. A., 1990, Combination of 16S rRNA-targeted oligonucleotide probes with flow cytometry for analyzing mixed microbial populations, *Appl. Environ. Microbiol.* **56:**1919–1925.
14. Pace, N. R., Stahl, D. A., Lane, D. J., and Olsen, G. J., 1986, The analysis of natural microbial populations by ribosomal RNA sequences, in: *Advances in Microbial Ecology*, Volume 9 (K. C. Marshall, R. Atlas, B. B. Jorgesen, and J. H. Slater, eds), Plenum Press, New York, pp. 1–55.
15. Weller, R., and Ward, D. M., 1989, Selective recovery of 16S rRNA sequences from natural communities in the form of cDNA, *Appl. Environ. Microbiol.* **55:**1818–1822.
16. Giovannoni, S. L., Britschgi, T. B., Moyer, C. L., and Field, K. G., 1990, Genetic diversity in Sargasso Sea bacterioplankton, *Nature* **345:**60–63.

17. Schmidt, T. M., DeLong, E. F., and Pace, N. R., 1991, Analysis of a marine picoplankton community by 16S rRNA gene cloning and sequencing, *J. Bacteriol.* **173:**4371–4378.
18. Stackebrandt, E., Liesack, W., and Goebel, B. M., 1993, Bacterial diversity in a soil sample from a subtropical Australian environment as determined by 16S rDNA analysis, *FASEB J.* **7:**232–236.
19. Rochelle, P. A., Fry, J. C., Parkes, R. J., and Weightman, A. J., 1992, DNA extraction for 16S rRNA gene analysis to determine genetic diversity in deep sediment communities, *FEMS Microbiol. Lett.* **100:**59–66.
20. Wilson, K. H., Blitchington, R. B., Frothingham, R., and Wilson, J. A. P., 1991, Whipple's disease associated with a bacterium related to *Streptomyces* and *Arthrobacterium*, Abstracts of the 31st Interscience Conference on Antimicrobial Agents and Chemotherapy 1092.
21. Solnick, J. V., O'Rourke, J., Lee, A., Paster, B. J., Dewhirst, F. E., and Tompkins, L. S., 1993, An uncultured gastric spiral organism is a newly identified *Helicobacter* in humans, *J. Infect. Dis.* **168:**379–385.
22. Angert, E. R., Clements, K. D., and Pace, N. R., 1993, The largest bacterium, *Nature* **362:** 239–241.
23. Duncan, A. J., Carman, R. J., Olsen, G. J., and Wilson, K. H., 1993, Assignment of the agent of Tyzzer's disease to *Clostridium piliforme* comb. nov. on the basis of 16S rRNA sequence analysis, *Int. J. Syst. Bacteriol.* **43:**314–318.
24. Lindenbaum, J., 1968, Small intestine dysfunction in Pakistanis and Americans resident in Pakistan, *Am. J. Clin. Nutr.* **21:**1023.
25. Raibaud, P., Ducluzeau, R., Dubos, F., Hudault, S., Bewa, H., and Muller, M. C., 1980, Implantation of bacteria from the digestive tract of man and various animals into gnotobiotic mice, *Am. J. Clin. Nutr.* **33:**2440–2447.
26. Hazenberg, M. P., Bakker, M., and Burggraaf, A. V., 1981, Effects of the human intestinal flora on germ-free mice, *J. Appl. Bacteriol.* **50:**95–106.
27. Moore, W. E. C., Cato, E. P., Good, I. J., and Holdeman, L. V. 1981, The effect of diet on the human fecal flora, in: *Branbury Report 7, Gastrointestinal Cancer: Endogenous Factors* (W. R. Bruce, P. Correa, M. Lipkin, S. R. Tannenbaum, and T. D. Wilkins, eds.), Cold Spring Harbor Laboratory, New York, pp. 11–24.
28. Finegold, S. M., Attebery, H. R., and Sutter, V. L., 1974, Effect of diet on human fecal flora: Comparison of Japanese and American diets, *Am. J. Clin. Nutr.* **27:**1456–1469.
29. Wilson, K. H., Sheagren, J. N., and Freter, R., 1985, Population dynamics of ingested *Clostridium difficile* in the gastrointestinal tract of the Syrian hamster, *J. Infect. Dis.* **151:** 355–361.
30. Hoskins, L. C., and Boulding, E. T., 1976, Degradation of blood group antigens in human colon ecosystems. I. In vitro production of ABH blood group-degrading enzymes by enteric bacteria, *J. Clin. Invest.* **1:**63–73.
31. Hoskins, L. C., and Boulding, E. T., 1976, Degradation of blood group antigens in human colon ecosystems. II. A gene interaction in man that affects the fecal population density of certain enteric bacteria, *J. Clin. Invest.* **1:**74–82.
32. Wilson, K. H., and Perini, F., 1988, Role of competition for nutrients in suppression of *Clostridium difficile* by the colonic microflora, *Infect. Immun.* **56:**2610–2614.
33. Bollard, J. E., Vanderwee, M. A., Smith, G. W., Tasman-Jones, C., Gavin, J. B., and Lee, S. P., 1986, Location of bacteria in the mid-colon of the rat, *Appl. Environ. Microbiol.* **51:**604–608.
34. Cohen, P. S., Rossoll, R., Cabelli, V. J., Yang, S.-L., and Laux, D. C., 1983, Relationship between the mouse colonizing ability of a human fecal strain and its ability to bind a specific mouse colonic mucous gel protein, *Infect. Immun.* **40:**62–69.
35. Boren, T., Falk, P., Roth, K. A., Larson, G., and Normark, S., 1993, Attachment of *Helicobacter pylori* to human gastric epithelium mediated by blood group antigens, *Science* **262:**1892–1895.

36. Midura, T. F., and Arnon, S. S., 1976, Infant botulism: Identification of *Clostridium botulinum* and its toxins in feces, *Lancet* **2:**934–936.
37. Viscidi, R., Willey, S., and Bartlett, J. G., 1981, Isolation rates and toxigenic potential of *Clostridium difficile* isolates from various populations, *Gastroenterology* **81:**5–9.
38. Chia, J. K., Clark, J. B., Ryan, C. A., and Pollack, M., 1986, Botulism in an adult associated with food-borne intestinal infection with *Clostridium botulinum*, *N. Engl. J. Med.* **315:**239–240.
39. Bohnhoff, M., Drake, B. L., and Miller, C. P., 1954, Effect of streptomycin on susceptibility of the intestinal tract to experimental salmonella infection, *Proc. Soc. Exp. Biol. Med.* **86:**132–137.
40. Freter, R., 1955, Fatal enteric cholera infection in the guinea pig achieved by inhibition of normal enteric flora, *J. Infect. Dis.* **97:**57–64.
41. Krause, W., Matheis, H., and Wulf, K., 1969, Fungaemia and funguria after oral administration of *Candida albicans*, *Lancet* **1:**598–599.
42. Buck, A. C., and Cooke, M., 1969, The fate of ingested *Pseudomonas aeruginosa* in normal persons, *J. Med. Microbiol.* **2:**521–525.
43. Wells, C. L., and Balish, E., 1983, *Clostridium tetani* growth and toxin production in the intestines of germfree rats, *Infect. Immun.* **41:**826–828.
44. Schimpff, S. C., Young, V. M., Greene, W. H., Vermeulen, G. D., Moody, M. R., and Wiernik, P. H., 1972, Origin of infection in acute nonlymphocytic leukemia: Significance of hospital acquisition of potential pathogens, *Ann. Intern. Med.* **77:**707–714.
45. Freter, R., and Abrams, G. D., 1972, Function of various intestinal bacteria in converting germfree mice to the normal state, *Infect. Immun.* **6:**119–126.
46. Wilson, K. H., and Freter, R., 1986, Interactions of *Clostridium difficile* and *E. coli* with microfloras in continuous-flow cultures and gnotobiotic mice, *Infect. Immun.* **54:**354–358.
47. Su, W. J., Bourlioux, P., Bournaud, M., Besnier, M. O., and Fourniat, J., 1986, Mise au point d'un modele experimental animal permettant de la microflore coecale du hamster, antagoniste de *Clostridium difficile*, *Ann. Inst. Pasteur* (Paris) **137A:**89–96.
48. Wilson, K. H., and Sheagren, J. N., 1983, Antagonism of toxigenic *Clostridium difficile* by nontoxigenic *C. difficile*, *J. Infect. Dis.* **147:**733–736.
49. Seal, D., Borriello, S. P., Barclay, F. E., Welch, A., Piper, M., and Bonnycastle, M., 1987, Treatment of relapsing *Clostridium difficile* diarrhea by administration of a nontoxigenic strain, *Eur. J. Clin. Microbiol.* **6:**51–53.
50. Shinefield, H. R., Ribble, J. C., Boris, M., and Eichenwald, H. F., 1963, Bacterial interference: Its effect on nursery-acquired infection with *Staphylococcus aureus*. I. Preliminary observations on artificial colonization of newborns, *Am. J. Dis. Child.* **105:**146–154.
51. Shinefield, H. R., Ribble, J. C., Eichenwald, H. F., Boris, M., and Sutherland, J. M., 1963, An analysis and interpretation, *Am. J. Dis. Child.* **105:**683–688.
52. Branche, W. C., Jr., Young, V. M., Robinet, H. G., and Massey, E. D., 1963, Effect of colicine production on *Escherichia coli* in the normal human intestine, *Proc. Soc. Exp. Biol. Med.* **114:** 198–201.
53. Suzuki, K., Benno, Y., Mitsuoka, T., Takebe, S., Kobashi, K., and Hase, J., 1979, Urease-producing species of intestinal anaerobes and their activities, *Appl. Environ. Microbiol.* **37:** 379–382.
54. Freter, R., 1974, Interactions between mechanisms controlling the intestinal microflora, *Am. J. Clinc. Nutr.* **27:**1409–1416.
55. Hentges, D. J., and Freter, R., 1962, In vivo and in vitro antagonism of intestinal bacteria against *Shigella flexneri*. I. Correlation between various tests, *J. Infect. Dis.* **110:**30–37.
56. Monod, J., 1950, La technique de cultur continue: theorie et applications, *Ann. Inst Pasteur* (Paris) **79:**390–410.
57. Novik, A., and Szilard, L., 1950, Experiments with the chemostat on spontaneous mutations of bacteria, *Proc. Natl. Acad. Sci. USA* **36:**708–719.

58. Freter, R., 1962, In vivo and in vitro antagonism of intestinal bacteria against *Shigella flexneri*. II. The inhibitory mechanism, *J. Infect. Dis.* **110:**38–46.
59. Freter, R., Stauffer, E., Cleven, D., Holdeman, L. V., and Moore, W. E. C., 1983, Continuous-flow cultures as in vitro models of the ecology of large intestinal flora, *Infect. Immun.* **39:** 666–675.
60. Veilleux, B. G., and Rowland, I., 1981, Simulation of the rat intestinal ecosystem using a two-stage continuous culture system, *J. Gen. Microbiol.* **123:**103–115.
61. Miller, T. L., and Wolin, M. J., 1981, Fermentation by the human large intestine microbial community in a semicontinuous culture system, *Appl. Environ. Microbiol.* **42:**400–407.
62. Ozawa, A., and Freter, R., 1964, Ecologic mechanism controlling growth of *Escherichia coli* in continuous-flow culture and in the mouse intestine, *J. Infect. Dis.* **114:**235–242.
63. Freter, R., Brickner, H., Botney, M., Cleven, D., and Aranki, A., 1983, Mechanisms that control bacterial populations in continuous-flow culture models of mouse large intestinal flora, *Infect. Immun.* **39:**676–685.
64. Guiot, H. F., 1982, Role of competition for substrate in bacterial antagonism in the gut, *Infect. Immun.* **38:**887–892.
65. Freter, R., Brickner, H., Fekete, J., Vickerman, M. M., and Carey, K. E., 1983, Survival and implantation of *Escherichia coli* in the intestinal tract, *Infect. Immun.* **39:**686–703.
66. Duvall-Iflah, Y., Raibaud, P., and Rousseau, M., 1981, Antagonisms among isogenic strains of *Escherichia coli* in the digestive tracts of gnotobiotic mice, *Infect. Immun.* **34:**957–969.
67. Frederickson, A. G., 1977, Behavior of mixed cultures of microorganisms, *Annu. Rev. Microbiol.* **31:**63–87.
68. Rolfe, R. D., 1984, Role of volatile fatty acids in colonization resistance to *Clostridium difficile*, *Infect. Immun.* **45:**185–191.
69. Hentges, D. J., and Maier, B. R., 1972, Inhibition of *Shigella flexneri* by the normal intestinal flora. III Interactions with *Bacteroides fragilis* in vitro, *Infect. Immun.* **6:**168–173.
70. Maier, B. R., and Hentges, D. J., 1972, Experimental shigella infection in animals, *Infect. Immun.* **6:**168–173.
71. Su, W. J., Waechter, M. J., Bourlioux, P., Dolegeal, M., Fourniat, J., and Mahuzier, G., 1987, Role of volatile fatty acids in colonization resistance to *Clostridium difficile* in gnotobiotic mice, *Infect. Immun.* **55:**1686–1691.
72. Meynell, G. G., 1963, Antibacterial mechanisms of the mouse gut, *Br. J. Exp. Pathol.* **44:** 209–219.
73. Krogfelt, K. A., Poulsen, L. K., and Molin, S., 1993, Identification of coccoid *Escherichia coli* BJ4 cells in the large intestine of streptomycin-treated mice, *Infect. Immun.* **61:**5029–5034.
74. McCormick, B. A., Laux, D. C., and Cohen, P. S., 1990, Neither motility nor chemotaxis plays a role in the ability of *Escherichia coli* F-18 to colonize the streptomycin-treated mouse large intestine, *Infect. Immun.* **58:**2957–2961.

2

Mucosal Immunity

Molecular and Cellular Aspects of Immune Protection to Enteric Infections

DAVID W. PASCUAL, HIROSHI KIYONO, and JERRY R. McGHEE

1. INTRODUCTION

Until recently, investigation into the mechanisms of mucosal immune protection has been overlooked as an integral component of immune surveillance. The failure to measure B- and T-cell responses in various mucosa-associated tissues and their secretions was complicated by difficulties in the isolation and characterization of mucosal lymphoid cells and in obtaining external secretions. The relevance of investigating mucosal immunity has been accentuated by recent developments, i.e., the failure of conventional vaccines to protect from mucosal pathogens when administered in parenteral sites and by the need for a mucosal vaccine to prevent the sexual transmission of HIV. In this review, we will discuss the nature of the gastrointestinal (GI) immune system and its subsequent manipulation, i.e., for vaccine development, to effect protection to enteric infectious agents. Furthermore, a better understanding of the GI immune system will contribute

DAVID W. PASCUAL • Veterinary Molecular Biology, Montana State University, Bozeman, Montana 59717-0360. HIROSHI KIYONO and JERRY R. McGHEE • The Immunobiology Vaccine Center, Departments of Oral Biology and Microbiology, University of Alabama at Birmingham, Birmingham, Alabama 35294-2170; and Department of Mucosal Immunology, Research Institute for Microbial Disease, Osaka University, Osaka 565, Japan.

Enteric Infections and Immunity, edited by Lois J. Paradise *et al.* Plenum Press, New York, 1996.

to the application of mucosal immunity for other tissues that are exposed to ubiquitous pathogens.

Because of its relative inaccessibility and size, studying the mucosal immune system (Fig. 1), particularly in the GI tract, has been a barrier for assessing vaccine efficacies. Not only does the system span large surface areas, but there also is a functional segregation of sites for antigen/vaccine processing where B and T lymphocytes are initially stimulated (referred to as inductive sites) and for immune cells involved in protection (referred to as effector sites).[1,2] Furthermore, it is important to understand the molecular pathogenesis of the infectious agent and how the GI immune system responds to it. Attempts to induce mucosal immunity to these agents via the

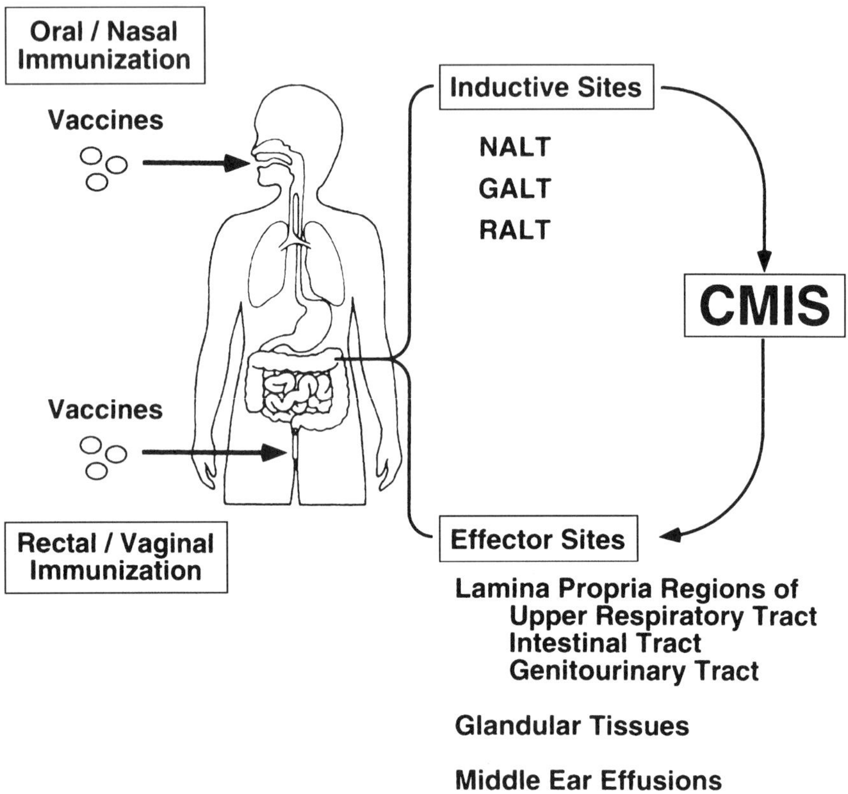

FIGURE 1. Involvement of the common mucosal immune system (CMIS) in vaccine delivery. This depiction emphasizes the segregation of inductive sites from effector sites during mucosal immune responses.

oral route have not always been successful. Thus, in this review we will discuss features of the GI immune system important for induction and expression of mucosal immunity. Specifically, we will examine the role of secretory IgA (S-IgA) antibodies, T cells and their subsets, cytokines and neuropeptides that promote IgA responses and those that promote cell-mediated immunity (CMI). In the case of T cells, emphasis will be on $CD4^+$ helper T (Th) cells including Th1 and Th2 subpopulations for the induction and regulation of IgA B-cell responses and on CD8+ cytotoxic T lymphocytes (CTLs) in antiviral immunity.

2. THE GASTROINTESTINAL IMMUNE SYSTEM

2.1. S-IgA Antibodies for Immune Protection in the GI Tract

Considering the surface area of the human intestinal tract, which can be equated to an area of greater than 300 m^2, effective immune protection for such an expansive area would require considerable amounts of antibody to be generated. In fact, IgA represents 60% of all Ig produced, of which >98% consists of S-IgA derived from plasma cells in mucosal sites such as the GI tract.[3,4] Over two-thirds of the IgA produced in humans is the product of plasma cells (IgA) and epithelial cells [secretory component (SC)] lining the lamina propria (LP) regions of the GI, upper respiratory, and genitourinary tracts and the acinar regions of the exocrine glands.[2,4] The GI tract alone contains $1–2 \times 10^{10}$ IgA^+ plasma cells/m intestine,[5] which underlines the strength of the host's commitment to protecting the GI tract.

Human IgA occurs as two subclasses: IgA1 and IgA2. Differences in their distribution and molecular form in serum and in various external secretions,[3,4,6] particularly in the GI tract, have posed a barrier for assessing vaccine efficacies. More than 90% of serum IgA is of the IgA1 subclass consisting predominantly of monomers, whereas various external secretions contain dimeric or polymeric forms of 50–70% IgA1 and 30–50% IgA2. The distribution of IgA1 and IgA2 plasma cells in various human mucosal effector sites supports the notion of separate mucosal inductive sites, e.g., bronchus-associated and nasopharyngeal-associated lymphoreticular tissues (BALT and NALT, respectively) versus gut-associated lymphoreticular tissues (GALT). These differences may result from differences in B-cell subsets present. For example, in the nasal mucosa, lacrimal glands and LP of the upper respiratory tract (URT) contain mainly $IgA1^+$ plasma cells, while fewer $IgA2^+$ cells are observed.[7,8] In contrast, higher numbers of IgA2 plasma cells are often seen in the LP of the lower intestinal tract.[7,8] If the tonsils represent an inductive site for NALT/BALT analogous to that of the appendix for GALT, then an interesting pattern emerges. For example, pokeweed mitogen-triggered tonsil B-cell cultures generate all major Ig isotypes, including IgA.[9] Greater than 90% of the IgA produced is IgA1, which supports the tenet that tonsils may represent the site for precursor IgA1^{+} B cells.[10] Likewise, IL-6-stimulated appendix B cells produce more IgA2 antibody.[10] This would suggest that the GALT is a major source for IgA precursor B cells which populate the effector regions of the small and large intestine with some tendency to IgA2 expression.

Our external secretions contain high proportions of IgA2 which are more resistant to bacterial proteases.[4,11] This higher degree of resistance of IgA2 to proteolytic attack over IgA1 is attributed to differences in the hinge region of the $\alpha 2$ H chain molecule. This difference in the $\alpha 2$ H chain is manifested by a deletion of 13 amino acids relative to the $\alpha 1$ H chain, which results in an unusual hinge structure. To this end, a number of bacteria produce proteases that can selectively cleave IgA1 into Fab and Fc fragments.[6,12] In addition to the altered hinge region, further resistance to proteolytic attack is conferred by SC.[13] Human SC (~80 kDa), a member of the immunoglobulin supergenefamily,[14] is a component of the poly-Ig receptor (~116 kDa) expressed on basolateral surfaces of mucosal epithelial cells capable of binding IgA and IgM polymeric antibody molecules. SC is disulfide-linked to IgA and after transcytosis to the luminal surface, the SC–polymeric IgA complex (S-IgA) is secreted.

2.2. Mucosal Inductive Sites of the GI Immune System

Inductive sites (Fig. 2) are where antigen is first encountered and processed and where initial induction of immune and memory B and T cells occurs. In the gut, the Peyer's patches (PP) and appendix represent inductive sites for GALT. A PP is an organized area of lymphoid tissue in the intestine with well-demarcated regions of B-cell follicles. In humans, PP are truly dynamic mucosal tissues manifested by an ability to fluctuate their numbers with age. At birth, infants have approximately 100 PP (>5 follicles/PP) in the small intestine. This number increases to 225–300 by late adolescence and then declines with increasing age until eventually it resembles levels observed at birth.[15,16] It is here where antigen/vaccine uptake into the PP leads to effective presentation to MHC class II-restricted CD4^{+} Th cells and, for intracellularly processed and class I-associated peptides, to CD8^{+} CTL precursors (Fig. 2). Furthermore, B cells in PP respond to antigen and undergo expansion and memory cell formation through the help provided by Th cells and emigrate to mucosal effector tissues such as the diffuse LP regions of the small and large intestine. On settling into these effector tissues, the B cells may undergo clonal expansion and differentia-

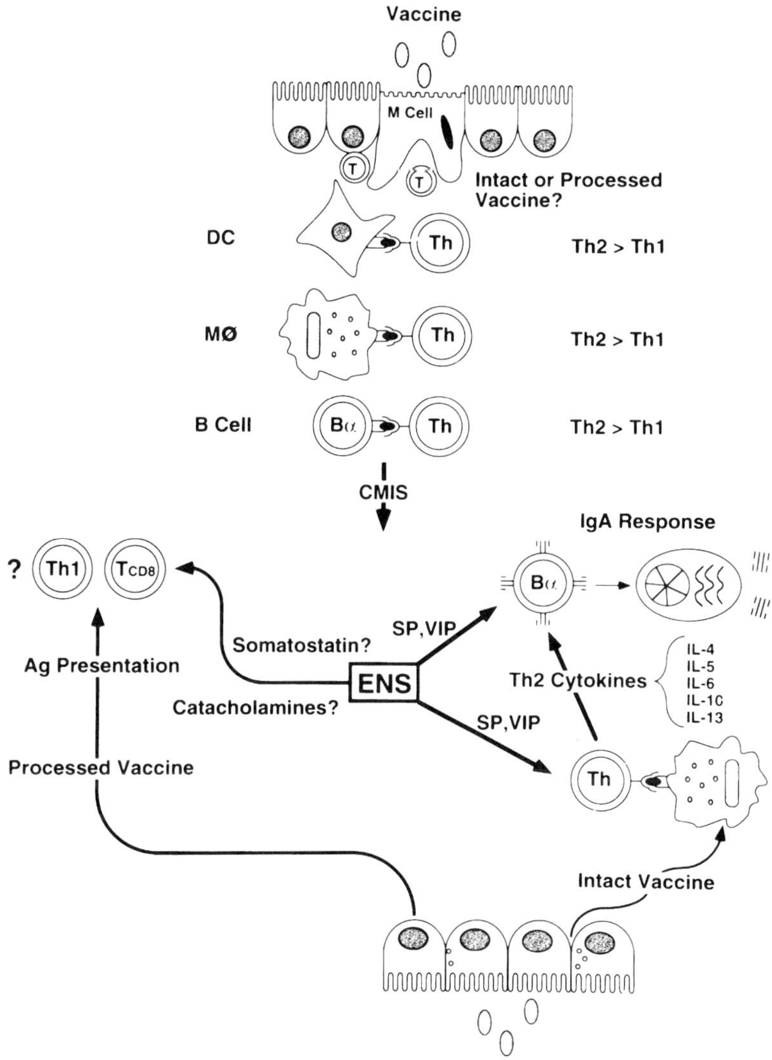

FIGURE 2. On antigen uptake in the Peyer's patch, antigen is transported or possibly processed and presented to B and T lymphocytes. The antigen-specific B cell travels through the CMIS for deposition in effector sites where it becomes terminally differentiated. Th1-derived cytokines and possibly neuropeptides promote the differentiation of the B cell into an IgA plasma cell.

tion into IgA plasma cells on reexposure to antigen and with help from Th-cell derived cytokines. The induction of immune B and Th cells in GALT (Fig. 2) is followed by migration of their progeny to effector sites such as LP regions of the GI tract for subsequent development of mucosal immunity via the common mucosal immune system (CMIS).[2,4,5]

The PP can be functionally and anatomically separated into three distinctive areas: (1) the dome with a unique lymphoepithelium; (2) the B-cell follicles which usually contain one or more germinal centers; and (3) the perifollicular or T-cell-dependent area (TDA).[4] The dome region is covered by an epithelium that contains antigen-sampling cells termed M cells.[17,18] Others have termed the epithelial layer a follicle-associated epithelium (FAE) to denote a relatively high frequency of cells that take up antigen (~10% in mice and humans)[19] and lymphocytes which apparently originate from the underlying dome area.[17–19] The M cells have membranelike surface structures that facilitate antigen uptake. M cells can take up soluble proteins[18–21] or viruses such as reovirus types 1 and 3,[22] HIV,[23] poliovirus,[24] and rotaviruses. Interestingly, all members of the Enterobacteriaceae family, e.g., *Salmonella, Escherichia coli,* and *Shigella,* as well as *Vibrio cholerae* are transported by M cells.[20] Consequently, many of the bacterial and viral pathogens actually exploit the M cell as a means of infecting the host. In this regard, it was shown that some of these microorganisms bind to specific receptors on M cells.[30] Such entry is facilitated by unique features of M cells. For example, the apical surface expresses little mucus and lesser amounts of brush border enzymes such as alkaline phosphatase than do adjacent enterocytes.[20] M cells contain few lysosomes and exhibit a well-organized microvesicular system. The apical membrane has binding sites for S-IgA and it is possible that S-IgA complexes (with, e.g., viruses or bacteria) could facilitate antigen presentation to underlying lymphoreticular cells for subsequent immune responses.[25] To this end, M cells have an extracellular space termed a hollow or pocket that is often occupied by one or more lymphocytes or even by plasma cells.[20] The combination of these properties complements M-cell function for antigen uptake and subsequent transport.[20]

The evidence presented thus far suggests that M cells serve primarily in antigen transport and to a lesser extent in antigen processing. There are currently two views on this subject. There is some debate as to whether M cells have the appropriate surface and intracellular mechanisms necessary to process antigen. One view is that epithelial cells in the dome region express class II MHC, but M cells do not.[26] The opposing view is that M cells are capable of antigen processing since class II MHC has been reported to be present on human[27] and rat[28] PP M cells. Second, there is contrasting evidence as to whether M cells can process antigen. However, in a recent study by Allan *et al.*,[27] rat M cells were shown to contain the necessary intracellular acidic endosomal–lysosomal compartments as well as to express class II MHC determinants in prelysosomal and lysosomal compartments. While it is clear that soluble antigen or infectious agents can bind to M cells, it remains to be determined if M cells do serve only for antigen transport or if they actually can contribute to antigen processing. Future studies will be required to address this at the molecular level. Adherence and transport do not necessarily involve antigen processing. For example, reovirus types 1 and 3, rotavirus, *E. coli*-RDEC, and *V. cholerae* bind to surfaces of M cells and are transported through them in nondegradative endosomal/lysosomal pathways.[20] Macromolecules may enter endosomes (possibly coated with specific S-IgA) which fuse with class II^+ secondary lysosomes and subsequently be processed. Antigen-presenting cells (APCs) do provide $CD4^+$ Th cells with the triggering signal by presenting the foreign peptide–MHC class II complex and this results in initial activation steps and induction of memory cells.

Distinct follicles (B-cell areas) occur just beneath the FAE enclosed dome region of PP. The follicles comprise one or two germinal centers where active B cells expressing membrane IgA [or surface IgA ($sIgA^+$)] are thought to be immediate precursors of IgA plasma cells.[29,30] The germinal centers, which contain 60–70% of $mIgA^+$ B cells present in the PP,[30] are considered to be the major sites where initial commitment of B cells to IgA is made. However, it is important to note that final differentiation of B cells into IgA-producing plasma cells does not occur in B-cell zones of the PP but instead in distant mucosal effector tissues via the CMIS as described above. The relationship between $sIgA^+$ B cells in the follicular germinal centers and IgA memory cells has been studied in detail.[30,31] B cells enriched for sIgA can be subdivided into at least two subsets based on binding to peanut agglutinin (PNA),[30] staining with anti-κ monoclonal antibody, and levels of mRNA for α-chain present.[31] The IgA memory population consists of small, resting, PNA^{low} and κ^{high} B cells which have low levels of α-chain mRNA.[31] It is now thought that the PNA^{low}, κ^{high} B cells are the immediate precursors of IgA plasma cells at mucosal sites.[31]

The TDA of the PP are adjacent to the follicles and some $CD4^+$ T cells actually reside in the germinal centers and presumably contribute help to ongoing B-cell responses. T cells are the first to populate the PP during development and >95% use the αβ form of the T-cell receptor (TCR). Approximately 60% of PP T cells are $CD4^+$ $CD8^-$ and exhibit properties of Th cells. However, significant numbers of $CD4^-$ $CD8^+$ T cells occur in PP and functional CTLs can be induced and are associated with this T-cell subset.[32] It should also be mentioned that all three major types of APCs (e.g., macrophages, dendritic cells, and sIg^+ B cells) occur in significant numbers in GALT. Macrophages, including both classical and tingible body types associated with uptake of apoptotic (dying) cells, are associated with the dome region and with germinal centers, respectively. Their perifollicular TDA is enriched for dendritic cells, which have been shown in functional assays to induce clusters of $CD4^+$ T cells which are particularly adept in supporting IgA synthesis.[33,34] It is intriguing that few studies have addressed APC function of B cells in PP, in spite of the presence of activated, dividing B cells which express class II MHC. It could be possible that activation signals provided by these three different kinds of APC may influence the outcome of different patterns of cytokine profiles induced by Th1/Th2-type $CD4^+$ T cells (see below).

2.3. Mucosal Effector Sites of the GI Immune System

Following antigen stimulation in GALT and presentation of the antigen to B and T cells, the antigen-induced B and T cells leave PP via efferent lymphatics and reach the systemic circulation through the thoracic duct. These lymphocytes then enter effector sites such as the LP of the respiratory, GI, and reproductive tracts and glandular tissues via the CMIS. These lymphoid cells are selectively retained in these tissues by mechanisms not yet fully understood, but clearly may involve the expression of adhesion molecules. B cells clonally expand under the influence of antigen, T cells, and cytokines and become mature IgA plasma cells in the IgA effector tissues.[2,4,5]

The LP of the GI tract is the major mucosal effector site and lymphocytes are the dominating cell type found and consist of 20–40% B cells (including plasma cells) and 40–60% T cells. Other important cell types include macrophages (~10%), eosinophils (~5%), and mucosal mast cells (~1–3%). Although large numbers of plasma cells are present, mostly of the IgA isotype, T cells are the most frequent cell type seen in these regions.[35,36] The majority of T cells are $CD3^+$ $CD4^+$ $CD8^-$ and exhibit helper functions.[35] Approximately one-third of the T cells in these effector regions are $CD3^+$ $CD4^-$ $CD8^+$ and may exhibit CTL (and possibly suppressor) functions. A unique feature of $CD3^+$ T cells in the LP is their resistance to activation via their αβ TCR.[37] This puzzling attribute has made it difficult to study the precise Th cell functions on isolated LP cells, and most studies thus far have measured spontaneous cytokine production.[38,39] It has been shown that relatively high numbers of Th2-type $CD4^+$ Th cells which produce cytokines, such as IL-5 and IL-6 which are optimal for stimulating $sIgA^+$ B cells to become IgA plasma cells, reside in the LP of the GI tract in addition to Th1-type cells. More studies are required to determine the precise functions of the two major effector T-cell subsets in orally immunized as well as in normal animals. Macrophages are found in significant numbers in the LP[36] and may be involved in antigen processing and presentation at this site. Only limited information is available at present on antigen-specific $CD4^+$ Th cell, $sIgA^+$ B cell, and macrophage interactions in effector tissues. It is important to understand the cellular and molecular interactions among these three subsets of mononuclear cells for induction of mucosal responses to vaccine antigens. Further, the induction of functional $CD8^+$ CTLs in effector sites must also be studied in more detail.

2.4. Mucosal Memory in the GI Immune System

Early studies addressing whether memory could be induced at mucosal effector sites had been a subject of some controversy since numerous studies had shown that IgA responses were sometimes short-lived and boosting failed to induce IgA responses faster or with increased affinity/avidity characteristic of memory T and B cells. However, these earlier studies assessed mucosal responses to polysaccharide antigens which do not elicit memory responses. Clearly, the nature of the antigen/vaccine will determine memory B-cell induction for anamnestic mucosal IgA responses.

Encounter with antigen leads to clonal expansion of antigen-specific B cells which develop in a germinal center. It is thought that memory B cells also arise in the germinal centers, including those present in the PP B-cell zones. As described above, these germinal center B cells include memory cells and acquire unique properties, including loss of sIgD and sIgM, and enhanced expression of PNA receptors ($sIgM^-$, $sIgD^-$, PNA^{high} B cells). It is usually assumed that a single B cell can give rise to the progeny for a germinal center and hybridoma analyses have revealed that the entire secondary antibody responses are generated by a small number of clones (~20).[40] This presumably should be sufficient for anamnestic IgA responses, yet this needs to be addressed experimentally. During germinal center responses and memory B-cell generation, switches to other isotypes occur, and for IgA, high frequencies of $sIgA^+$ B cells occur in PP germinal centers.[30] Such switches to IgA would entail $\mu \rightarrow \alpha$ or $\mu \rightarrow \gamma \rightarrow \alpha$ switching, but evidence is lacking to substantiate that antigen-specific switching transpires in the PP. It is known that PP germinal center B cells are $sIgM^+$, PNA^{high}, and κ^{low},[31]; however, significant numbers of $sIgA^+$, PNA^{high} B cells are also located in PP germinal centers.[30,31] The preference for switching to IgA in GALT is thought to be intrinsic to the gut microenvironment.[41]

The immediate sequential series of events following induction of IgA memory cells and $sIgA^+$ B cells in the PP has not been completely studied. It is known that Th cells assist in the induction process for memory B-cell development.[42–44] Consequently, $CD4^+$ T cells are essential for germinal center formation in the PP.(45) Thus, it is plausible to suggest that memory B cells exit germinal centers into T-cell areas enriched in $CD4^+$ T cells.(44) It is known that memory cells do not begin to recirculate until 2–3 weeks after antigen priming and memory B cells (and T cells) may remain in IgA inductive sites such as the PP for longer durations. It has been more difficult to study isolated B- and T-cell populations from the LP of the GI tract. However, recent evidence clearly suggests that the majority of B cells as well as $CD4^+$ and $CD8^+$ T cells which reside in LP exhibit a memory phenotype.(37–39)

2.5. Neuropeptides and Mucosal Immunity

Adhering to current tenets of immunology, humoral and cellular immune responses are self-regulating. The immune system has its own regulatory pathway to govern B- and T-cell responses. While these regulatory pathways are important for delineating the fundamental mechanisms of mucosal immunity, additional mechanisms outside the immune circuit also impact the regulation of the GI immune system, more specifically, the enteric nervous system (ENS).(46) To appreciate this additional means of immune regulation, it is important to understand the extent to which the ENS permeates the GI tract and the expression of neuropeptide receptors on lymphoid cells. The ENS encompasses the same immense area as the GI immune system and in addition to its role in GI physiology the ENS can affect both inductive and effector arms of the mucosal immune system (Fig. 2). This additional regulatory arm suggests that the GI immune system can integrate with other nonimmune components to further regulate mucosal immune function. Although summarizing the roles of more than 18 neuromodulators on immune function is beyond the scope of this discussion, we will briefly address the two predominant neuropeptides found in the GI tract, substance P (SP) and vasoactive intestinal polypeptide (VIP), which permeate throughout the GI tract and exhibit contrasting effects on mucosal lymphocytes.

2.5.1. *Roles for Substance P in the GI Immune System*

SP is a highly conserved, 11-amino-acid neuropeptide(47) belonging to the family of neuropeptides referred to as tachykinins.(48) Each peptide shares a similar C-terminus sequence (Phe-X-Gly-Leu-Met-NH_2) distinguished by amidation of its C-terminus methionine. SP is encoded by preprotachykinin A (PPT-A) gene, which generates three SP mRNAs by alternate RNA splicing mechanisms.(49) Expression of the separate PPT-A mRNAs is both species and tissue dependent.(50) Interestingly, outside the brain, SP is found in greatest concentrations in the gut.(51) SP is a product of the sensory ganglion cells and is transported to peripheral sites where it is stored and released on noxious stimulation.(51) The SP receptor (NK-1R), one of three tachykinins receptors (NK-R), is a G-protein-coupled receptor(52) with seven transmembrane domains and is expressed on lymphocytes.(53) Recent studies demonstrated that the lymphoid and brain NK-1R cDNA are identical.(54,55) Studies from our laboratory demonstrate that the second and third extracellular regions are important for SP binding.(56)

The PP is innervated with SP-containing nerve fibers.(57) These fibers infiltrate the perifollicular T-cell zones and are in close proximity with APCs.(58) Interestingly, these fibers are absent in B-cell follicular areas, suggesting that such avoidance may represent one form of regulation since PP B cells have been shown to express high numbers of NK-1R (~1000/cell), 5-fold higher than levels found on splenic B cells.(59) Evasion of B-cell follicles in PP may prevent the premature differentiation of B cells, since various studies have shown that SP can directly affect B-cell differentiation.(60–62) On the other hand, fiber contact with PP T cells would suggest that SP may be directly involved in cell activation since these lymphocytes also express NK-1R. Such may be the case in that enhanced IgA production (300% increase) was observed when PP mononuclear cells were stimulated with concanavalin A (Con A).(63) Thus, SP may directly stimulate mucosal B cells to promote IgA synthesis and indirectly may influence IgA production through the direct stimulation of Th2-type cells by SP.(53) LP of the small and large intestines are also innervated with SP-containing nerve fibers.(64) The density and distribution of SP-containing fibers varies among mammals.(64) Such anatomical differences may be important when analyzing the type and location of the lymphocytes examined. An interesting experiment is currently ongoing in our group, where we are examining whether SP and lymphoid NK-1R levels in the GI tract change on oral immunization with protein vaccine.

2.5.2. *Roles for Vasoactive Intestinal Polypeptide in the GI Immune System*

VIP is a 28-amino-acid neuropeptide with a C-terminal amidated asparagine,(65) whose sequence, like SPs, is conserved. It is translated as a preprohormone and it undergoes posttranslational processing to obtain VIP.(66) The rat VIP-R has recently been cloned and belongs to the family of G-protein-coupled receptors having seven membrane-spanning domains.(67) Recently, various related receptors which bind VIP with lesser affinity have been cloned.(68) Thus, it appears that there is also a family of VIP-related receptors. While lymphocytes can bind radiolabeled VIP, it is currently unclear which VIP receptor(s) are expressed on lymphocytes. VIP-containing nerve fibers have been shown to enter into the T-cell areas of the PP[69] and presumably they affect $CD4^+$ T cells since $CD8^+$ T cells do not express VIP-R.[70] In contrast to the stimulatory role SP exerts, VIP inhibits mitogen-induced proliferation of murine T cells[63,71,72] apparently through the reduction of IL-2 and not as the result of stimulating $CD8^+$ T cells.[70] Consequently, decreased IgA production is observed in PP cultures containing VIP.[63]

Like SP-containing nerve fibers, VIP-containing fibers infiltrate to varying degrees throughout the LP of the small and large intestines.[64] LP T cells have been shown to respond to VIP stimulation evident by increased IL-5 production[73] and to inhibit Con A-mediated T-cell proliferation.[74] With the recent cloning of the NK-1R and VIP-R, future studies should provide further insight on how SP and VIP may cross-regulate mucosal immune responses.

3. ANTIGEN-SPECIFIC IMMUNE RESPONSES IN THE GI TRACT

3.1. Immune Responses to Soluble Proteins

In order to focus on the induction of antigen-specific Th cell responses and to determine which cytokines are responsible for promoting these responses, antigen priming of mucosal inductive sites was performed. Initial studies with sheep erythrocytes (SRBC) revealed that oral immunization resulted in preferential induction of Th2 cells in PP whereas systemic immunization resulted in primarily Th1 cells in the spleen.[75] Experiments were performed in order to ascertain whether the finding that Th responses are predominantly Th2-dependent in mucosal inductive sites in contrast to systemic responses, which are both Th1- and Th2-dependent, is universal. Oral immunization with cholera toxin (CT) and tetanus toxoid (TT) was tested.[76] In addition, the role of CT as a mucosal adjuvant was assessed. After three oral immunizations with CT or TT at 1-week intervals, the numbers and isotypes of antigen-specific, antibody-producing cells [spot-forming cells (SFC)] were determined from lymphoid suspensions of spleen, PP, and LP lymphocytes by the ELISPOT method. When levels of secreted antigen-specific antibodies were examined, high CT- and TT-specific IgA responses were detected in fecal and serum samples from orally immunized mice.[76,77] High levels of serum anti-CT and anti-TT IgG antibodies were also induced and were predominantly of the IgG1 isotype.[77] TT-specific as well as CT subunit B (CT-B)-specific IgA SFC responses were elevated in the LP of orally immunized mice in contrast to systemically immunized mice.[76–78] Orally administered CT enhanced TT-specific IgA

SFC responses in the LP and secreted IgA in fecal samples.[78] In terms of antigen-specific Th1 and Th2 responses, $CD4^+$ Th cells from PP of mice orally immunized with CT and restimulated *in vitro* with CT-B resulted in enhanced Th2-cell responses as became evident by elevated numbers of IL-4 and IL-5 SFC[76] in comparison with Th1-type cells producing IFN-γ and IL-2. Likewise, increased frequencies of Th2-type responses were detected for splenic $CD4^+$ T-cell cultures from orally immunized mice. However, splenic $CD4^+$ T-cell cultures from mice systemically immunized demonstrated increased numbers of both Th1- and Th2-type cells. Thus, it appears that CT may have a propensity to stimulate Th2-type cells. Increased numbers of TT-specific IL-5 SFC were also induced in PP $CD4^+$ T cells isolated from orally immunized mice with TT and CT as adjuvant when compared to IFN-γ SFC.[76] Thus, these studies show that oral immunization results in enhanced Th2-cell responses in IgA-inductive sites (e.g., PP) resulting in elevated IgA anti-CT and anti-TT SFC responses in GI effector tissues (e.g., LP), whereas splenic SFC responses to CT and TT were largely of IgG and IgA isotypes. Further, CT, when coadministered as adjuvant, may specifically influence the predominant production of Th2-type cells.

3.2. Immune Responses to Infectious Agents

3.2.1. *IgA Antibody-Mediated Protection*

Induction of localized IgA production or passive transfer of immune IgA has been shown to protect mucosal tissues from subsequent infection. The classical manifestation and application of mucosal immune protection is vaccination with the live attenuated Sabin oral polio vaccine highlighting the importance of oral delivery for immune protection.[79] In general, coadministration of appropriate mucosal adjuvant is necessary for the induction of vaccine antigen-specific immune response in mucosa-associated tissues. Thus, as we previously discussed, CT acts as a potent adjuvant for oral immunization with soluble antigen. Likewise, induction of protective IgA to *Toxoplasma gondii* was seen on oral immunization with a cell sonicate when coadministered with CT.[80] The regimen of three oral immunizations over 20 days protected mice against oral challenge when compared to mice immunized without CT or with CT-B. Adjuvant effect of CT could be explained by our new finding that CT may specifically activate Th2 pathway in mucosa-associated tissues, which leads to the induction of high IgA response.

An alternative form of effective antigen delivery is the use of a live vector system. Induction of appropriate immunity could be afforded by the oral administration of live attenuated *S. typhimurium* expressing the fragment C gene of TT.[81] On administration of 500 LD_{50} units of TT, 75% of the immune mice resisted challenge. Subsequent oral boosting induced complete protection when challenged with toxin.[81] Examination of PP Th responses induced by *Salmonella* expression systems has been limited. In most cases, splenocyte responses were examined and from them, mucosal immune responses were extrapolated. Thus, our current effort is focused on the characterization of antigen-specific Th1 and Th2 cells in both IgA inductive (e.g., PP) and effector (e.g., LP of the GI tract and salivary gland) sites of orally immunized mice with TT fragment C gene expressed in *S. typhimurium.* As in the case of expressing *Leishmania major* gp63 surface protein in live attenuated *Salmonella,* a Th1-like response was indicated by the production of IFN-γ and IL-2 and lack of IL-4 in culture supernatants from antigen-stimulated immune spleen cells.[82] A similar Th1 profile is indicated, but not conclusively, with the generation of IgG2a (generally IFN-γ-dependent) antibodies specific for hepatitis B core antigen expressed in *Salmonella.*[83] Future studies using these live attenuated *Salmonella in vivo* expression systems to immunize mucosal tissues should reveal the Th profiles as well as isotype of antigen-specific Ig responses for both systemic and mucosal inductive sites.

Some of the best evidence of passive mucosal immune protection was seen with the "hybridoma backpack" model.[84,85] The subcutaneous implantation of hybridoma cells secreting monoclonal IgA anti-*V. cholerae* LPS antibody protected mice from oral challenge with 100 LD_{50} of *V. cholerae*[84] and similar results were obtained for mice implanted with hybridoma cells producing a monoclonal IgA anti-*S. typhimurium* antibody when they were orally challenged with *S. typhimurium.*[85] While these antibodies were not locally produced, it does show that IgA antibodies in the gut (monoclonal IgA antibodies were detected in gut washings) can protect the host against lethal bacterial infection. Taken together, it is now clear that the induced antigen-specific IgA antibody in external secretions can provide a first line of defense, which is tightly regulated by $CD4^+$ Th-cell subsets and their derived cytokines.

3.2.2. CTL-Mediated Protection

Induction of CTL responses is important for the elimination of virus-infected cells[86,87] and presumably occurs during the course of infection with endogenous processing of viral peptides by its association with MHC class I.[88] CTLs (mostly $CD4^-$ $CD8^+$ T cells) can be induced on immunization with either killed or live influenza virus vaccine.[89,90] These CTLs can be found in mucosal effector tissues such as the epithelium and LP of the GI tract.[91–93] Virus-specific CTLs in the PP and other mucosa-associated tissues can be induced following oral immunization with virus.[94–96] The oral administration of vaccinia virus to rats resulted in the induction of virus-specific CD4⁻ CD8⁺ CTLs in PP and mesenteric lymph nodes within 1 week of oral immunization.[94] This evidence suggests that following enteric immunization, antigen-stimulated CTLs disseminated from PP into mesenteric lymph nodes via the lymphatic drainage.[32] Similar induction was observed following oral administration of reovirus and rotavirus.[95,96] Six days after subsequent oral immunization, a high frequency of virus-specific CTLs was induced in PP and also resulted in the induction of CTL formation in systemic lymphoid tissues, such as the spleen.[96] Collectively, these studies suggest that induction of mucosal CTLs is important for virus clearance and protection of mucosal tissues may also protect against subsequent systemic infection. One mechanism for CTL protection involves lysis of infected cells by the formation of perforin, which forms pores in target cell membrane, and the release of lytic granzymes,[97,98] but there may be other forms of lysis.[99,100] IFN-γ appears to be important for CTL antiviral immunity. $CD8^+$ T cells from influenza virus-infected subjects generate more IFN-γ following *in vitro* stimulation with influenza virus antigens[101,102] and this enhanced IFN-γ production may enhance class I MHC on infected target cells.[103] Furthermore, it was shown that IFN-γ production by $CD4^+$ Th cells is also increased. Recently, the possible relationship between IFN-γ and perforin synthesis has been shown, since preeffector CTL blasts contained perforin mRNA and became functional CTLs in the presence of IFN-γ.[104] Future studies will have to address this relationship for virus-specific immunity. However, although an exact role of IFN-γ in virus-specific T-cell immunity is still not known, it is clear that mRNA expression and secretion of IFN-γ are increased in virus-specific $CD8^+$ CTLs and $CD4^+$ Th1-type cells.

4. SUMMARY

An advantage of mucosal immunization via the oral–GI tract is that one can induce both mucosal IgA and systemic IgG responses. The induction and regulation of IgA responses are tightly controlled by defined groups of cytokines which are products of Th1 and Th2 cells. In addition, another layer of regulation may be provided by numerous neuropeptides (e.g., SP and VIP) which are also expressed in both mucosal inductive and effector tissues. In order to optimize mucosal IgA responses in the GI tract as well as other mucosal effector tissues, one must consider the tripartite regulation offered by Th1/Th2 cells, cytokines, and neuropeptides. In future mucosal vaccine studies, it will be essential to characterize how these three regulatory mechanisms can influence the outcome of mucosal and systemic immune responses induced by oral immunization at both molecular and cellular levels.

ACKNOWLEDGMENTS. Work summarized in this review was supported by U.S. Public Health Service Grants CA 54430, AI 18958, DE 04217, DE 09837, AI 35932, AI 35544, and DE 08228. We thank Sheila D. Shaw for the preparation of this manuscript and Dr. Masafumi Yamamoto for preparation of the figures. (Note: This article was submitted in July, 1994 and may not include references in this area since that time.)

REFERENCES

1. McGhee, J. R., Mestecky, J., Dertzbaugh, M. T., Eldridge, J. H., Hirasawa, M., and Kiyono, H., 1992, The mucosal immune system: From fundamental concepts to vaccine development, *Vaccine* **10:**75–88.
2. Kiyono, H., Bienenstock, J., McGhee, J. R., and Ernst, P. B., 1992, The mucosal immune system: Features of inductive and effector sites to consider in mucosal immunization and vaccine development, *Reg. Immunol.* **4:**54–62.
3. Conley, M. E., and Delacroix, D. L., 1987, Intravascular and mucosal immunoglobulin A: Two separate but related systems of immune defense? *Ann. Intern. Med.* **106:**892–899.
4. Mestecky, J., and McGhee, J. R., 1987, Immunoglobulin A (IgA): Molecular and cellular interactions involved in IgA biosynthesis and immune response, *Adv. Immunol.* **40:**153–245.
5. Brandtzaeg, P., 1989, Overview of the mucosal immune system, *Curr. Top. Microbiol. Immunol.* **146:**13–25.
6. Mestecky, J., and Russell, M. W., 1986, IgA subclasses, *Monogr. Allergy* **19:**277–301.
7. Brandtzaeg, P., 1994, Distribution and characteristics of mucosal immunoglobulin-producing cells, in: *Handbook of Mucosal Immunology* (P. L. Ogra, J. Mestecky, M. E. Lamm, W. Strober, J. R. McGhee, and J. Bienenstock, eds.), Academic Press, San Diego, pp. 251–262.
8. Kett, K., Brandtzaeg, P., Radl, J., and Haaijman, J. F., 1986, Different subclass distribution of IgA-producing cells of human lymphoid organs and various secretory tissues, *J. Immunol.* **136:**3631–3635.
9. Brandtzaeg, P., 1984, Immune functions of human nasal mucosa and tonsils in health and disease, in: *Immunology of the Lung and Upper Respiratory Tract* (J. Bienenstock, ed.), McGraw–Hill, New York, pp. 28–95.
10. Fujihashi, K., McGhee, J. R., Lue, C., Beagley, K. W., Taga, T., Hirano, T., Kishimoto, T., Mestecky, J., and Kiyono, H., 1991, Human appendix B cells naturally express receptors for and respond to interleukin 6 with selective IgA1 and IgA2 synthesis, *J. Clin. Invest.* **88:**248–252.
11. Kilian, M., Mestecky, J., and Russell, M. W., 1988, Defense mechanisms involving Fc-dependent functions of immunoglobulin A and their subversion by bacterial immunoglobulin A proteases, *Microbiol. Rev.* **52:**296–303.
12. Putnam, F. W., Liu, Y. S., and Low, T. L. K., 1979, Primary structure of human IgA1 immunoglobulin. IV. Streptococcal IgA1 protease digestion: Fab and Fc fragment and the complete amino acid sequence of the a1 heavy chain, *J. Biol. Chem.* **254:**2865–2874.
13. Tomasi, T. B., and Czerwinski, D. S., 1968, The secretory IgA systems, in: *Immunological Deficiency Disease in Man* (D. Bergsma, ed.), Volume 4, The National Foundation, New York, pp. 270–275.
14. Krajci, P., Solberg, R., Sanberg, M., Oyen, O., Jahnsen, T., and Brandtzaeg, P., 1989, Molecular cloning of the human transmembrane secretory component (poly-Ig receptor) and its mRNA expression in human tissues, *Biochem. Biophys. Res. Commun.* **158:**783–789.
15. Cornes, J. S., 1965, Number, size, and distribution of Peyer's patches in the human small intestine. Part 1: The development of Peyer's patches, *Gut* **6:**225–229.

16. Cornes, J. S., 1965, Number, size, and distribution of Peyer's patches in the human small intestine. Part 2: The effect of age on Peyer's patches, *Gut* **6:**230–233.
17. Owen, R. L., and Jones, A. L., 1974, Epithelial cell specialization within human Peyer's patches: An ultrastructural study of intestinal lymphoid follicles, *Gastroenterology* **66:**189–203.
18. Owen, R. L., 1977, Sequential uptake of horseradish peroxidase by lymphoid follicle epithelium of Peyer's patch in the normal unobstructed mouse intestine: An ultrastructural study, *Gastroenterology* **72:**440–451.
19. Bockman, D. E., and Cooper, M. D., 1973, Pinocytosis by epithelium associated with lymphoid follicles in the bursa of Fabricius, appendix, and Peyer's patches. An electron microscopic study, *Am. J. Anat.* **136:**455–477.
20. Kato, T., and Owen, R. L., 1994, Structure and function of intestinal mucosal epithelium, in: *Handbook of Mucosal Immunology* (P. L. Ogra, J. Mestecky, M. E. Lamm, W. Strober, J. R. McGhee, and J. Bienenstock, eds.), Academic Press, San Diego, pp. 11–26.
21. Neutra, M. R., Phillips, T. L., Mayer, E. L., and Fishkind, D. J., 1987, Transport of membrane-bound macromolecules by M cells in follicle-associated epithelium of rabbit Peyer's patch, *Cell Tissue Res.* **247:**537–546.
22. Wolf, J. L., Kauffman, R. S., Finberg, R., Dambrauskas, R., Fields, B. N., and Trier, J. S., 1983, Determinants of reovirus interaction with the intestinal M cells and absorptive cells of murine intestine, *Gastroenterology* **85:**291–300.
23. Amerongen, H. M., Weltzin, R., Farnet, C. M., Michetti, P., Haseltine, W. A., and Neutra, M. R., 1991, Transepithelial transport of HIV-1 by intestinal M cells: A mechanism for transmission of AIDS, *J. AIDS* **4:**760–765.
24. Sicinki, P., Rowinski, J., Warchol, J. R., Jarzabek, Z., Gut, W., Szczgiel, B., Bielecki, K., and Koch, G., 1990, Poliovirus type 1 enters the human host through intestinal M cells, *Gastroenterology* **98:**56–58.
25. Weltzin, R. A., Lucia-Jandris, P., Michetti, P., Fields, B. N., and Kraehenbuhl, J. P., 1989, Binding and transepithelial transport of immunoglobulins by intestinal M cells: Demonstration using monoclonal IgA antibodies against enteric viral proteins, *J. Cell Biol.* **108:** 1673–1685.
26. Nagura, H., Ohtani, H., Masuda, T., Kimura, M., and Nakamura, S., 1991, HLA-DR expression on M cells overlying Peyer's patches is a common feature of human small intestine, *Acta Pathol. Jpn.* **41:**818–823.
27. Allan, C. H., Mendrick, D. L., and Trier, J. S., 1993, Rat intestinal M cells contain acidic endosomal–lysosomal compartments and express class II major histocompatibility complex determinants, *Gastroenterology* **104:**698–708.
28. Brandtzaeg, P., and Bjerke, K., 1990, Immunomorphological characteristics of human Peyer's patches, *Digestion* **46:**262–273.
29. Jones, P. P., and Cebra, J. J., 1974, Restriction of gene expression in B lymphocytes and their progeny. III. Endogenous IgA and IgM on the membranes of different plasma cell precursors, *J. Exp. Med.* **140:**966–976.
30. Butcher, E. C., Rouse, R. V., Coffman, R. L., Nottenburg, C. N., Hardy, R. R., and Weissman, I. L., 1982, Surface phenotype of Peyer's patch germinal center cells: Implication for the role of germinal centers in B cell differentiation, *J. Immunol.* **129:**2698–2707.
31. Lebman, D. A., Griffin, P. M., and Cebra, J. J., 1987, Relationship between expression of IgA by Peyer's patch cells and functional IgA memory cells, *J. Exp. Med.* **166:**1405–1418.
32. McGhee, J. R., Mestecky, J., Elson, C. O., and Kiyono, H., 1989, Regulation of IgA synthesis and immune responses by T cells and interleukins, *J. Clin. Immunol.* **9:**175–199.
33. Spalding, D. M., Koopman, W. J., Eldridge, J. L., McGhee, J. R., and Steinman, R., 1983, Accessory cells in murine Peyer's patch: I. Identification and enrichment of functional dendritic cells, *J. Exp. Med.* **157:**1646–1659.

34. Spalding, D. M., Williamson, S. I., Koopman, W. J., and McGhee, J. R., 1984, Preferential induction of polyclonal IgA secretion by murine Peyer's patch dendritic cell–T cell mixtures, *J. Exp. Med.* **160:**941–946.
35. Kanof, M. E., Strober, W., Fiocchi, C., Zeitz, M., and James, S. P., 1988, CD4 positive Leu-8 negative helper-inducer T cells predominate in the human intestinal lamina propria, *J. Immunol.* **141:**3029–3036.
36. Bull, D. M., and Bookman, M. A., 1977, Isolation and functional characterization of human intestinal mucosal lymphoid cells, *J. Clin. Invest.* **59:**966–974.
37. Qiao, L., Schürman, G., Betzler, M., and Meuer, S. C., 1991, Functional properties of human lamina propria T lymphocytes assessed with mitogenic monoclonal antibodies, *Immunol. Res.* **10:**218–225.
38. James, S. P., and Zeitz, M., 1994, Human gastrointestinal mucosal T cells, in: *Handbook of Mucosal Immunology* (P. L. Ogra, J. Mestecky, M. E. Lamm, W. Strober, J. R. McGhee, and J. Bienenstock, eds.), Academic Press, San Diego, pp. 275–285.
39. James, S. P., Kwan, W. C., and Sneller, M. C., 1990, T cells in the inductive and effector compartments of the intestinal mucosal immune system of nonhuman primates differ in lymphokine mRNA expression, lymphokine utilization, and regulatory function, *J. Immunol.* **144:**1251–1256.
40. Blier, P., and Bothwell, A., 1987, A limited number of B cell lineages generates the heterogeneity of a secondary immune response, *J. Immunol.* **139:**3996–4006.
41. Weinstein, P. D., and Cebra, J. J., 1991, The preference for switching to IgA expression by Peyer's patch germinal center B cells is likely due to the intrinsic influence of their microenvironment, *J. Immunol.* **147:**4126–4135.
42. Liu, Y.-J., Johnson, G. D., Gordon, J., and McLennan, I. C. M., 1992, Germinal centres in T-cell-dependent antibody responses, *Immunol. Today* **13:**17–21.
43. Rouse, R., Ledbetter, J., and Weissman, I. L., 1982, Mouse lymph node germinal centers contain a selected subset of T cells: The helper phenotype, *J. Immunol.* **128:**2243–2246.
44. Krall, W. J., and Braun, J., 1992, *In vivo* retroviral marking of antigen-specific B lymphocytes, *Semin. Immunol.* **4:**19–28.
45. Mega, J., Bruce, M. G., Beagley, K. W., McGhee, J. R., Taguchi, T., Pitts, A. M., Bucy, R. P., Eldridge, J. H., Mestecky, J., and Kiyono, H., 1991, Regulation of mucosal responses by $CD4^+$ T lymphocytes: Effect of anti-L3T4 treatment on the gastrointestinal immune system, *Int. Immunol.* **3:**793–805.
46. Furness, J. B., and Costa, M., 1980, Types of nerves in the enteric nervous system, *Neuroscience* **5:**1–20.
47. Chang, M. M., and Leeman, S. E., 1971, Amino acid sequence of substance P, *Nature New Biol.* **232:**86–87.
48. Erspamer, V., 1981, The tachykinin peptide family, *Trends Neurosci.* **4:**267–269.
49. Nawa, H., Kotani, H., and Nakanishi, S., 1984, Tissue-specific generation of two preprotachykinin mRNAs from one gene by alternative RNA splicing, *Nature* **312:**729–734.
50. Helke, C. J., Krause, J. E., Mantyh, P. W., and Bannon, M. J., 1990, Diversity in mammalian tachykinin peptidergic neurons: Multiple peptides, receptors, and regulatory mechanisms, *FASEB J.* **4:**1606–1615.
51. Pernow, B., 1983, Substance P, *Pharmacol. Rev.* **35:**85–140.
52. Hershey, A. D., and Krause, J. E., 1990, Molecular characterization of a functional cDNA encoding the rat substance P receptor, *Science* **247:**958–962.
53. Pascual, D. W., Stanisz, A. M., and Bost, K. L., 1994, Functional aspects of the peptidergic circuit in mucosal immunity, in: *Handbook of Mucosal Immunology* (P. L. Ogra, J. Mestecky, M. E. Lamm, W. Strober, J. R. McGhee, and J. Bienenstock, eds.), Academic Press, San Diego, pp. 203–216.

54. Bost, K. L., 1993, Expression of the mRNA encoding substance P receptors by activated macrophages, *J. Immunol.* **150**(Part II):54A, #297.
55. Cook, G. A., Elliott, D., Metwali, A., Blum, A. M., Sandor, M., Lynch, R., and Weinstock, J. V., 1994, Molecular evidence that granuloma T lymphocytes in murine Schistosomiasis mansoni express an authentic substance P (NK-1) receptor, *J. Immunol.* **152:**1830–1835.
56. van Ginkel, F. W., Bost, K. L., Kiyono, H., McGhee, J. R., and Pascual, D. W., 1994, IL-1α production by the C6 astrocyte cell line is regulated through its substance P (SP) receptor, *FASEB J.* **8:**A247, #1426.
57. Felten, D. L., Felten, S. Y., Bellinger, D. L., Carlson, S. L., Ackerman, K. D., Madden, K. S., Olschowki, J. A., and Livnat, S., 1987, Noradrenergic sympathetic neural interactions with the immune system: Structure and function, *Immunol. Rev.* **100:**225–260.
58. Stead, R. H., Bienenstock, J., and Stanisz, A. M., 1987, Neuropeptide regulation of mucosal immunity, *Immunol. Rev.* **100:**333–359.
59. Stanisz, A. M., Scicchitano, R., Dazin, P., Bienenstock, J., and Payan, D. G., 1987, Distribution of substance P receptors on spleen and Peyer's patch T and B cells, *J. Immunol.* **139:** 749–754.
60. Pascual, D. W., Xu-Amano, J., Kiyono, H., McGhee, J. R., and Bost, K. L., 1991, Substance P acts directly upon cloned B lymphoma cells to enhance IgA and IgM production, *J. Immunol.* **146:**2130–2136.
61. Pascual, D. W., McGhee, J. R., Kiyono, H., and Bost, K. L., 1991, Neuroimmune modulation of lymphocyte function: I. Substance P enhances immunoglobulin synthesis in LPS activated murine splenic B cells, *Int. Immunol.* **3:**1223–1229.
62. Pascual, D. W., Bost, K. L., Beagley, K. W., Kiyono, H., and McGhee, J. R., 1995, Substance P promotes Peyer's patch B and splenic B cell differentiation, in: *Advances in Mucosal Immunology* (J. Mestecky *et al.*, eds.), Plenum Press, New York, pp. 55–59.
63. Stanisz, A. M., Befus, D., and Bienenstock, J., 1986, Differential effects of vasoactive intestinal peptide, substance P, and somatostatin on immunoglobulin synthesis and proliferation by lymphocytes from Peyer's patches, mesenteric lymph nodes, and spleen, *J. Immunol.* **136:**152–156.
64. Keast, J. R., Furness, J. B., and Costa, M., 1985, Distribution of certain peptide-containing nerve fibres and endocrine cells in the gastrointestinal mucosa in five mammalian species, *J. Comp. Neurol.* **236:**403–422.
65. Said, S. I., and Mutt, V., 1970, Polypeptide with broad biological activity: Isolation from small intestine, *Science* **169:**1217–1218.
66. Linder, S., Barkhem, T., Torberg, A., Persson, H., Schalling, M., Hökfelt, T., and Magnusson, G., 1987, Structure and expression of the gene encoding the vasoactive intestinal peptide precursor, *Proc. Natl. Acad. Sci. USA* **84:**605–609.
67. Ishibara, T., Shigemoto, R., Mori, K., Takahashi, K., and Nagata, S., 1992, Functional expression and tissue distribution of a novel receptor for vasoactive intestinal polypeptide, *Neuron* **8:**811–819.
68. Christophe, J., 1993, Type I receptors for PACAP (a neuropeptide even more important than VIP?), *Biochim. Biophys. Acta* **1154:**183–199.
69. Ottaway, C. A., Lewis, D. L., and Asa, S. L., 1987, Vasoactive intestinal peptide-containing nerves in Peyer's patches, *Brain Behav. Immun.* **1:**148–158.
70. Ottaway, C. A., 1987, Selective effects of vasoactive intestinal peptide on the mitogenic response of murine T cells, *Immunology* **62:**291–297.
71. Ottaway, C. A., and Greenberg, F. R., 1984, Interaction of vasoactive intestinal peptide with mouse lymphocytes: Specific binding and the modulation of mitogen responses, *J. Immunol.* **132:**417–423.
72. Krco, C. J., Gores, A., and Go, V. L. W., 1986, Gastrointestinal regulatory peptides

modulate in vitro immune reactions of mouse lymphoid cells, *Clin. Immunol. Immunopathol.* **39:**308–318.
73. Blum, A. M., Mathew, R., Cook, G. A., Metwali, A., Felman, R., and Weinstock, J. V., 1992, Murine mucosal T cells have VIP receptors functionally distinct from those on intestinal epithelial cells, *J. Neuroimmunol.* **39:**101–108.
74. Elitsur, Y., and Luk, G. D., 1990, Gastrointestinal neuropeptides suppress human colonic lamina propria lymphocyte DNA synthesis, *Peptides* **11:**879–884.
75. Xu-Amano, J., Aicher, W. K., Taguchi, T., Kiyono, H., and McGhee, J. R., 1992, Selective induction of Th2 cells in murine Peyer's patches by oral immunization, *Int. Immunol.* **4:**433–445.
76. Xu-Amano, J., Kiyono, H., Jackson, R., Staats, H. F., Fujihashi, K., Burrows, P. D., Elson, C. O., Pillai, S., and McGhee, J. R., 1993, Helper T cell subsets for immunoglobulin A responses: Oral immunization with tetanus toxoid and cholera toxin as adjuvant selectively induces Th2 cells in mucosa associated tissues, *J. Exp. Med.* **178:**1309–1320.
77. Jackson, R. J., Fujihashi, K., Xu-Amano, J., Kiyono, H., and McGhee, J. R., 1993, Optimizing oral vaccines: Induction of systemic and mucosal B cell and antibody responses to tetanus toxoid by use of cholera toxin as adjuvant, *Infect. Immun.* **61:**4272–4279.
78. Xu-Amano, J., Fujihashi, K., Jackson, R., Kiyono, H., and McGhee, J. R., 1994, Helper Th1 and Th2 cell responses following mucosal or systemic immunization with cholera toxin, *Vaccine* **12:**903–911.
79. Ogra, P. L., and Karzon, D. T., 1969, Distribution of poliovirus antibody in serum, nasopharynx, and alimentary tract following segmental immunization of lower alimentary tract with polio vaccine, *J. Immunol.* **102:**1423–1430.
80. Bourguin, I., Chardes, T., and Bout, D., 1993, Oral immunization with *Toxoplasma gondii* antigens in association with cholera toxin induces enhanced protective and cell-mediated immunity in C57BL/6 mice, *Infect. Immun.* **61:**2082–2088.
81. Fairweather, N. F., Chatfield, S. N., Makoff, A. J., Strugnell, R. A., Bester, J., Maskell, D. J., and Dougan, G., 1990, Oral vaccination of mice against tetanus by use of a live attenuated *Salmonella* carrier, *Infect. Immun.* **58:**1323–1326.
82. Yang, D. M., Fairweather, N., Button, K. L., McMaster, W. R., Kahl, L. P., and Liew, F. Y., 1990, Oral *Salmonella typhimurium* (AroA$^-$) vaccine expressing a major leishmanial surface protein (gp63) preferentially induces T helper 1 cells and protective immunity against leishmaniasis, *J. Immunol.* **145:**2281–2285.
83. Schodel, F., Milich, D. R., and Will, H., 1990, Hepatitis B virus nucleocapsid/pre-S2 fusion proteins expressed in attenuated *Salmonella* for oral vaccination, *J. Immunol.* **145:**4317–4321.
84. Winner, L., Mack, J., Weltzin, R., Mekalanos, J. J., Kraehenbuhl, J. P., and Neutra, M. R., 1991, New model for analysis of mucosal immunity: Intestinal secretion of specific monoclonal immunoglobulin A from hybridoma tumors protects against *Vibrio cholerae* infection, *Infect. Immun.* **59:**977–982.
85. Michetti, P., Mahan, M. J., Slauch, J. M., Mekalanos, J. J., and Neutra, M. R., 1992, Monoclonal secretory immunoglobulin A protects mice against oral challenge with the invasive pathogen *Salmonella typhimurium*, *Infect. Immun.* **60:**1786–1792.
86. Yap, K. L., Ada, G. L., and McKenzie, I. F. C., 1978, Transfer of specific cytotoxic T lymphocytes protects mice inoculated with influenza virus, *Nature* **273:**238–239.
87. Zinkernagel, R. M., and Doherty, P. C., 1979, MHC-restricted cytotoxic T cells: Studies on the biological role of polymorphic major transplantation antigens determining T-cell restriction-specificity, function, and responsiveness, *Adv. Immunol.* **27:**51–177.
88. Marrack, P., and Kappler, J., 1987, The T cell receptor, *Science* **238:**1073–1079.
89. Ennis, F. A., Yi-Hua, Q., and Schild, G. C., 1982, Antibody and cytotoxic T lymphocyte responses of humans to live and inactivated influenza vaccines, *J. Gen. Virol.* **58:**273–281.

90. McMichael, A. J., and Askonas, B., 1978, Influenza virus-specific cytotoxic T cell in man: Induction and properties of the cytotoxic cell, *Eur. J. Immunol.* **8**:705–711.
91. Davies, M. D. J., and Parrott, D. M. W., 1981, Cytotoxic T cells in small intestine epithelial, lamina propria and lung lymphocytes, *Immunology* **44**:367–371.
92. Nauss, K. M., Pavlina, T. M., Kumar, V., and Newberne, P. M., 1984, Functional characteristics of lymphocytes isolated from the rat large intestine: Response to T-cell mitogens and natural killer cell activity, *Gastroenterology* **86**:468–475.
93. Ernst, P. B., Befus, A. D., and Bienenstock, J., 1985, Leukocytes in the intestinal epithelium: An unusual immunological compartment, *Immunol. Today* **6**:50–55.
94. Issekutz, T. B., 1984, The response of gut-associated T lymphocytes to intestinal viral immunization, *J. Immunol.* **133**:2955–2960.
95. London, S. D., Rubin, D. H., and Cebra, J. J., 1987, Gut mucosal immunization with reovirus serotype 1/L stimulated virus-specific cytotoxic T cell precursors as well as IgA memory cells in Peyer's patches, *J. Exp. Med.* **165**:830–847.
96. Offit, P. A., and Dudzik, K. I., 1989, Rotavirus-specific cytotoxic T lymphocytes appear at the intestinal mucosal surface after rotavirus infection, *J. Virol.* **63**:3507–3512.
97. Podack, E. R., and Kupef, A., 1991, T cell effector functions: Mechanisms for delivery of cytotoxicity and help, *Annu. Rev. Cell Biol.* **7**:479–504.
98. Young, L. H. Y., Klavinskis, L. S. D., Oldstone, M. B. A., and Young, J. D. E., 1989, *In vivo* expression of perforin by $CD8^+$ lymphocytes during an acute viral infection, *J. Exp. Med.* **169**:2159–2171.
99. Liu, C.-C., Steffen, M., King, F., and Young, J. D. E., 1987, Identification, isolation and characterization of a novel cytotoxin in murine cytolytic lymphocytes, *Cell* **51**:393–403.
100. Trenn, G., Takayama, H., and Sitkovsky, M. V., 1987, Exocytosis of cytolytic granules may not be required for target cell lysis by cytotoxic T-lymphocytes, *Nature* **330**:72–74.
101. Taylor, P. M., Wraith, D. C., and Askonas, B. A., 1985, Control of immune interferon release by cytotoxic T-cell clones specific for influenza, *Immunology* **54**:607–614.
102. Yamada, Y. K., Meager, A., Yamada, A., and Ennis, F. A., 1986, Human interferon alpha and gamma production by lymphocytes during the generation of influenza virus-specific cytotoxic T lymphocytes, *J. Gen. Virol.* **67**:2325–2334.
103. Wong, G. H., Bartlett, P. F., Lewis Clark, I., and McKimm-Breschkin, J.-L., 1985, Interferon-gamma induces the expression of H-2 and Ia antigens on brain cells, *J. Neuroimmunol.* **7**:255–278.
104. Nelson, P. J., Geller, R. L., Podack, E., and Bach, F. H., 1992, Molecular events in late stage of T-cell functional maturation, *Scand. J. Immunol.* **35**:311–320.

3

Vibrio cholerae

Molecular Pathogenesis, Immune Response, and Vaccine Development

MATTHEW K. WALDOR and JOHN J. MEKALANOS

1. INTRODUCTION

The curved, gram-negative bacterium *Vibrio cholerae* is the cause of cholera, a severe and potentially lethal diarrheal disease. The disease has been recognized since ancient times and descriptions of cholera can be found in Sanskrit texts dating back to 500 B.C.[1] Besides its often dramatic clinical presentation, two epidemiologic features distinguish cholera from other diarrheal diseases. These are the tendency of cholera to appear in explosive outbreaks and its predilection for pandemic spread. The explosiveness of a cholera outbreak was sadly witnessed in 1994 with the massive death counts in the refugee camps of Goma, Zaire.

History has recorded seven cholera pandemics since 1817. All of the cholera pandemics began in Asia and subsequently spread to other continents. The first six pandemics of cholera were probably caused by the classical biotype of *V. cholerae*. The El Tor biotype of *V. cholerae* has given rise to the seventh pandemic of cholera. This pandemic began on the Indonesian island of Celebes in 1961 and continues today. The seventh pandemic reached Peru in 1991 and subsequently has given rise to more than a million cases of cholera throughout South America. All seven of the cholera pandemics are believed to have been caused by the O1 serogroup of *V. cholerae*. This has been the only *V. cholerae* serogroup associated with epidemic chol-

MATTHEW K. WALDOR and JOHN J. MEKALANOS • Department of Microbiology and Molecular Genetics, Harvard Medical School, Boston, Massachusetts 02115.

Enteric Infections and Immunity, edited by Lois J. Paradise *et al.* Plenum Press, New York, 1996.

era. However, in late 1992, massive cholera outbreaks occurred in India and Bangladesh that were caused by a previously unrecognized serogroup of *V. cholerae.* This novel serogroup has been designated *V. cholerae* O139[2] and has now spread outside of the Indian subcontinent giving rise to the eighth cholera pandemic.

This chapter will focus on the molecular pathogenesis of cholera with particular emphasis on the bacterial virulence factors that are crucial for infection and disease. The application of some of the knowledge gained from studies of the virulence factors of *V. cholerae* and the host immune response to *V. cholerae* toward the development of live attenuated *V. cholerae* vaccines will also be discussed. More comprehensive reviews can been found in two recent books on cholera.[3,4]

2. SEROLOGIC AND BIOTYPIC CLASSIFICATION

V. cholerae are oxidase-positive, gram-negative, facultative anaerobes. A variety of selective media and biochemical tests can be used to identify *V. cholerae.*[5] *V. cholerae* cells bear a single sheathed polar flagellum and are highly motile in wet mount preparations.

A serotyping scheme for *V. cholerae* has been developed by Shimada and colleagues.[6] There are now over 140 designated serogroups of *V. cholerae*[6] The chemical basis for the serogroup antigens is the O side chain of lipopolysaccharide (LPS) molecules (the O-antigen). Prior to the identification of *V. cholerae* O139, only O1 serogroup strains were associated with epidemic cholera. Non-O1 strains (other than *V. cholerae* O139) have given rise to sporadic cases of diarrhea as well as to extraintestinal diseases including wound infections but not to epidemic cholera.[7] Except for *V. cholerae* O139, the great majority of non-O1 strains do not contain the genes encoding cholera toxin, the principal virulence factor responsible for the secretory diarrhea that is the hallmark of cholera.

The O1 serogroup has been subdivided into three serotypes designated Inaba, Ogawa, and Hikojima based on the epitopes found on the O-antigen of the LPS. Serotype conversion has been shown to occur during the course of epidemics. Thus, the initial isolates in the current Latin American epidemic were all Inaba strains, but in less than a year after the outbreak began, most of the isolates were of the Ogawa serotype.[8] Manning and his coworkers cloned the *V. cholerae* operon encoding O-antigen biosynthesis (the *rfb* operon) and found that Inaba strains contain mutations in the *rfbT* gene resulting in truncated, inactive rfbT proteins.[9] Generation of an Inaba-to-Ogawa serotype shift therefore requires a mutation that restores the complete reading frame of the *rfbT* gene.

V. cholerae O1 are divided into two biotypes, classical and El Tor. Although the first six cholera pandemics were probably caused by classical strains of *V. cholerae*, in the past decade classical strains have only been isolated in Bangladesh.(10) The two biotypes have been traditionally distinguished on the basis of five tests: hemolysis of sheep erythrocytes, hemagglutination of chicken erythrocytes, polymyxin B sensitivity, bacteriophage sensitivity, and the Voges–Proskauer reaction. Various restriction fragment polymorphisms have now been identified which can also be used to distinguish the two *V. cholerae* biotypes.(11,12) Many studies have now demonstrated that *V. cholerae* O139 is closely related to El Tor O1 strains.(13,16) This new epidemic serogroup arose from an El Tor O1 strain by recombination event(s) which resulted in at least a partial deletion of the O1 *rfb* region along with an insertion of DNA encoding the novel serogroup antigen into this locus.(17–19)

3. PATHOGENESIS OF CHOLERA

Contaminated food or water are the sources of *V. cholerae* that give rise to cholera in the human host.(20) The ingested dose of *V. cholerae* must be large enough to allow the bacterium to pass through the gastric acid barrier to reach the small intestine where it multiples. Host factors (e.g., gastric acidity and O blood group, see below) as well as bacterial factors (e.g., ingested dose) are important in the development of the disease. Little is known about bacterial factors that may be important in the spread of *V. cholerae* in populations. Such factors might include gene products that facilitate the bacterium's survival in the environment or that make the organism more infectious. In contrast, there is considerably more known about the *V. cholerae* virulence factors that lead to bacterial colonization of the small intestine and the genesis of diarrhea. This review will focus on these bacterial virulence factors.

4. CHOLERA TOXIN AND OTHER TOXINS

The most important bacterial factor leading to the development of the severe secretory diarrhea that is the hallmark of cholera is cholera toxin (CT). The central role CT plays in the genesis of diarrhea was dramatically demonstrated in human volunteer studies conducted by Levine and colleagues. These workers fed purified CT to volunteers and found that as little as 5 μg of purified toxin led to severe diarrhea in the volunteers.(21)

CT is an A–B-type toxin. The B subunit binds the toxin to the eukaryotic cell receptor, the ganglioside G_{M1}, and the A subunit functions intracellularly as an enzyme to catalyze the transfer of the ADP-ribose moiety from NAD to the $G_{s\alpha}$ protein.[22] The ADP-ribosylation of $G_{s\alpha}$ leads to the activation of adenylate cyclase with subsequent increases in intracellular cAMP concentrations. Elevated cAMP levels in turn lead to increased Cl^- secretion by intestinal crypt cells and decreased absorption by villus cells. The precise intracellular pathway whereby raised cAMP levels lead to diarrhea is not completely understood. Interestingly, the CFTR gene, which is mutated in cystic fibrosis patients, is a cAMP-regulated chloride channel and this protein may be one of the targets of the cascade of events triggered by CT. In support of this possibility, it has recently been shown that mice that contain a truncated CFTR protein do not secrete fluid in response to CT.[23] All of the secretory effects of CT may not be attributable to increased levels of cAMP; rather, there is evidence suggesting that prostaglandins and other products of arachidonic acid metabolism play a role in CT-induced diarrhea.[24] Also, CT may influence the function of the autonomic nervous system which innervates the intestine and can regulate intestinal secretion.[25]

The genes encoding the A and B subunits of CT are part of the *ctxAB* operon.[26–28] This operon is part of a larger genetic element which has been termed the CTX genetic element.[29] The CTX genetic element consists of a core region flanked by one or more directly repeated copies of RS sequences. The core region is composed of the *ctxAB* operon along with *zot*, which encodes the zonula occludens toxin,[30] *ace*, which encodes accessory cholera enterotoxin,[31] and *cep*, which encodes a pilinlike intestinal colonization factor[29] (see below). The RS sequences encode a site-specific recombination system that mediates the recombination between the CTX genetic element and an 18-bp target sequence on the *V. cholerae* chromosome termed attRS1.[29] Thus, as a whole the CTX genetic element can be thought of as a site-specific virulence transposon. The fact that the genes encoding CT are part of a mobile genetic element has important implications for vaccine development (see below).

Although CT plays a critical role in the pathogenesis of the diarrhea induced by *V. cholerae*, it does not explain all of the diarrhea or associated symptoms that accompany infection with *V. cholerae*. This was clearly demonstrated in studies of prototype live oral *V. cholerae* vaccine strains that contained deletions of *ctxA* or *ctxAB*. In these studies, volunteers who were administered these oral vaccine candidates experienced a high frequency of "reactogenicity," namely, symptoms of mild diarrhea, nausea, malaise, and anorexia.[32] The residual reactogenicity of these *ctx* deletion strains is not completely understood. It has led to the hypotheses that *V. cholerae* may possess additional toxic activities or alternatively that some aspect of the bacterial colonization process may by itself lead to symptoms.[33]

Besides CT, other *V. cholerae* toxins have been described. Two of these other toxins are encoded in the core region of the CTX genetic element. Fasano and colleagues described Zot, a toxin that alters the structure of epithelial intercellular tight junctions and thereby may increase intestinal permeability.[(30)] Zot is encoded by a gene located immediately upstream from *ctxAB* and the sequence of *zot* shows some homology to an ATPase,[(34)] but the molecular mechanism of Zot remains to be determined. Trucksis and colleagues described Ace, a toxin that increases short-circuit current in Ussing chambers perhaps by stimulating chloride secretion and leads to fluid secretion in rabbit ileal loops.[(31)] The gene encoding Ace is located immediately upstream of *zot* in the core region of the CTX element. The sequence of *ace* showed significant homology to eukaryotic ion-transporting ATPases and to the δ toxin of *Staphylococcus aureus*.[(31)] This homology led Trucksis *et al.* to propose that multimers of Ace might insert into eukaryotic membranes and establish an ion channel,[(31)] but this interesting hypothesis awaits verification. However, the toxic activities encoded by *zot* and *ace* cannot explain the residual reactogenicity that was observed in humans who were fed live prototype *V. cholerae* vaccine strains containing deletions of *ctxAB*. Vaccine constructs that contained deletions of the entire CTX genetic element including all of the core-encoded toxins still led to significant reactogenicity in volunteers.[(35,36)]

V. cholerae also produce a hemolysin.[(37)] The purified hemolysin can induce fluid accumulation in rabbit ileal loops.[(38)] However, strains that are deleted for *hlyA*, which encodes the hemolysin, along with *ctx*, still lead to reactogenicity in human volunteers,[(32)] indicating that this hemolysin, like Zot and Ace, is not the cause of the residual reactogenicity seen in *ctx* deletion strains. *V. cholerae* has also been reported to produce other toxins including a Shiga-like toxin,[(39)] and "new cholera toxin."[(40)] Molecular characterization of the latter two toxins has not been accomplished.

5. COLONIZATION FACTORS

A broad definition of *V. cholerae* colonization factors includes all bacterial gene products that specifically facilitate either the survival or the multiplication of the bacterium in the small intestine. This definition includes many bacterial factors and properties that include but are not restricted to those that mediate the attachment of vibrios to the small intestinal epithelium. Most of the bacterial properties, products, and metabolic pathways that impart on *V. cholerae* the unusual capability to colonize the small intestine have not yet been delineated. CT itself may play some role in facilitating intestinal colonization.[(41)] Early studies of colonization suggested that bacterial motility and chemotaxis played a role in facilitating penetration of the intestinal mucus by *V. cholerae*.[42] A number of specific bacterial gene products have now been described that play important roles in colonization.

TCP, the toxin co-regulated pili, are the best-characterized colonization factors of *V. cholerae*.[43] The pilus is composed of a 20.5-kDa pilin subunit which is encoded by *tcpA*. Sequence analysis of *tcpA* shows that it is a member of the type IV family of bacterial pilins. At least 13 genes are required for the secretion and assembly of TCP.[44] Most of these genes together with *tcpA* are linked on the *V. cholerae* chromosome in the *tcp* gene cluster.[43,45,46] Another unlinked locus *tcpG*, which encodes an enzyme that catalyzes disulfide bond formation, is required for the elaboration of functional pili.[17] Classical *V. cholerae* strains that do not elaborate normal TCP are more than 500-fold reduced in their capacity to colonize the infant mouse[43] and showed drastically reduced colonization of human volunteers as well.[48] There has been some controversy as to whether TCP are colonization factors for El Tor strains; however, two groups have now constructed El Tor strains with knockouts of *tcpA* and have clearly demonstrated that TCP are essential colonization factors for this biotype as well as the classical biotype.[49,50]

The mechanism by which TCP mediate colonization of the small intestine by *V. cholerae* has not been elucidated. Although it seems likely that some part of the pili function as an adhesin for enterocytes or to intestinal mucus, to date no such binding has been clearly established. It remains a possibility that TCP act as a colonization factor without being an adhesin *per se*. For example, TCP are known to aggregate into "bundles" and the expression of TCP causes autoagglutination of *V. cholerae* cultures. This clumping behavior might reflect a similar role for TCP in the establishment of microcolonies on the intestinal epithelium.[43] It is also known that strains that do not make TCP are exceptionally sensitive to the bactericidal effects of serum.[51] Thus, it is possible that TCP function to enhance bacterial resistance to some complement-like bactericidal factor expressed in the intestine.[51]

Other *V. cholerae* gene products coordinately regulated with TCP and CT are known to effect intestinal colonization. For example, Peterson and Mekalanos identified a cluster of four genes linked to the *tcp* cluster that encode accessory colonization factors (*acfA*, *acfB*, *acfC*, and *acfD*).[52,53] Strains with mutations in *acf* genes are not as severely defective in colonization of the suckling mouse intestine as strains with mutations in *tcpA*.[53] The exact function of the *acf* genes in colonization is not known. AcfD has been shown to be a lipoprotein.[51] Sequence analysis of *acfB* demonstrated that this gene bears significant homology to genes that encode methyl-accepting chemotaxis proteins. Members of this class of proteins act as sensors in signal transduction pathways that regulate bacterial chemotaxis. This observation lends credence to the notion that has been proposed and explored by Gardel and Mekalanos that bacterial chemotaxis is important for colonization while motility *per se* is not required for colonization.[54] Waldor and

Mekalanos have recently identified other genes coordinately regulated with CT and TCP which encode additional intestinal colonization factors but these genes have not been fully characterized (unpublished observations).

In addition to the toxins encoded in the core of the CTX genetic element, the core region also encodes an 82-amino-acid protein, designated the core-encoded pilus (cep), which shows striking homology to the flexible pilus of *Aeromonas hydrophila.*[29] It has been proposed that this protein may be a subunit of a pilus colonization factor,[29] although this putative pilus structure has so far eluded detection.

At least four hemagglutinins (HA) have been identified in *V. cholerae* and these have been implicated in the colonization process. The cell-associated D-mannose-, L-fucose-resistant HA (MFRHA) has been cloned and one strain with an insertional mutation in the gene encoding this HA showed markedly reduced colonization of the infant mouse.[55] As is the case for TCP, the precise role of the MFRHA during the course of infection is not known. Cloning of the gene encoding a *V. cholerae* mannose-sensitive hemagglutinin, *mshA*, has recently been reported.[56] MshA, like TcpA, is a type 4 pilin.[56] Colonization studies of strains with mutations in *mshA* have not yet been reported. The gene (*hap*) for a soluble hemagglutinin/protease (HA/protease) has also been cloned, but strains with null mutations in *hap* exhibited wild-type virulence in the rabbit model.[57] It has been suggested that the HA/protease might function as a "detachase" to enable vibrios to become free of the intestinal epithelium during the course of infection.[57]

Iron scavenging mechanisms are presumably important for the growth of *V. cholerae* within the small intestine. Payne and colleagues have constructed a strain with a knockout in both iron siderophore synthesis and heme utilization and found that this double mutant was severely defective in colonization.[58] Low concentrations of iron are known to increase the transcription of several genes in *V. cholerae* and at least one of these iron-regulated genes, *irgA*, apparently encodes a colonization factor.[59] The function of the outer membrane protein IrgA is not known and recent evidence suggests that the iron-regulated promoter controlling *irgA* expression may not be fully induced during gastrointestinal infections in animals.[60]

Rough mutants of *V. cholerae* O1 which do not synthesize the O side chain of LPS are severely defective in the colonization of the suckling-mouse intestine.[9,18] The O-antigen itself may be important in colonization through interaction with host lectins or by conferring resistance to intestinal bactericidal activities or, alternatively, by facilitating the assemble and/or secretion of bacterial surface structures like TCP. Unlike *V. cholerae* O1, *V. cholerae* O139 expresses a capsular polysaccharide, and it has recently been demonstrated that this capsule is also a colonization factor.[18]

Other possible colonization factors that have not yet been characterized with defined knockout mutations include the major ToxR-regulated outer membrane protein OmpU, the genes required for the secretion of CT,[(61)] other *V. cholerae* fimbria besides TCP,[(62)] and biosynthetic genes involved in the synthesis of compounds that are presumably hard to scavenge in the intestine (e.g., purine biosynthetic genes).[(63)]

Table I lists some of the factors and properties of *V. cholerae* (discussed above) that are required for survival and multiplication of the organism in the small intestine. Certainly this is not a complete list. New colonization factors will continue to be discovered. Future research must define in more molecular terms how most of these colonization factors work. This includes determining when and where in the infection cycle a particular colonization factor is expressed. The use of the genetic method recently described by Camilli and Mekalanos,[(60)] which employs the activity of resolvase to irreversibly mark expressed genes, should facilitate this study. Also, definition of the eukaryotic molecules that interact with the *V. cholerae* colonization factors must be identified and the interactions characterized.

6. REGULATION OF VIRULENCE GENE EXPRESSION

V. cholerae was one of the initial bacterial species in which the regulation of virulence gene expression was observed. Many of the virulence genes in *V. cholerae* including *ctx*, *tcp*, and *acf* are regulated by ToxR, a transmembrane DNA binding protein.[(64)] ToxR acts as transcriptional activator for *ctxAB* as well as *toxT*, another transcriptional activator. ToxT activates the transcription of most of the other genes in the ToxR regulon. There is therefore a regulatory cascade controlling the expression of many of the virulence factors in *V. cholerae.*[(65)] The genes in this regulon are coordinately regulated

TABLE I
Gene Products and Properties of *Vibrio cholerae* Required for Survival and Growth in the Small Intestine

Toxin co-regulated pilus
Accesory colonization factor
Mannose fucose resistant hemagglutinin
IrgA
LPS/capsular polysaccharide
Iron scavenging/utilization system
Chemotaxis

by similar environmental signals in the laboratory. It remains to be demonstrated whether or not ToxR is actually the "sensor" in this signal transduction pathway leading from environmental conditions external to the bacterium to the activation of transcription of virulence genes. Other regulatory systems such as the maltose regulon[66] have been shown to have some influence over the expression of the ToxR regulon. Besides *toxR*, *fur*, and *hlyU*, two other transcriptional regulators are known to influence virulence gene expression in *V. cholerae*.[67,68]

7. HOST SUSCEPTIBILITY

Host factors play an important role in susceptibility both to *V. cholerae* infection and to the expression of disease (reviewed in Ref. 69). In endemic regions it is believed the majority of people who develop transient *V. cholerae* infections do not develop symptoms of cholera. For the human host, susceptibility factors have not been well defined, but epidemiologic data and experimental data from volunteer studies indicate that the following factors play a role in susceptibility to infection and disease.

1. Blood type O has been found to be correlated with the development of more severe cholera both in endemic areas and in volunteer studies.[70,71] This correlation may only be valid for the El Tor biotype and not the classical biotype of *V. cholerae*.[71] The molecular basis for this correlation is not known but we can speculate that it might reflect increased adherence of *V. cholerae* to the intestinal tract of O blood group individuals.
2. Elevated stomach pH with the concomitant lowering of the gastric acid barrier to infection. Since *Helicobacter pylori* can result in chronic gastritis with hypochlorhydria, chronic infection with this organism, which is extremely common in areas where cholera is endemic, may turn out to be an important risk factor for the development of cholera.[69]

Studies by Richardson and co-workers have demonstrated a genetic control over intestinal fluid accumulation in mice in response to CT. The phenotype of enhanced sensitivity to CT has been shown to correlate with enterocyte ganglioside expression, and the gene(s) regulating this expression have been mapped to chromosome 17 near *H-2*, the murine major histocompatibility complex.[69]

8. IMMUNE RESPONSE

In pioneering studies, Levine and colleagues have demonstrated that volunteers experimentally infected with *V. cholerae* generate long-lasting protective immunity against cholera. Volunteers experimentally infected with wild-type strains of either biotype of *V. cholerae* showed 90–100% protection against subsequent challenge with the same biotype organism[70] up to 3 years after the initial experimental infection.[70] It is conceivable that conclusions concerning immunity to *V. cholerae* from experimental *V. cholerae* infections would not apply to endemically acquired *V. cholerae* infections. The epidemiologic data with regard to immunity to cholera following infection in endemic regions are not as clear as the data from volunteer studies. Glass *et al.* concluded from studies of data from Bangladesh that an initial clinically apparent case of cholera led to an approximately 90% reduction in the risk of a second clinically apparent case of cholera.[72] In another study of data from Bangladesh, Clemens *et al.* found that clinically significant cholera was only associated with a 61% reduction in the rate of subsequent cholera and an initial infection with an El Tor strain engendered significantly less protection than did an initial infection with a classical strain.[73]

The targets and mechanisms of the protective immune response against *V. cholerae* are incompletely understood. It is generally agreed that the antibacterial response is a more important component of the protective immune response to *V. cholerae* than antitoxic immunity. The major component of the antibacterial response is believed to be against the O-antigen of LPS.[70] This notion is supported by the emergence of the *V. cholerae* O139 epidemic. This epidemic occurred in the adult population in a cholera-endemic region where there is significant antitoxic and antibacterial immunity to *V. cholerae* O1 in adults. Although the O-antigen of LPS seems to be the major target of the antibacterial immune response, other *V. cholerae* antigens are also targets. Other antigens that may be important in the antibacterial response include outer membrane proteins, hemagglutinins, flagellar proteins, heat-shock proteins and antigens which are only expressed *in vivo* (reviewed in Ref. 74).

Not all colonization factors are immunogenic. For instance, TCP, which is probably the most important colonization factor of *V. cholerae*, does not engender a significant antibody response following infection.[75] The relative lack of immunogenicity of TCP may constitute a strategy of the bacterium to avoid clearance by the host immune response. Although TCP is not particularly immunogenic after natural infection, antiserum specifically raised against purified TCP is protective in the suckling mouse model.[76] Thus, the lack of immunogenicity of a particular protein during natural infection does not mean that the protein cannot be the target of a protective immune response. Hopefully, future studies will define the rules that govern antigenicity for the enteric mucosal immune response and then rational manipulation of this response may be achievable.

The most significant mechanism of protective immunity against *V. cholerae* is believed to be secretory IgA (sIgA) molecules.[77] Some experimental verification for this has come from studies with "backpack" tumors. In this model, neonatal mice that were subcutaneously injected with a hybridoma secreting a monoclonal IgA anti-Ogawa LPS antibody were protected from *V. cholerae* infection.[78] The mechanisms of protection by sIgA are not certain but probably include bacterial agglutination with concomitant prevention of bacterial adherence to enterocytes. Antibody-dependent cellular cytotoxicity may also play some role.[78] Although *V. cholerae* infection is limited to the intestine, serum vibriocidal antibodies have been shown to be a useful surrogate for predicting protective immunity to *V. cholerae*.[70] Some intestinal colonization seems to be crucial for effective immunization against cholera. Thus, administration of a noncolonizing *tcpA* mutant to human volunteers did not result in the generation of a vibriocidal antibody response.[48] Future studies elucidating the molecular mechanisms of immunity to cholera should provide important and much needed basic understanding of the mechanisms whereby the enteric immune response protects the intestinal surface.

Although the immune response to CT does not seem to be the most important component of the protective immune response against cholera, there is evidence of synergy between the antitoxic and antibacterial immune responses to *V. cholerae* (reviewed in Ref. 70). CT has also been shown to exert a powerful regulatory effect on the host immune response. CT is one of the most potent oral adjuvants known (reviewed in Ref. 79). Co-oral administration of CT with a variety of other antigens has been shown to augment substantially the sIgA response to the coadministered antigen. There is some controversy as to whether the catalytic activity of the A subunit is required for this adjuvant effect. It has been shown that a single amino acid substitution in the A subunit that destroys the toxic activity of the molecule also abrogates its adjuvant activity for coadministered KLH.[80] However, it has also been shown that chemical coupling of an antigen to the nontoxic B subunit of CT can lead to enhanced mucosal immune response to the coupled antigen.[81] This effect is presumably related to the targeting of the coupled antigen to the G_{M1} receptors for the B subunit found on enterocytes and M cells, the specialized intestinal antigen-sampling cells.[82] Recently, it has also been shown that oral administration of antigens coupled to the B subunit can induce suppression of the systemic delayed-type hypersensitivity response to the coupled antigen.[83] The molecular bases for the immunoregulatory effects are not clear but CT has been shown to promote isotype switching toward IgA,[79] to enhance antigen presentation,[79] and to induce a Th2 T-cell subset response in Peyer's patches.[84]

9. VACCINE DEVELOPMENT

The basic and clinical research that has elucidated many aspects of cholera pathogenesis has made the goal of developing a safe and effective cholera vaccine realizable. Work toward the development of cholera vaccines began more than a century ago. The early parenteral killed whole-cell (WC) vaccine candidates have been largely abandoned because these vaccines induce only weak and short-lived immunity. Presumably this is because parenteral immunization does not generate an adequate local immune response where it is required, that is, on the intestinal mucosal surface. Attention has now focused on the development of oral cholera vaccines to take advantage of the strong mucosal immune response that is elicited in the intestine following exposure to *V. cholerae*. Both live attenuated *V. cholerae* strains and killed whole bacterial cells are being pursued as oral vaccines.

A killed WC oral vaccine formulation has been subjected to a large placebo-controlled trial in rural Bangladesh.[85] Participants in this large field trial received killed whole cells alone or in combination with the B subunit of CT (BS-WC) or placebo. In the first 6 months, the BS-WC vaccine provided a substantially greater rate of protection from cholera (85%) compared to the WC vaccine (58%). However, this advantage disappeared after the initial 6 months of the trial. At the 3-year follow-up point, both the BS-WC and WC vaccine formulation showed approximately equal protection rates of 50%.[86] The BS-WC and WC vaccine formulations were not nearly as effective in children. Limitations of the killed WC oral vaccine include: the requirement for multiple vaccine doses, a relatively expensive production process, lower protection in children and individuals of the blood group O, the reduced uptake by M cells of killed bacteria compared to live bacteria,[87] and the inability of killed bacteria to engender immune responses to bacterial antigens that are only expressed in the intestine.

Recombinant DNA technology eventually led to the construction of genetically stable, live attenuated *V. cholerae* vaccines about a decade ago.[27,28] The goal of this approach was to reproduce as much as possible the highly immunizing effects of natural infection without its toxic consequences. Since CT is largely responsible for the diarrheal symptoms caused by *V. cholerae* infection, deletion or alteration of the gene encoding the toxic A subunit of CT was believed to be the fundamental genetic alteration of *V. cholerae* that was required to attenuate the bacterium.

The first recombinant live oral vaccines tested in volunteers (JBK70, which expresses neither the A nor B subunit of CT, and CVD101, which only expresses the nontoxic B subunit of CT) were both effective immunogens. However, some of the volunteers who received these vaccines developed mild to severe diarrhea, cramps, nausea, and anorexia.[70] The explanation for this "reactogenicity" is not known. Searches for residual toxic activities in these *V. cholerae* strains that can account for the reactogenicity have not been successful (see above).

The first live attenuated cholera vaccine that was well tolerated but immunogenic was CVD-103, a *ctxA* deletion in the classical strain 569B.[70] A derivative of this vaccine candidate, CVD-103-HgR, has undergone extensive trials (reviewed in Ref. 70) where it has proved to be nonreactogenic and to induce a high rate of protective immunity against subsequent challenge with the homologous strain. However, this vaccine strain was not as effective when volunteers were challenged with a heterologous El Tor strain providing only 63% protection from diarrhea.[70] The reason why this vaccine strain does not exhibit reactogenicity is not known.

Since El Tor O1 strains and *V. cholerae* O139 have replaced classical strains as the predominant causes of cholera worldwide, recent efforts have focused on generating vaccine constructs in these strain backgrounds. Properties of an ideal live oral attenuated *V. cholerae* include: (1) the development of long-lasting protective immunity against El Tor O1 and O139 strains of *V. cholerae* following a single dose; (2) safety for vaccine recipients and their close contacts; (3) safety from possible reversion to enterotoxigenicity by reacquisition of the transposable CTX genetic element; and (4) inexpensive production and ease of administration.

Recent strides toward achieving these ideals have been made in El Tor O1 and O139 strain backgrounds. Properties of two live attenuated vaccine candidates in these strain backgrounds are listed in Table II. Both of the vaccine constructs were derived from current epidemic *V. cholerae* strains. Peru-14 is a derivative of strain C6709, an El Tor O1 clinical isolate from the 1991 Peruvian cholera outbreak. Bengal-15 is a derivative of strain MO10, an O139 clinical isolate from the 1992 O139 outbreak in India. Deletion of the CTX genetic element in these constructs results in the loss of *ctxAB* as well as the other virulence genes found in the core region as well as RS and attRS1 sequences. This should prevent these strains from being able to reacquire the CTX genetic element through site-specific recombination. Since *V. cholerae* are capable of gene transfer by transduction, conjugation, and transformation, *recA* was also deleted in these strains. This combination of mutations should confer an unprecedented level of safety from possible reversion to enterotoxigenicity in these strains. The *recA* gene was deleted in these strains by marker exchange which results in the insertion of *ctxB* under control of the powerful heat shock promoter.[88] It is hoped that the overex-

TABLE II
Properties of Peru-14 and Bengal-15, Two Live Attenuated Oral Cholera Vaccine Candidates

Deletion of all of the CTX genetic element-encoded virulence genes (*cep*, *zot*, *ace*, and *ctxAB*)
Deletion of the recombination systems that could potentially allow the strains to revert to enterotoxicity (RS, attRS1, and recA)
Expression of the presumed major protective antigens (the O1 serogroup antigen for Peru-14 and the O139 serogroup antigen for Bengal-15)
Deletion of a locus to render the strain less fit to survive in the environment (*recA*)
Overexpression of a protein that should enhance the protective response of the vaccine against both cholera and enterotoxic *E. coli* (B subunit of CT)
Ability to colonize the small intestine
Motility deficiency

pressed nontoxic B subunit of CT will have an adjuvant role for the vaccine as well as contribute to immunity to both cholera and enterotoxigenic *E. coli.*[89] The initial studies of both the safety as well as the efficacy of these vaccine candidates have been very encouraging. In volunteer studies, neither strain led to significant reactogenicity while both engendered an approximately 90% protection from subsequent challenge with wild-type *V. cholerae* strains of the same serogroup as the vaccine strains.[90]

These strains are both motility-deficient mutants of the motile parental strains which contained deletions of the CTX genetic element (see above). The motility deficiency may have contributed to the remarkable lack of reactogenicity that these vaccine strains exhibited in volunteers. It is possible that adherence of *V. cholerae* to the epithelium of the small intestine leads to a local inflammatory response which accounts for the clinical symptoms (reactogenicity) which vaccine recipients have previously experienced. Nonmotile *V. cholerae* mutants may not be able to traverse the mucus barrier that covers enterocytes and therefore may not be able to colonize certain areas of the epithelium such as the intravillous spaces and crypts; however, such motility mutants should be taken up by antigen-sampling M cells (which are covered by relatively little mucus), and therefore be effective immunogens. Motility-deficient mutants of *ctx* deletion strains that had previously caused reactogenicity in volunteers have now been constructed and will be used to further test this hypothesis (J.J.M., unpublished data).

The current progress in the development of a safe and effective oral live attenuated cholera vaccine suggests that such a vaccine will be become available for public health use in this decade. Continued progress should also allow the development of *V. cholerae* as a bacterial vector to deliver heterologous antigens to the mucosal immune system. Unraveling the molecular mechanisms that lead to protective immunity to cholera will advance our understanding of the general physiology of mucosal immunity.

ACKNOWLEDGMENTS. Our collaborators and the volunteers at Fort Detrick, Maryland, and at the Virus Research Institute, Cambridge, Massachusetts, made our vaccine studies possible. We gratefully acknowledge the helpful comments on the manuscript from Drs. A. Camilli and Eric J. Rubin. M.K.W. is a Howard Hughes postdoctoral fellow.

REFERENCES

1. Barua, D., 1992, History of cholera, in: *Cholera* (D. Barua and W. B. Greenough III, eds.), Plenum Press, New York, pp. 1–36.
2. Shimada, T., Nair, G. B., Deb, B. C., Albert, M. J., Sack, R. B., and Takeda, Y., 1993, Outbreak of *Vibrio cholerae* non-O1 in India and Bangladesh, *Lancet* **341:**1347.
3. Wachsmuth, I. K., Blake, P. A., and Olsvik, Ø. (eds.), 1994, *Vibrio cholerae and Cholera: Molecular to Global Perspectives,* ASM Press, Washington, D.C.
4. Barua, D., and Greenough, W. B., 1992, *Cholera,* Plenum Medical, New York.
5. Kaper, J. B., Morris, J. G., and Levine, M. M., 1995, Cholera, *Clin. Microbiol. Rev.* **8:**48–86.
6. Shimada, T., Arakawa, E., Itoh, K., Okitsu, T., Matsushima, A., Asai, Y., Yamai, S., Nakazato, T., Nair, G. B., Albert, M. J., and Takeda, Y., 1994, Extended serotyping scheme for *Vibrio cholerae, Curr. Microbiol.* **28:**175–178.
7. Morris, J. G., 1990, Non-O group 1 *Vibrio cholerae*: A look at the epidemiology of an occasional pathogen, *Epidemiol. Rev.* **12:**179–191.
8. Wachsmuth, I. K., Evins, G. M., Fields, P. I., Olsvik, O., Popovic, T., Bopp, C. A., Wells, J. G., Carrillo, C., and Blake, P. A., 1993, The molecular epidemiology of cholera in Latin America, *J. Infect. Dis.* **167:**621–626.
9. Manning, P. A., Stroeher, U. H., and Morona, R., 1994, Molecular basis for O-antigen biosynthesis in *Vibrio cholerae* O1: Ogawa–Inaba switching, in: *Vibrio cholerae and Cholera: Molecular to Global Perspectives* (I. K. Wachsmuth, P. A. Blake, and Ø. Olsvik, eds.), ASM Press, Washington, D.C., pp. 77–94.
10. Siddique, A. K., Baqui, A., Eusof, A., Haider, K., Hossain, M., Bashir, I., and Zaman, K., 1991, Survival of classic cholera in Bangladesh, *Lancet* **337:**1125–1127.
11. Popovic, T., Bopp, C., Olsvik, Ø., and Wachsmuth, I. K., 1993, Epidemiologic application of a standardized ribotype scheme for *Vibrio cholerae* O1, *J. Clin. Microbiol.* **31:**2474–2482.
12. Taylor, R., Shaw, C., Peterson, K., Spears, P., and Mekalanos, J., 1988, Safe, live *Vibrio cholerae* vaccines? *Vaccine* **6:**151–154.
13. Waldor, M. K., and Mekalanos, J. J., 1994, ToxR regulates virulence gene expression in non-O1 strains of *Vibrio cholerae* that cause epidemic cholera, *Infect. Immun.* **62:**72–78.
14. Calia, K. E., Murtagh, M., Ferraro, M. J., and Calderwood, S. B., 1994, Comparison of *Vibrio cholerae* O139 with *V. cholerae* O1 classical and El Tor biotypes, *Infect. Immun.* **62:**1504–1506.
15. Cholera Working Group, 1993, Large epidemic of cholera-like disease in Bangladesh caused by *Vibrio cholerae* O139 synonym Bengal, *Lancet* **342:**387–390.
16. Hall, R. H., Khambaty, F. M., Kothary, M., and Keasler, S. D., 1993, Non-O1 *Vibrio cholerae, Lancet* **342:**430.

17. Waldor, M. K., and Mekalanos, J. J., 1994, *Vibrio cholerae* O139 specific gene sequences, *Lancet* **343:**1366.
18. Waldor, M. K., Colwell, R., and Mekalanos, J. J., 1994, The *Vibrio cholerae* O139 serogroup antigen includes O-antigen capsule and lipopolysaccharide virulence determinants, *Proc. Natl. Acad. Sci. USA* **91:**11388–11392.
19. Bik, E. M., Bunschoten, A. E., Gouw, R. D., and Mooi, F., 1995, Genesis of the novel epidemic *Vibrio cholerae* O139 strain: Evidence for horizontal transfer of genes involved in polysaccharide synthesis, *EMBO J.* **14:**209–216.
20. Mintz, E. D., Popovic, T., and Blake, P. A., 1994, Transmission of *Vibrio cholerae* O1, in: *Vibrio cholerae and Cholera: Molecular to Global Perspectives* (I. K. Wachsmuth, P. A. Blake, and Ø. Olsvik, eds.), ASM Press, Washington, D. C., pp. 345–356.
21. Levine, M. M., Kaper, J. B., Black, R. E., and Clements, M. L., 1983, New knowledge on pathogenesis of bacterial enteric infections as applied to vaccine development, *Microbiol. Rev.* **47:**510–550.
22. Finkelstein, R. A., 1992, Cholera enterotoxin (choleragen): A historical perspective, in: *Cholera* (D. Barua and W. B. Greenough III, eds.), Plenum Press, New York, pp. 155–187.
23. Gabriel, S. E., Brigman, K. N., Koller, B. H., Boucher, R. C., and Stutts, M. J., 1994, Cystic fibrosis heterozygote resistance to cholera toxin in the cystic fibrosis mouse model, *Science* **266:**107–109.
24. Peterson, J. W., and Ochoa, G. L., 1989, Role of prostaglandins and cAMP in the secretory effects of cholera toxin, *Science* **245:**857–859.
25. Cassuto, J., Jodal, M., Tuttle, R., and Lundgren, O., 1981, On the role of intramural nerves in cholera toxin induced intestinal secretion, *Scand. J. Gastroenterol.* **16:**377–384.
26. Pearson, G. D., and Mekalanos, J. J., 1982, Molecular cloning of *Vibrio cholerae* enterotoxin genes in Escherichia coli K-12, *Proc. Natl. Acad. Sci. USA* **79:**2976–2980.
27. Mekalanos, J. J., Swartz, D. J., Pearson, G. D., Harford, N., Groyne, F., and deWilde, M., 1983, Cholera toxin genes: Nucleotide sequence, deletion analysis and vaccine development, *Nature* **306:**551–557.
28. Kaper, J. B., Lockman, H., Balini, M., and Levine, M. M., 1984, Recombinant nontoxinogenic *Vibrio cholerae* strains as attenuated cholera vaccine candidates, *Nature* **308:** 655–658.
29. Pearson, G. D. N., Woods, A., Chiang, S. L., and Mekalanos, J. J., 1993, CTX genetic element encodes a site-specific recombination system and an intestinal colonization factor, *Proc. Natl. Acad. Sci. USA* **90:**3750–3754.
30. Fasano, A., Baudry, B., Pumplin, D. W., Wasserman, S. S., Tall, B. D., Ketley, J. N., and Kaper, J. B., 1991, *Vibrio cholerae* produces a second enterotoxin which affects intestinal tight junctions, *Proc. Natl. Acad. Sci. USA* **88:**5242–5246.
31. Trucksis, M., Galen, J., Michalski, J., Fasano, A., and Kaper, J. B., 1993, Accessory cholera enterotoxin (Ace), the third toxin of a *Vibrio cholerae* virulence cassette, *Proc. Natl. Acad. Sci. USA* **90:**5267–5271.
32. Levine, M. M., Kaper, J. B., Herrington, D., Losonsky, G., Morris, J. G., Clements, M., Black, R. E., Tall, B., and Hall, R., 1988, Volunteer studies of deletion mutants of *Vibrio cholerae* O1 prepared by recombinant techniques, *Infect. Immun.* **56:**161–167.
33. Mekalanos, J. J., and Sadoff, J. C., 1994, Cholera vaccines: Fighting an ancient scourge, *Science* **265:**1387–1389.
34. Koonin, E. V., 1992, The second cholera toxin, Zot, and its plasmid-encoded and phage-encoded homologues constitute a group of putative ATP-ases with an altered purine NTP-binding motif, *FEBS Lett.* **312:**3–6.
35. Taylor, D. N., Killeen, K. P., Hack, D. C., Kenner, J. R., Coster, T. S., Beattie, D. T., Ezzell, J., Hyman, T., Troga, A., Sjogren, M. H., Friedlander, A., Mekalanos, J. J., and Sadoff, J. C.,

1994, Development of a live, oral, attenuated vaccine against El Tor cholera, *J. Infect. Dis.* **170:**1518–1523.

36. Tacket, C. O., Losonsky, G., Nataro, J. P., Cryz, S. J., Edelman, R., Fasano, A., Michalski, J., Kaper, J. B., and Levine, M. M., 1993, Safety and immunogenicity of live oral cholera vaccine candidate CVD 110, a Δ*ctxA* Δ*zot* Δ*ace* derivative of El Tor Ogawa *Vibrio cholerae*, *J. Infect. Dis.* **168:**1536–1540.
37. Honda, T., and Finkelstein, R. A., 1979, Purification and characterization of a hemolysin produced by *Vibrio cholerae* biotype El Tor: Another toxic substance produced by cholera vibrios, *Infect. Immun.* **26:**1020–1027.
38. Ichinose, Y., Yamamoto, K., Nakasone, N., Tanabe, M. J., Takeda, T., Miwatani, T., and Iwanaga, M., 1987, Enterotoxicity of El Tor-like hemolysin of non-O1 *Vibrio cholerae*, *Infect. Immun.* **55:**1090–1093.
39. O'Brien, A. D., Chen, M. E., Holmes, R. K., Kaper, J., and Levine, M. M., 1984, Environmental and human isolates of *Vibrio cholerae* and *Vibrio parahaemolyticus* produce a *Shigella dysenteriae* 1 (Shiga)-like cytotoxin, *Lancet* **1:**77–78.
40. Saha, S., and Sanyal, S. C., 1988, Cholera toxin gene-positive *Vibrio cholerae* O1 Ogawa and Inaba strains produce the new cholera toxin, *FEMS Microbiol. Lett.* **50:**113–116.
41. Pierce, N. F., Kaper, J. B., Mekalanos, J. J., and Cray, W. J., 1985, Role of cholera toxin in enteric colonization by *Vibrio cholerae* O1 in rabbits, *Infect. Immun.* **50:**813–816.
42. Freter, R., O'Brien, P. C. M., and Macsai, M. M. S., 1981, Role of chemotaxis in the association of motile bacteria with intestinal mucosa in vivo studies, *Infect. Immun.* **34:** 234–240.
43. Taylor, R. K., Miller, V. L., Furlong, D. B., and Mekalanos, J. J., 1987, Use of *phoA* gene fusions to identify a pilus colonization factor coordinately regulated with cholera toxin, *Proc. Natl. Acad. Sci. USA* **84:**2833–2837.
44. Kaufman, M. R., and Taylor, R. K., 1994, The toxin-coregulated pilus: Biogenesis and function, in: *Vibrio cholerae and Cholera: Molecular to Global Perspectives* (I. K. Wachsmuth, P. A. Blake, and Ø. Olsvik, eds.), ASM Press, Washington, D.C., pp. 187–202.
45. Ogierman, M. A., Zabihi, S., Mourtzios, L., and Manning, P. A., 1993, Genetic organization and sequence of the promoter-distal region of the *tcp* gene cluster of *Vibrio cholerae*, *Gene* **126:** 51–60.
46. Kaufman, M. R., Shaw, C. E., Jones, I. D., and Taylor, R. K., 1993, Biogenesis and regulation of the *Vibrio cholerae* toxin-coregulated pilus: Analogies to other virulence factor secretory systems, *Gene* **126:**43–49.
47. Peek, J. A., and Taylor, R. K., 1992, Characterization of a periplasmic thiol:disulfide interchange protein required for the functional maturation of secreted virulence factors of *Vibrio cholerae*, *Proc. Natl. Acad. Sci. USA* **89:**6210–6214.
48. Herrington, D. A., Hall, R. H., Losonsky, G., Mekalanos, J. J., Taylor, R. K., and Levine, M. M., 1988, Toxin, toxin-coregulated pili, and the *toxR* regulon are essential for *Vibrio cholerae* pathogenesis in humans, *J. Exp. Med.* **168:**1487–1492.
49. Rhine, J. A., and Taylor, R. K., 1994, TcpA pilin sequences and colonization requirements for O1 and O139 *Vibrio cholerae*, *Mol. Microbiol.* **13:**1013–1020.
50. Attridge, S. R., Voss, E., and Manning, P. A., 1993, The role of toxin-coregulated pili in the pathogenesis of *Vibrio cholerae* O1 El Tor, *Microb. Pathog.* **15:**421–431.
51. Parsot, C., Taxman, E., and Mekalanos, J. J., 1991, ToxR regulates the production of lipoproteins and the expression of serum resistance in *Vibrio cholerae*, *Proc. Natl. Acad. Sci. USA* **88:**1641–1645.
52. Everiss, K. D., Hughes, K. J., Kovach, M. E., and Peterson, K. M., 1994, The *Vibrio cholerae acfB* colonization determinant encodes an inner membrane protein that is related to a family of signal-transducing proteins, *Infect. Immun.* **62:**3289–3298.

53. Peterson, K. M., and Mekalanos, J. J., 1988, Characterization of the *Vibrio cholerae* ToxR regulon: Identification of novel genes involved in intestinal colonization, *Infect. Immun.* **56:**2822–2829.
54. Gardel, C. L., and Mekalanos, J. J., 1994, Modus operandi of *Vibrio cholerae*: Swim to arrive, stop to kill. The relationship among chemotaxis, motility and virulence, *J. Cell Biochem.* **18A:**65.
55. Franzon, V. L., Barker, A., and Manning, P. A., 1993, Nucleotide sequence encoding the mannose-fucose resistant hemagglutinin of *Vibrio cholerae* O1 and construction of a mutant, *Infect. Immun.* **61:**3032–3037.
56. Jonson, G., Lebens, M., and Holmgren, J., 1994, Cloning and sequencing of *Vibrio cholerae* mannose-sensitive haemagglutinin pilin gene: Localization of *mshA* within a cluster of type 4 pilin genes, *Mol. Microbiol.* **13:**109–118.
57. Finkelstein, R. A., Boesman-Finkelstein, M., Chang, Y., and Hase, C. C., 1992, *Vibrio cholerae* hemagglutinin/protease, colonial variation, virulence, and detachment, *Infect. Immun.* **60:** 472–478.
58. Henderson, D. P., and Payne, S. M., 1994, *Vibrio cholerae* iron transport systems: Roles of heme and siderophore iron transport in virulence and identification of a gene associated with multiple iron transport systems, *Infect. Immun.* **62:**5120–5125.
59. Goldberg, M. B., DiRita, V. J., and Calderwood, S. B., 1990, Identification of an iron-regulated virulence determinant in *Vibrio cholerae*, using Tn*phoA* mutagenesis, *Infect. Immun.* **58:**55–60.
60. Camilli, A., Beattie, D., and Mekalanos, J., 1994, Use of genetic recombination as a reporter of gene expression, *Proc. Natl. Acad. Sci. USA* **91:**2634–2638.
61. Overbye, L. J., Sandkvist, M., and Bagdasarian, M., 1993, Genes required for extracellular secretion of enterotoxin are clustered in *Vibrio cholerae*, *Gene* **132:**101–106.
62. Hall, R. H., Vial, P. A., Kaper, J. B., Mekalanos, J. J., and Levine, M. M., 1988, Morphological studies on fimbriae expressed by *Vibrio cholerae* O1, *Microb. Pathog.* **4:**257–65.
63. Baselski, V. S., Medina, R. A., and Parker, C. D., 1979, In vivo and in vitro characterization of virulence deficient mutants of *Vibrio cholerae*, *Infect. Immun.* **24:**111–116.
64. Ottemann, K. M., and Mekalanos, J. J., 1994, Regulation of cholera toxin expression, in: *Vibrio cholerae and Cholera: Molecular to Global Perspectives* (I. K. Wachsmuth, P. A. Blake, and Ø. Olsvik, eds.), ASM Press, Washington, D.C., pp. 177–185.
65. DiRita, V. J., 1992, Co-ordinate expression of virulence genes by ToxR in *Vibrio cholerae*, *Mol. Microbiol.* **6:**451–458.
66. Lang, H., Jonson, G., Holmgren, J., and Palva, E. T., 1994, The maltose regulon of *Vibrio cholerae* affects production and secretion of virulence factors, *Infect. Immun.* **62:**4781–4788.
67. Litwin, C. M., and Calderwood, S. B., 1994, Analysis of the complexity of gene regulation by Fur in *Vibrio cholerae*, *J. Bacteriol.* **176:**240–248.
68. Williams, S. G., Attridge, S. R., and Manning, P. A., 1993, The transcriptional activator HlyU of *Vibrio cholerae*: Nucleotide sequence and role in virulence gene expression, *Mol. Microbiol.* **9:**751–760.
69. Richardson, S. H., 1994, Host susceptibility, in: *Vibrio cholerae and Cholera: Molecular to Global Perspectives* (I. K. Wachsmuth, P. A. Blake, and Ø. Olsvik, eds.), ASM Press, Washington, D.C., pp. 273–292.
70. Levine, M. M., and Pierce, N. F., 1992, Immunity and vaccine development, in: *Cholera* (D. Barua and W. B. Greenough III, eds.), Plenum Press, New York, pp. 285–328.
71. Clemens, J. D., Sack, D. A., Harris, J. R., Chakraborty, J., Khan, M., Huda, S., Ahmed, F., Goes, J., Rao, M., Svennerholm, A., and Holmgren, J., 1989, ABO blood groups and cholera: New observations on specificity of risk and modification of vaccine efficacy, *J. Infect. Dis.* **159:**770–773.

72. Glass, R. I., Svennerholm, A. M., Khan, M. R., Huda, S., Huq, M. I., and Holmgren, J., 1985, Seroepidemiological studies of El Tor cholera in Bangladesh: Association of serum antibody levels with protection, *J. Infect. Dis.* **151:**236–242.
73. Clemens, J. D., Van Loon, F., Sack, D., Rao, M., Ahmed, F., Chakraborty, J., Kay, B., Khan, M., Yunus, M., Harris, J., Svennerholm, A., and Holmgren, J., 1991, Biotype as determinant of natural immunising effect of cholera, *Lancet* **337:**883–884.
74. Svennerholm, A.-M., Jonson, G., and Holmgren, J., 1994, Immunity to Vibrio cholerae infection, in: *Vibrio cholerae and Cholera: Molecular to Global Perspectives* (I. K. Wachsmuth, P. A. Blake, and Ø. Olsvik, eds.), ASM Press, Washington, D.C., pp. 257–272.
75. Hall, R. H., Losonsky, G., Silveira, A. P., Taylor, R. K., Mekalanos, J. J., Witham, N. D., and Levine, M. M., 1991, Immunogenicity of *Vibrio cholerae* O1 toxin-coregulated pili in experimental and clinical cholera, *Infect. Immun.* **59:**2508–2512.
76. Sun, D. X., Mekalanos, J. J., and Taylor, R. K., 1990, Antibodies directed against the toxin-coregulated pilus isolated from *Vibrio cholerae* provide protection in the infant mouse experimental cholera model, *J. Infect. Dis.* **161:**1231–1236.
77. Cash, R. A., Music, S. I., Libonati, J. P., Criag, J. P., Pierce, N. F., and Hornick, R. B., 1974, Reponse of man to infection with *Vibrio cholerae.* II. Protection from illness afforded by previous disease and vaccine, *J. Infect. Dis.* **130:**325–333.
78. Winner, L. D., Mack, J., Weltzin, R., Mekalanos, J. J., Kraehenbuhl, J. P., and Neutra, M. R., 1991, New model for analysis of mucosal immunity: Intestinal secretion of specific monoclonal immunoglobulin A from hybridoma tumors protects against *Vibrio cholerae* infection, *Infect. Immun.* **59:**977–982.
79. Holmgren, J., Lycke, N., and Czerkinsky, C., 1993, Cholera toxin and cholera B subunit as oral-mucosal adjuvant and antigen vector systems, *Vaccine* **11:**1179–1184.
80. Lycke, N., Tsuji, T., and Holmgren, J., 1992, The adjuvant effect of *Vibrio cholerae* and *Escherichia coli* heat-labile enterotoxins is linked to their ADP-ribosyltransferase activity, *Eur. J. Immunol.* **22:**2277–2281.
81. McKenzie, S., and Halsey, J., 1984, Cholera toxin B subunit as a carrier protein to stimulate a mucosal immune response, *J. Immunol.* **133:**1818–1822.
82. Kraehenbuhl, J. P., and Neutra, M. R., 1992, Molecular and cellular basis of immune protection of mucosal surfaces, *Physiol. Rev.* **72:**853–879.
83. Sun, J.-B., Holmgren, J., and Czerkinsky, C., 1994, Cholera toxin B subunit: An efficient transmucosal carrier-delivery system for induction of peripheral immunological tolerance, *Proc. Natl. Acad. Sci. USA* **91:**10795–10799.
84. Xu-Amano, J., Kiyono, H., Jackson, R., Staats, H., Fujihashi, K., Burrows, P., Elson, C. O., Pillai, S. and McGhee, J. R., 1993, Helper T cell subsets for immunoglobulin A responses: Oral immunization with tetanus toxoid and cholera toxin as adjuvant selectively induces Th2 cells in mucosa associated tissues, *J. Exp. Med.* **178:**1309–1320.
85. Clemens, J. D., Harris, J. R., Sack, D. A., Chakraborty, J., Ahmed, F., Stanton, B. F., Khan, M. U., Kay, B. A., Huda, N., and Khan, M. R., 1988, Field trial of oral cholera vaccines in Bangladesh: Results of one year of follow-up, *J. Infect. Dis.* **158:**60–69.
86. Clemens, J. D., Sack, D. A., Harris, J. R., Loon, F. V., Chakraborty, J., Ahmed, F., Rao, M. R., Khan, M. R., Yunus, M. and Huda, N., 1990, Field trial of oral cholera vaccines in Bangladesh: Results from three-year follow-up, *Lancet* **335:**270–273.
87. Owen, R. L., Pierce, M. F., Apple, R. T., and Cray, W. C., 1986, M cell transport of *Vibrio cholerae* from the intestinal lumen into Peyer's patches: A mechanism for antigen sampling and for microbial transepithelial migration, *J. Infect. Dis.* **153:**1108–1118.
88. Waldor, M. K., and Mekalanos, J. J., 1994, Emergence of a new cholera pandemic: Molecular analysis of virulence determinants in *Vibrio cholerae* O139 and development of a live vaccine prototype, *J. Infect. Dis.* **170:**278–283.

89. Clemens, J. D., Sack, D. A., Harris, J. R., Chakraborty, J., Neogy, P. K., Stanton, B., Huda, N., Khan, M. U., Kay, B. A., and Khan, M. R., 1988, Cross-protection by B subunit-whole cell cholera vaccine against diarrhea associated with heat-labile toxin-producing enterotoxigenic *Escherichia coli*: Results of a large-scale field trial, *J. Infect. Dis.* **158:**372–377.
90. Coster, T. S., Killeen, K. P., Waldor, M. K., Beattie, D., Spriggs, D., Kenner, J. R., Trofa, A., Sadoff, J., Mekalanos, J. J., and Taylor, D. N., 1995, Safety, immunogenicity and efficacy of a live attenuated *Vibrio cholerae* O139 vaccine protype, Bengal-15, *Lancet* **345:**949–952.

4

Immunity to Salmonella Infections

TOBY K. EISENSTEIN, DUAN HUANG, and MARTIN G. SCHWACHA

1. INTRODUCTION

Immunity to typhoid fever has been of interest since the turn of the century, principally because of a desire to develop optimal vaccines for use by military personnel. The British were the first to try vaccination under the direction of Wright and Semple in 1897.(1) The United States Army adopted compulsory vaccination in 1911, using the British method, which involved three injections of heat-killed organisms, spaced 1 week apart.(2) The results were a stunning success, with the number of cases falling from over 350/100,000 to less than 25/100,000 once compulsory vaccination was instituted.(2) At that time the causative organism was designated as *Eberthella typhosa*, the name being changed later to *Salmonella typhosa*, and finally to *Salmonella typhi*. Initially, the major antigens on the organism were determined to be the O (somatic antigens) and the H antigens (flagellar). In 1934 the Vi capsular antigen was discovered by Felix and Pitt(3) and shown to be a major virulence factor of *S. typhi*. Considerable research was invested in trying to assess the relative contribution of the O and Vi antigens to the protection conferred by vaccination. Since the Vi antigen was preserved better by alcohol than by heat, alcohol-preserved whole-cell vaccines were also tested. In 1953 Landy introduced acetone-killed and dried cells. These various whole killed-cell formulations were compared in extensive double-blind field trials carried

---

TOBY K. EISENSTEIN, DUAN HUANG, and MARTIN G. SCHWACHA • Department of Microbiology and Immunology, Temple University, School of Medicine, Philadelphia, Pennsylvania 19140.

Enteric Infections and Immunity, edited by Lois J. Paradise *et al.* Plenum Press, New York, 1996.

out in the late 1950s and 1960s in British Guyana, Yugoslavia, Poland, and Russia under the auspices of the World Health Organization (WHO).[5,6] It was found that many of these nonviable preparations conferred high levels of relatively long-lasting protection (at least 7 years) on vaccinees.[7] Thus, vaccines consisting of 10^9 whole killed cells have consistently demonstrated protection in humans against typhoid fever.

However, attempts to understand the antigens involved in the protection yielded data that were confusing. One approach was to correlate levels of antibody titers to the O, H, and Vi antigens induced by the various vaccine preparations with protective potency. Although poor correlations were found between protection and anti-O or anti-Vi titers, it should be appreciated that the antibody assays were agglutination tests which only measured IgM.[4] A second approach was to try to assess vaccine potency with the capacity of vaccines to immunize and protect mice from challenge with *S. typhi* or with their ability to induce immune serum that would passively protect naive animals. Twenty laboratories in twelve countries engaged in these tests. A lack of concordance in the findings between laboratories emerged, with no clear correlations between vaccine efficacy in field trials and in laboratory tests in animals.[8] Nonetheless, the field trial data showed that the heat-killed phenol-preserved vaccine and the acetone-killed vaccine were highly efficacious, and they have been used not only by the military, but also for vaccination of civilians in typhoid-endemic areas. However, all parenteral whole killed-cell vaccines have a major drawback in that they have considerable toxicity. Their endotoxin content causes reactogenicity, which includes systemic symptoms, such as fever and chills, and local symptoms, including pain, erythema, and swelling at the injection site. These problems with nonviable vaccines provided a major impetus for development of alternative vaccination strategies, which will be discussed in detail later in this chapter. One of these approaches is the use of live attenuated Salmonella as vaccines. Various mutants or genetically engineered strains have been tested and found to be efficacious. Studies in mice using these organisms, as well as sublethal doses of virulent Salmonella, have been instrumental in formulation of hypotheses regarding mechanisms of immunity to Salmonella.

2. ANIMAL MODELS

Vaccine potency testing for the WHO trials was carried out in mice that were challenged with *S. typhi* that had been coated with hog gastric mucin. In the absence of mucin, this organism was not virulent for mice, requiring greater than 10^7 organisms to achieve an LD_{50}. An alternative mouse model, which is thought to be more physiologic, has been used extensively to investigate mechanisms of immunity to typhoid fever. This model uses one of many inbred strains of mice that are highly susceptible to a naturally occurring typhoidlike disease caused by strains of *Salmonella typhimurium* (which means mouse typhoid). Human isolates of *S. typhimurium*, as well as certain other strains of Salmonella with different O antigens, such as *S. enteritidis* (O:9,12), are also virulent in this model. To avoid Salmonella epidemics in mouse colonies, Swiss mice have actually been bred for resistance to infection with Salmonella. A substantial body of literature has accumulated using the *S. typhimurium* murine model of typhoid fever to dissect, by analogy, the mechanisms of immunity to typhoid in the human. In human typhoid fever, organisms enter the body by ingestion and become systemic by penetrating the intestinal epithelium and disseminating to the lymphatics and subsequently to the bloodstream. The major symptoms of the disease result from bacteremia and proliferation of the organisms in the liver and spleen. Diarrhea is not a prominent feature of typhoid fever. Consequently, inoculation of mice with *S. typhimurium* to study mechanisms of immunity has most commonly been carried out by intravenous (i.v.) or intraperitoneal (i.p.) injection, not by oral inoculation, the rationale being that the enteric phase of the disease is only of interest for studying the pathogenesis of breaching the intestinal epithelial barrier, but not particularly relevant to understanding systemic immune responses. Thus, most of our knowledge on mechanisms of host defense to typhoid is derived from the *S. typhimurium* challenge model using systemic inoculation (i.p. or i.v.) in the mouse.

3. KILLED VACCINES AND HUMORAL IMMUNITY IN MURINE MODELS OF TYPHOID FEVER

The literature cited above on human vaccination shows empirically that nonviable *S. typhi* vaccines protect humans against typhoid fever. Although there was poor correlation between human and mouse data, the difficulty could be attributed to the use of *S. typhi* for challenge, a poor model for the human disease. From mouse models of immunization and infection with *S. typhimurium* or *S. enteritidis*, the conclusion that killed cells and bacterial extracts are protective was supported by numerous studies from different laboratories around the world.

Jenkin and Rowley, in Australia, reported that either acetone-killed or alcohol-killed cells provided solid protection against Salmonella infection for Swiss Webster mice.[9] Additionally, Herzberg *et al.* found that heat-killed cells, as well as a deoxycholate extract of live cells, gave 100% protection against 100 LD_{50} doses of virulent Salmonella, also in Swiss mice.[10] Work at

Stanford University in the United States by Ornellas *et al.* demonstrated protection against 10,000 LD_{50} doses using multiple inoculations of heat-killed cells in mice.[11] Angerman and Eisenstein at Temple University showed that acetone-killed cells, ribosomal extracts of Salmonella, and purified lipopolysaccharide (LPS) protected against challenge doses of 100 to 1000 LD_{50}s in CD-1 mice.[12] Studies by Lindberg's group in Sweden demonstrated unequivocally that O antigens were sufficient for protection in mice by showing that a vaccine made from the synthetic disaccharide, tyvelose-mannose (O:9), chemically coupled to bovine serum albumin (BSA), protected mice against challenge with *S. enteritidis*.[13] *S. enteritidis* has the O:9 antigen, unlike *S. typhimurium*, which has the O:4 determinant. In further studies, this group purified an octasaccharide from *S. typhimurium* by chemical fractionation after enzymatic digestion of the LPS, which contained the O antigens. The octasaccharide was then chemically coupled to diphtheria toxin or to porin proteins of Salmonella. A variety of these semisynthetic vaccines protected mice against challenge with 20 to 100 LD_{50} doses of virulent Salmonella.[14]

There is also strong evidence that outer membrane protein antigens of Salmonella are protective in mice. Makela's group showed that porins purified from rough Salmonella (which lack O antigens) act as vaccines.[15] Eisenstein's laboratory reported that extraction of a rough strain of Salmonella to yield "endotoxin protein," a mixture of several polypeptides that are closely associated with lipid A in the Salmonella outer membrane, protected against 27 LD_{50} doses of virulent Salmonella.[16] The presence of O antigen or of rough polysaccharide significantly enhanced the protective capacity of the porin and of the endotoxin protein preparations.[16,17] These studies, along with others not cited, provide unequivocal evidence that nonviable vaccines, including purified or synthetic O antigens and protein antigens of Salmonella, are protective in the mouse model of Salmonella infection.

Another approach to assessing the nature of the protective antigens and the mechanisms of protection in Salmonella is to raise antibody to various Salmonella vaccines in rabbits or mice and to test the passive protective capacity of serum in mice challenged with virulent organisms. Rowley's group showed that immune serum raised to attenuated Salmonella in mice could passively protect naive recipient mice.[9,18,19] Roantree's laboratory reported that hyperimmune mouse serum raised to heat-killed organisms conferred protection against 1000 to 10,000 LD_{50} challenge doses.[11] Blanden *et al.* found that immune serum raised to whole heat-killed cells was protective.[20] Eisenstein *et al.* observed that murine hyperimmune serum raised to acetone-killed cells passively protected some strains of mice, but not others.[21] Svenson *et al.* reported that rabbit serum raised to the semisynthetic O antigen octasaccharide conjugated to various protein carriers protected mice against 100 LD_{50} doses.[14] Kuusi *et al.* showed that antiporin antibodies raised in rabbits protected mice against lethal challenge with Salmonella.[15,17] These studies clearly established that antibody alone was sufficient to protect some mouse strains from mortality caused by naturally mouse-virulent strains of Salmonella. The evidence shows definitively that anti-O antibodies protect in the mouse model and also supports a role for anti-outer membrane protein antibodies in host defense against Salmonella in mice.

4. THE CONCEPT OF CELLULAR IMMUNITY AND IMMUNITY TO SALMONELLA INFECTION

Some investigators working on typhoid vaccine models in mice using *S. typhimurium* challenge made observations that did not fit with those cited in the section above. Hobson had reported in 1957 that administration of a sublethal dose of live Salmonella gave high levels of protection against challenge with the virulent strain, but heat-killed vaccines provided poor protection.[22] Furthermore, protection by the live organisms did not correlate with induction of anti-O antibodies.[22] Dr. Ushiba's laboratory in Japan reported better protection using a live attenuated strain of Salmonella that was rough and lacked O antigens than was observed with multiple injections of heat-killed cells.[23] Necropsy of challenged mice showed that animals receiving the live vaccine were able to control replication of the virulent challenge strain in the spleen and liver, whereas mice vaccinated with the killed cells were unable to control replication. Passive serum extended the mean time to death, but was not protective. Ushiba formulated the hypothesis that there were two mechanisms of immunity in mouse typhoid, humoral and cellular, and that the cellular response was crucial to controlling microbial replication.[24] In 1966, the group at the Trudeau Institute, Drs. Mackaness, Collins, and Blanden, published three papers in the *Journal of Experimental Medicine* that altered the thinking about mechanisms of immunity to Salmonella in mice.[20,25,26] In agreement with Hobson's observation, they also found that heat-killed or alcohol-killed Salmonella were poorly protective compared with a sublethal dose of a virulent strain. However, more importantly, they found that live, but not killed, Salmonella activated macrophages to kill Salmonella. Furthermore, mice infected with live, but not killed, Salmonella were found to be resistant to challenge with the antigenically unrelated organism, *Listeria monocytogenes.* A theory was proposed to explain the superior protection of viable versus nonviable cells.[27] It was hypothesized that living but not killed Salmonella induced "cellular immunity" which was manifested as activated macrophages that had enhanced bactericidal capacity against Salmonella and other facultative intracellular pathogens.[20] At certain times after immunization with live Salmonella as vaccines, and using a particular range of challenge doses, enhanced killing of virulent Salmonella could also be demonstrated in the livers and spleens of vaccinated animals.[26] Cellular immunity was correlated with the appearance of delayed-type hypersensitivity (DTH), measured in the mouse by footpad swelling 24–48 hr after injection of a Salmonella eliciting antigen.[28,29] The developing understanding of the T-cell arm of the immune system as an anatomically distinct entity from the B-cell arm provided further clarification of the model. Important to the theory was the inclusion of Salmonella with other organisms for which immunity was also proposed to be cellular. Among these were *Mycobacterium bovis*, strain bacillus Calmette–Guérin (BCG), *Listeria monocytogenes*, and *Brucella abortus*.[27] These three organisms all are facultative intracellular pathogens of macrophages. Importantly, for BCG and Listeria, it was shown that T cells could transfer immunity. It is noteworthy, however, that this group never demonstrated adoptive transfer of Salmonella immunity.[29] They did show that macrophages of live-cell-vaccinated mice, exposed to Salmonella *in vivo* and tested *ex vivo*, could kill the organisms better than macrophages from killed-cell-vaccinated mice under similar conditions.[20] The theory that T cells secrete substances that activate macrophages to kill intracellular pathogens that parasitize them emerged from these studies.[27] The theory has proven immensely powerful, particularly for understanding host defense against Mycobacteria and Listeria. However, hindsight indicates that this theory has been somewhat overinterpreted by the host defense community in its application to Salmonella. The prevailing view became a certainty that immunity to Salmonella required sensitized T cells and cellular immunity, even though the evidence for Salmonella was largely circumstantial. Conversely, the protective capacity of nonviable Salmonella vaccines and antibody was dismissed because, in the mouse strain used at the Trudeau Institute, killed cells and passive serum performed poorly compared with living Salmonella as vaccines.[30] The disregard for the efficacy of nonviable vaccines occurred in spite of the rather impressive literature that was available before and after the publication of the 1966 papers.

5. GENETIC DETERMINANTS OF RESISTANCE TO SALMONELLA

Innate resistance to *S. typhimurium* in mice is genetically determined. The first gene identified was mapped to chromosome 1 and is designated *Ity*.[31] Many of the strains of mice commonly used in laboratory research, such as BALB/c and C57BL/6, have the susceptible allele.[31] This gene also controls resistance to Leishmania and to BCG.[32] Recently the gene has been cloned and shown to be expressed exclusively in macrophages,[33] which would fit with the observation that its phenotypic expression effects host defense to facultative intracellular pathogens of macrophages. Lissner *et al.* published evidence that *Ity* regulates the ability of host macrophages to kill Salmonella and other intracellular pathogens,[34] but Briles concluded that this susceptibility allele controls the rate of growth of Salmonella, rather than the rate of killing.[35] A second gene controlling innate resistance to Salmonella infection is *Xid*, and CBA/N mice with the defective allele are more susceptible to Salmonella infection.[36] Since the deficit in these animals involves poor antibody formation, the influence of this locus on Salmonella susceptibility implies a role for the capacity to mount a humoral immune response as important in the outcome of a Salmonella infection. Still a third locus involved in resistance to Salmonella is the gene determining responsiveness to gram-negative bacterial lipopolysaccharide (*Lps*). The *Lps* locus was shown to mediate enhanced susceptibility to Salmonella infection in C3H/HeJ mice, which have the defective allele (Lps^{d}), in contrast to other mouse strains in the C3H lineage which are innately resistant to Salmonella and are lipopolysaccharide (LPS) responders (Lps^{r}).[37] However, the role of Lps^{d} in Salmonella susceptibility is not clear-cut. It was shown by Eisenstein *et al.* that C3HeB/FeJ mice are normal responders to LPS, but are hypersusceptible to Salmonella infection, implying that there is yet another gene controlling resistance to Salmonella in these mice in the C3H lineage.[38] The existence of such a gene was later demonstrated using classical genetic mouse breeding experiments by O'Brien and Rosenstreich, but the gene was never named.[39] Crosses between C3H/HeJ and C3HeB/FeJ mice have not been done to show definitively that the susceptibility of the C3H/HeJ animals is related only to Lps^{d} and not to the FeJ locus gene. Interestingly, all mice in the C3H lineage are Ity^{r}. Therefore, the unnamed FeJ gene must override Ity^{r}, which suggests that this gene is even more important than *Ity* for the host. The genetic control of natural resistance to Salmonella infection has additional complexities, as it appears that there are different genes responsible for controlling early and late replication in the liver and spleen,[40] and the late genes are not yet mapped. Furthermore, there remain undefined factors that govern interactions between the virulence of the strain of Salmonella and the genetic resistance of the mouse strain in determining the outcome of an infection.[41,42] In the literature on host defense to Salmonella infection, many investigators only cite the work on *Ity* and appear to be unfamiliar with studies on the other genetic parameters that influence innate resistance to Salmonella. This is unfortunate, for the genetic background of the mouse is crucial to the ability of various types of Salmonella vaccines to protect, as will be demonstrated in the following section.

6. THE INFLUENCE OF GENETICALLY DETERMINED INNATE RESISTANCE OR SUSCEPTIBILITY TO SALMONELLA AND MECHANISMS OF ACQUIRED HOST RESISTANCE

The studies discussed above on innate susceptibility to Salmonella infection did not consider genetic influences on acquired immunity. Several groups have shown that there are differences in the capacity of various strains of mice to be vaccinated by different Salmonella vaccines.[21,43,44] One reason postulated for these differences is that mice differ in immune response genes to particular antigens, and thus in the capacity to mount an immune response.[31] It has been suggested that immune response genes might govern the ability to develop DTH to Salmonella antigens,[31] but there are examples of mouse strains where the correlation between induction of DTH by Salmonella and innate resistance or susceptibility is not found.[45–47] Alternatively, Eisenstein's laboratory[12,21,48] compared the capacity of LPS, acetone-killed cells, endotoxin protein, and ribosomal extracts of Salmonella to vaccinate C3H/HeJ, C3HeB/FeJ, C3H/HeN, and CD-1 mice against virulent Salmonella challenge. The first two mouse strains are Salmonella susceptible and the last two are innately resistant. It was found that all four strains made adequate antibody responses to the nonviable vaccines. However, the innately susceptible strains were poorly protected by the nonviable vaccines, but the innately resistant strains survived Salmonella challenge. Interestingly, in the resistant mice, some of the nonviable vaccines, including the acetone-killed cells, were almost as effective as the live cells.[48] It was also shown that hyperimmune serum raised to acetone-killed cells could passively protect the innately resistant C3H/HeN strain against Salmonella challenge, but not the innately susceptible strains.[21] From these results it was postulated that the disparities in the literature concerning the efficacy of nonviable Salmonella vaccines could be explained by the use of strains of mice that differ in their innate susceptibilities to Salmonella infection.[21] Review of previous studies did show that where the data were available, groups that found little protection by nonviable vaccines were using innately hypersusceptible mouse strains, whereas groups that found good to excellent protection by nonviable vaccines used innately resistant animals.[21] This observation led to the suggestion that the mouse does not present a single model of resistance to Salmonella infection, but rather a spectrum of models.[6] In the most hypersusceptible mice, only live cells can protect, with cellular immunity apparently being necessary for resistance, as passive antibody is ineffective. In resistant mice, antibody alone is sufficient, as even purified LPS or passive antibody alone is protective. Live cells can also protect innately resistant animals, so cellular immunity as well as humoral immunity is probably effective in these strains. There are also mice of intermediate susceptibility to Salmonella infection,[38] and their capacity to be immunized has not been systematically assessed.

7. IMPLICATIONS OF MOUSE STUDIES FOR HUMAN VACCINATION AND HOST DEFENSE TO SALMONELLA

The question has been raised as to whether the human is more like the innately susceptible or resistant mouse in terms of the need for cellular or humoral immunity.[6] Eisenstein concluded that because humans can be effectively immunized with killed vaccines, as documented above, they are more like the inherently resistant rodents.[6,21] It is not easily explicable why the research community has focused so intently on a perceived need to induce cellular immunity for superior protection against Salmonella infection in humans, when there is so much evidence against this proposition. The fallacy in this line of thinking was finally demonstrated definitively and empirically by Robbins and colleagues, who showed in two field trials that purified Vi polysaccaride antigen, whose mechanism of action is by induction of anti-Vi antibodies, gave excellent levels of protection in humans against typhoid fever.[49,50] It has been argued that the reason nonviable vaccines protect humans is because field trials are carried out in typhoid-endemic areas of the world where people may have had prior exposure to living, virulent organisms and recovered. The efficacy of nonviable vaccines in humans has been postulated to result from "boosting" the immune response already engendered by live organisms. There is actually no evidence for this theory. In fact, the opposite argument can be made. Most cases of typhoid fever in endemic areas occur in children or young adults and appear to be primary infections.[4,51] Furthermore, in humans, recovery from active infection, which is argued to confer cellular immunity that is long-lasting and stronger than that induced by killed cells, does not confer complete protection from reinfection, as there are documented case reports of second infections with typhoid after laboratory-confirmed primary infections.[52,53]

8. FURTHER STUDIES ON MECHANISMS OF HOST RESISTANCE TO SALMONELLA INFECTION

One approach to evaluating the mechanisms of immunity to Salmonella is to examine the protective capacity of live versus killed vaccines, with the assumption that the killed cells protect via antibody. Another approach is to use passive serum transfer to test directly the role of antibody. Studies of these types were reviewed in previous sections and shown to yield disparate results depending on the mouse strain used. Other approaches to determining the relative role of the cellular or humoral arm of the immune system in host defense to Salmonella infection are either transfer of defined lymphoid cell populations from Salmonella-immune mice to naive mice or depletion of specific cell subsets from mice before or after immunization to assess the effects on host resistance. A supporting role for B cells in induction of resistance to Salmonella in mice comes from the work of Hochadel and Keller in which adoptive transfer of B cells into nonirradiated recipients gave higher levels of protection than adoptive transfer of T cells.(54) Furthermore, Morris *et al.* found that cyclophosphamide, which primarily depresses B-cell function but not antithymocyte serum, which affects T cells, sensitized mice to outgrowth of a nonlethal dose of an avirulent strain of Salmonella.(55) A similar conclusion was reached by Davies and Kotlarski in studies of thymectomized, irradiated, bone marrow-reconstituted mice. Elimination of T cells did not affect the capacity of these animals to control replication of an attenuated strain of Salmonella.(56) However, lack of T cells did abolish the ability of the attenuated vaccine to protect mice against virulent challenge.(56) In contrast, Nauciel found that depletion of $CD4^+$ cells did result in a modest effect on clearance of an attenuated strain of Salmonella.(57) Nauciel also found that spleen cells from mice receiving the attenuated vaccine could transfer the ability to clear attenuated Salmonella more rapidly than naive mice, and anti-CD4 treatment of the donor cells abolished their protective capacity.(57) In agreement with Davies and Kotlarski, anti-CD4 treatment of mice receiving attenuated Salmonella abolished protection against challenge 50 days after immunization.(57) Mixter *et al.*(58) reported that depletion of T cells expressing the $\alpha\beta$ or the $\gamma\delta$ T-cell receptor sensitized mice to oral infection with Salmonella. Using transfer of enriched cell populations from mice receiving a live irradiated vaccine, Chander *et al.* showed a modest increase in mean time to death in mice receiving T-cell-enriched spleen cells taken from vaccinated mice; B-cell-enriched cell fractions were not effective.(59) A role for both $CD4^+$ and $CD8^+$ T cells in mice infected with *S. abortusovis* has been demonstrated by Guilloteau *et al.* using selective depletion of each T-cell subset prior to transfer of immune spleen cells into normal recipients.(60) Hougen and Jensen found that in rats, purified $CD4^+$ T cells taken from spleens of animals inoculated with a sublethal dose of live Salmonella protected recipients against challenge; B cells and $CD8^+$ cells were not effective.(61) Hormaeche's group found that both immune T cells and immune serum were needed to transfer resistance to oral infection with Salmonella.(62) One problem with studies using primary immune cells for transfer studies is that live Salmonella or Salmonella antigen may also be transferred with the protective cell populations. To circumvent this possibility, Paul *et al.* developed T-cell lines from lymph node cells or peritoneal exudate cells obtained from mice immunized with a culture supernatant antigen of an attenuated strain of Salmonella administered in Freund's complete adjuvant and showed that these cells were able to transfer protection to Salmonella challenge.[63,64] A difficulty with interpretation of these results is that any protein antigen incorporated into Freund's adjuvant will induce a T-cell response, but such a response does not necessarily represent what happens during infection. Nonetheless, these experiments show that Salmonella has antigens capable of inducing protective T cells. In contrast, Killar and Eisenstein were unable to transfer immunity to Salmonella using nonadherent or T-cell-enriched spleen cell fractions from mice receiving attenuated Salmonella. They found that macrophage-enriched fractions transferred immunity to lethal challenge.[65] A role for macrophages is supported by studies of Maskell *et al.*, who also found that macrophages and not T cells transferred early resistance to Salmonella.[66] These studies raise the possibility that macrophages may actually be unrecognized as the active cells in transferred T-cell-enriched populations, unless they are carefully eliminated. Hof *et al.* had also shown that depletion of macrophages by dextran sulfate sensitized mice to Salmonella infection,[67] and O'Brien *et al.* observed the same result after depleting macrophages with silica.[68] In Hof's studies the capacity of immune serum or immune peritoneal exudate cells to transfer protection was also abolished by the prior treatment of recipients with dextran sulfate.[67]

In aggregate, these studies show that it has been difficult to find a population of immune T cells from live-cell-vaccinated mice that will reliably transfer high levels of immunity to recipients, or to show consistently that depletion of T cells is crucial for host defense. These results stand in contrast to easier demonstrations of a role for T cells in immunity to Listeria or Mycobacteria. Furthermore, there has been no attempt to relate the results of T-cell transfer and depletion experiments to the innate resistance of the mouse strain used, so the contribution of the genetic background of the mice to the interpretation of the experimental outcome is yet to be evaluated. In most cases, investigators have not taken into account the possibility that $CD4^+$ cells may function not only in cellular immunity via macrophage activation, but also as helper cells in antibody formation. Hence, the possibility that transferred cells may actually contribute to humoral rather than cellular immunity has not been rigorously tested.

Another approach to implicating T cells in defense against Salmonella is to examine proliferation of T cells from Salmonella-immunized mice to specific Salmonella antigens *in vitro*, and DTH responses *in vivo*, to indicate the presence of sensitized T cells. These assays have been used to obtain evidence for a T-cell response to attenuated vaccines.[28,65] However, the requirement for expression of DTH in order to have resistance to Salmonella infection is brought into question by the observation that high levels of resistance to virulent challenge can exist in the absence of a positive footpad reaction.[47,69]

Still another complication of understanding cellular mechanisms of host defense to Salmonella is that early resistance in mice appears to be mediated by different mechanisms than late-onset immunity. O'Brien and Metcalf[70] and Nauciel[57] found that control of growth of Salmonella early in infection does not require T cells, as nude mice were no more sensitive to infection than euthymic heterozygotes. Hormaeche *et al.* showed that irradiated, bone marrow-reconstituted mice could control bacterial growth in the early phase of infection without T cells.[71] Schafer and Eisenstein provided evidence that natural killer (NK) cells play a role in early immunity to Salmonella infection, as depletion of NK cells abrogated the capacity of live attenuated Salmonella vaccines to protect against subsequent challenge with a virulent strain.[72] Polymorphonuclear leukocytes are also present in lesions of the livers and spleens of Salmonella-infected mice.[73] Early in infection, polymorphonuclear leukocytes may provide a "safe site" for the organisms to escape the microbicidal activity of the host.[74]

9. CYTOKINE RESPONSES AND HOST RESISTANCE TO SALMONELLA

As our understanding of the crucial role of cytokines in regulating cellular immune responses and macrophage activation has increased, researchers have investigated the role of various cytokines in resistance to Salmonella infection, either by giving the cytokine or by giving antibodies to the cytokine. Since the macrophage is thought to have a central role in control of Salmonella replication, cytokines that activate macrophages or that are produced by activated macrophages have been a focus of study. Nakano's group found that administration of recombinant murine interferon-γ (IFN-γ) or recombinant tumor necrosis factor-α (TNF-α), prior to i.p. challenge with Salmonella, protected against lethal infection and markedly decreased the number of bacteria that could be found in the peritoneal cavities 24 hr later.[75,76] Muotiala and Makela[77] showed that IFN-γ restricted the rate of bacterial multiplication in the liver and spleen. Nauciel and Espinasse-Maes reported that a monoclonal antibody against IFN-γ or a rabbit antiserum to TNF-α sensitized mice to Salmonella infection and increased the number of organisms retrieved from the liver and spleen after an i.v. infection; administration of both antibodies had an additive effect.[78] The source of the IFN-γ did not appear to be T cells as depletion of $CD4^+$ or

CD8⁺ cells with monoclonal antibodies did not sensitize the animals to infection.[78] The sensitizing effect of anti-IFN-γ was confirmed by Muotiala and Makela,[79] but it was shown to influence microbial growth rates only in the early phase of the infection in both normal and previously vaccinated mice. Mastroeni *et al.* examined the time at which anti-TNF-α antibodies were effective and found that they potentiate Salmonella growth in both the early and the plateau phase of the infection.[80] Furthermore, this group found that $CD4^+$ cells, IFN-γ, and TNF-α were all important in the recall of immunity in mice vaccinated orally with an *aro*A strain of Salmonella.[81] Klimpel's laboratory compared IFN-γ responses of spleen cells from Ity^r with cells from Ity^s mice and reported that the former produce higher levels of IFN-γ in response to Salmonella *in vitro.*[82] Based on earlier studies, it was concluded that splenic NK cells are the source of the IFN-γ. The augmented response of Ity^r mice appeared to be related to production of an inducing factor for NK cells to make IFN-γ by the adherent cells. The *Ity* locus has also been found by Morrissey and Charrier to control responses of mice to IL-1 with respect to modulation of resistance to Salmonella infection.[83] Mice with the Ity^r phenotype that were treated with IL-1 16 hr before lethal infection with Salmonella survived, but Ity^s mice did not. Recently, a chemokine, monocyte-chemotactic and -activating factor (MCAF), has also been shown to protect mice against Salmonella challenge by enhancing the bactericidal capacity of macrophages.[84]

The relatively recent data reviewed in this section support the concept that early defense against Salmonella involves macrophages and NK cells and may not require T cells. The dichotomy in the types of mediators involved in protective responses against Salmonella infection in mice with differing innate resistance to this class of organisms is also reinforced by these studies.

10. IMMUNE RESPONSES TO ORAL INFECTION

The natural route of Salmonella infection is via the gastrointestinal tract, in both humans and rodents. Organisms invade in the small intestine at the Peyer's patches and disseminate through the draining lymphatics to the bloodstream and subsequently colonize and multiply in the liver and spleen.[85] As the workings of the secretory immune system unfolded following the discovery of IgA, it became clear that local application of antigen leads to optimal IgA responses. A logical consequence of this new understanding of secretory immune responses was the hypothesis that immunity to Salmonella could best be achieved by oral vaccination with a live attenuated strain, which could be expected to stimulate secretory IgA in the intestinal tract. Such a strategy would also have several practical advantages. An oral vaccine would be easier to administer in parts of the world where typhoid is endemic and it would be less likely to produce the adverse reactions associated with injecting nonviable organisms by parenteral routes. Germanier developed an attenuated strain of *S. typhi,* Ty21a by chemical mutagenesis and showed in field trials among schoolchildren in Egypt that it was protective.(86) Ty21a is a rough strain of the galE type, which means it has a lesion in the capacity to make UDP-galactose epimerase, a necessary enzyme for formation of galactose from glucose to make the core and O polysaccharides of Salmonella LPS. Subsequent field trials in Chile confirmed the protective capacity of this vaccine,(87) although initial trials did not show the same high levels of protection seen in Egypt. The field trials in Chile showed that different formulations and dosing regimens of the vaccine result in different efficacy.(51,87) It also appears that the vaccine may not be protective against high challenge doses that can occur in a typhoid epidemic.(51,87) Understanding of the mechanism of protection, whether by IgA, serum IgG, or cellular immunity, has been hampered by the need to carry out the studies using humans. One of the most interesting findings, however, is that Ty21a does not protect unless it is grown in exogenous galactose, presumably because the organism must be smooth to induce anti-O antibody.(88) Furthermore, the serum response of vaccinees to O antigen correlates with protection, yet anti-O antibody is still not considered important for protection by the investigators carrying out the studies.(51) Experiments in volunteers using Ty21a showed that it induced IgA antibodies to LPS detectable in feces.(89) A novel test for cellular immunity was used to assess stimulation of this effector mechanism. Mononuclear peripheral blood cells were obtained from vaccinees and shown to have enhanced bactericidal activity against *S. typhi.*(89) This activity was thought to be mediated by T cells armed with cytophilic serum IgA.(89) In 1981 Hoiseth and Stocker showed that mouse-virulent *Salmonella typhimurium* strains, rendered avirulent by introduction of a lesion in the *aro*A gene, could be used as oral vaccines.(90) This approach was preferable to that of chemical mutagenesis because the lesion was defined and nonreverting, because of either a deletion or inversion of the *aro*A gene. Salmonella attenuated in this way gave high levels of protection in mice against oral and systemic challenge with virulent Salmonella of the same serotype.(90) The concept of using these and other attenuated strains to carry antigens of other pathogens was developed further in order make an oral vaccine that would be easy to deliver and could induce cellular immunity against a variety of microbial antigens including malaria circumsporozoite protein and M protein of streptococci.(91) Mice vaccinated with these strains have been shown to develop DTH and cytotoxic

T cells against the passenger antigen,[92,93] as well as antibody, suggesting that live attenuated Salmonella do induce cell-mediated immunity.

The studies on secretory immune responses still had many unanswered questions, which were being pursued in the laboratory simultaneously with the more practical aspects of vaccine development. One question was an assessment of the actual contribution of secretory IgA to host defense in the intestinal tract. Neutra's group showed that subcutaneous implantation of a hybridoma producing monoclonal IgA against a carbohydrate epitope on *S. typhimurium* protected BALB/c mice against oral challenge with the same strain of Salmonella.[94] This study was definitive in showing that IgA alone could protect hypersusceptible mice against oral Salmonella challenge. Interestingly, however, the monoclonal did not protect against i.p. challenge, suggesting that IgA has a specific protective role in the gut. Another question that emerged was whether use of the same carrier Salmonella for a variety of different passenger antigens would be detrimental if there were subsequent immunizations with the same or different passenger antigens. It was feared that oral tolerance might develop to the Salmonella and thus prevent booster responses or responses to new passenger antigens given at a later time. Bao and Clements carried out systematic studies to address this question and showed that mucosal but not serum antibody responses against foreign antigens increased if the mouse had first been primed with the carrier strain of homologous serotype to the subsequent challenge.[95] Thus, oral tolerance did not develop in mice to the Salmonella carrier, but instead the initial immunization primed the mouse for an anamnestic response on subsequent exposure. Both of these studies, along with the experiments showing induction of cellular immunity in mice, provided experimental evidence for the validity of the approach using attenuated live Salmonella orally as vaccines. A potential problem with these vaccines is that when given intraperitoneally to mice, they have been shown to induce transient immunosuppression as measured by hyporesponsiveness of spleen cells to mitogens[96] and suppressed capacity of spleen cells to make antibody to sheep red blood cells and tetanus toxoid.[97,98] Whether a similar suppression occurs in the secretory system is not known.

11. SUMMARY

The literature on immunity to Salmonella is vast, and only a small portion of the published papers could be reviewed in this chapter. The citations chosen address specific points about the theory of protective immune responses to Salmonella or provide definitive empirical evidence of strategies that induce protection, from which mechanisms might be inferred. It would appear that theoretical considerations have not been highly predictive of assessment of efficacy of potential Salmonella vaccines or vaccine strategies. A major reason for this situation has been the systematic failure of investigators to take into account diverse findings in the literature. Among the conclusions that emerge from survey of the literature are that antibody alone is protective in mice, using *S. typhimurium* challenge, and in humans, as deduced from vaccine trials. Thus, passive antibody protects many strains of mice against systemic Salmonella challenge and against oral challenge.[21,94] Vi antigen protects humans against typhoid fever as assessed in field trials.[49,50] Purified polysaccharide Vi antigen is thought to act only by inducing antibody. Clearly, humoral immunity alone is sufficient for protection against typhoid fever in humans. Mouse studies show that living Salmonella can induce cellular immunity.[99] However, the antigens against which the cellular immune responses are directed during natural infection are not precisely defined. Furthermore, attenuated vaccines can induce high levels of protection in the absence of DTH.[65] In humans, it has been much more difficult to detect responses that are clearly cellular. Skin tests to measure DTH have not been developed. The mononuclear cell bactericidal test is not on a firm theoretical basis. Some studies have suggested that positive proliferative responses of T cells of human vaccinees are directed against the O antigens of LPS. This conclusion does not fit with the current theory that T-cell responses are restricted to protein antigens. The role of T cells in host defense to Salmonella is still an open question. Part of the difficulty in developing evidence for T-cell involvement may be confounding events that occur early in infection. Recent studies in mice point toward a model in which NK cells, polymorphonuclear leukocytes, and macrophages are important in host defense in the absence of T cells. It may be possible to detect sensitized T cells only later in the infection or later after immunization with attenuated vaccines, when the early inflammatory events have subsided.

In spite of a lack of a firm theoretical foundation, live attenuated Salmonella have been shown to be effective as vaccines. Interestingly, in humans they are not more efficacious than Vi antigen, showing that if live vaccines do work by inducing cellular immunity, they are not substantially more protective than nonviable vaccines which confer humoral immunity. Live vaccines, like whole-cell vaccines, require multiple doses, which detracts somewhat from their appeal for use in developing areas of the world. They do, however, have the advantage of a much simpler mode of administration than vaccines that require parenteral injection.

The studies reviewed in this chapter reveal the complexity of host responses to Salmonella. Future directions for typhoid vaccines may rest on empirical and practical considerations rather than theoretical ones.

REFERENCES

1. Wright, A., and Semple, D., 1897, Remarks on vaccination against typhoid fever, *Br. Med. J.* **1:**256.
2. Siler, J. F., *et al.*, 1941, *Immunization to Typhoid Fever*, Johns Hopkins Press, Baltimore.
3. Felix, A., and Pitt, R. M., 1951, The pathogenic and immunogenic activity of *Salmonella typhi* in relation to its antigenic constituents, *J. Hyg.* **49:**92–110.
4. Robbins, J. D., and Robbins, J. B., 1984, Reexamination of the protective role of the capsular polysaccharide (Vi antigen) of *Salmonella typhi* [Review], *J. Infect. Dis.* **150:**436–449.
5. Joo, I., 1971, Bacterial diseases: The problem, and control by vaccines. Present status and perspectives of vaccination against typhoid fever, in: *International Conference on the Application of Vaccines Against Viral, Rickettsial, and Bacterial Diseases of Man*, Scientific Publication No. 226, Pan-American World Health Organization, Washington, D.C., pp. 329–341.
6. Eisenstein, T. K., and Sultzer, B. M., 1983, Immunity to Salmonella infection, *Adv. Exp. Med. Biol.* **162:**261–296.
7. Ashcroft, M. T., Ritchie, J. M., and Nicholson, C. C., 1964, Controlled field trial in British Guiana school children of heat-killed phenolized and acetone-killed lyophilized typhoid vaccines, *Am. J. Hyg.* **79:**196–206.
8. Spaun, J., and Uemura, K., 1964, International reference preparations of typhoid vaccine. A report on international collaborative laboratory studies, *Bull. WHO* **31:**761–791.
9. Jenkin, C. R., and Rowley, D., 1965, Partial purification of the "protective" antigen of *Salmonella typhimurium* and its distribution amongst various strains of bacteria, *Aust. J. Exp. Biol. Med. Sci.* **43:**65–78.
10. Herzberg, M., Nash, P., and Hino, S., 1972, Degree of immunity induced by killed vaccines to experimental salmonellosis in mice, *Infect. Immun.* **5:**83–90.
11. Ornellas, E. P., Roantree, R. J., and Steward, J. P., 1970, The specificity and importance of humoral antibody in the protection of mice against intraperitoneal challenge with complement-sensitive and complement-resistant Salmonella, *J. Infect. Dis.* **121:**113–123.
12. Angerman, C. R., and Eisenstein, T. K., 1978, Comparative efficacy and toxicity of a ribosomal vaccine, acetone-killed cells, lipopolysaccharide, and a live cell vaccine prepared from *Salmonella typhimurium*, *Infect. Immun.* **19:**575–582.
13. Lindberg, A. A., Rosenberg, L. T., Ljunggren, A., Garegg, P. J., Svensson, S., and Wallin, N. H., 1974, Effect of synthetic disaccharide–protein conjugate as an immunogen in Salmonella infection in mice, *Infect. Immun.* **10:**541–545.
14. Svenson, S. B., Nurminen, M., and Lindberg, A. A., 1979, Artificial Salmonella vaccines: O-antigenic oligosaccharide–protein conjugates induce protection against infection with *Salmonella typhimurium*, *Infect. Immun.* **25:**863–872.
15. Kuusi, N., Nurminen, M., Saxen, H., Valtonen, M., and Makela, P. H., 1979, Immunization with major outer membrane proteins in experimental salmonellosis of mice, *Infect. Immun.* **25:**857–862.
16. Phillips, M., Castagna, R., Sultzer, B. M., and Eisenstein, T. K., 1989, Immunogenic endotoxin associated protein from a rough strain of Salmonella, *FEMS Microbiol. Immunol.* **1:** 485–490.

17. Kuusi, N., Nurminen, M., Saxen, H., and Makela, P. H., 1981, Immunization with major outer membrane protein (porin) preparations in experimental murine salmonellosis: Effect of lipopolysaccharide, *Infect. Immun.* **34:**328–332.
18. Jenkin, C. R., Rowley, D., and Auzins, I., 1965, The basis for immunity to mouse typhoid. I. The carrier state, *Aust. J. Exp. Biol. Med. Sci.* **42:** 215–228.
19. Rowley, D., Auzins, I., and Jenkin, C. R., 1968, Further studies regarding the question of cellular immunity in mouse typhoid, *Aust. J. Exp. Biol. Med. Sci.* **46:**447–463.
20. Blanden, R. V., Mackaness, G. B., and Collins, F. M., 1966, Mechanisms of acquired resistance in mouse typhoid, *J. Exp. Med.* **124:**585–600.
21. Eisenstein, T. K., Killar, L. M., and Sultzer, B. M., 1984, Immunity to infection with *Salmonella typhimurium*: Mouse-strain differences in vaccine- and serum-mediated protection, *J. Infect. Dis.* **150:**425–435.
22. Hobson, D., 1957, Resistance to reinfection in experimental mouse typhoid, *J. Hyg.* **55:**334–343.
23. Ushiba, D., Saito, K., Akiyama, T., Nakano, M., Sugiyama, T., and Shirono, S., 1959, Studies on experimental typhoid: Bacterial multiplication and host cell response after infection with *Salmonella enteritidis* in mice immunized with live and killed vaccines, *Jpn. J. Microbiol.* **3:**231–242.
24. Ushiba, D., 1965, Two types of immunity in experimental typhoid; "Cellular immunity" and "humoral immunity," *Keio J. Med.* **14:**45–60.
25. Collins, F. M., Mackaness, G. B., and Blanden, R. V., 1966, Infection-immunity in experimental salmonellosis, *J. Exp. Med.* **124:**601–619.
26. Mackaness, G. B., Blanden, R. V., and Collins, F. M., 1966, Host–parasite relations in mouse typhoid, *J. Exp. Med.* **124:**573–583.
27. Mackaness, G. B., 1971, Resistance to intracellular infection, *J. Infect. Dis.* **123:**439–445.
28. Collins, F. M., and Mackaness, G. B., 1968, Delayed hypersensitivity and Arthus reactivity in relation to host resistance in Salmonella-infected mice, *J. Immunol.* **101:**830–845.
29. Collins, F. M., 1978, Cellular antimicrobial immunity [Review], *Crit. Rev. Microbiol.* **7:**27–91.
30. Collins, F. M., 1969, Effect of specific immune mouse serum on the growth of *Salmonella enteritidis* in mice preimmunized with living or ethyl alcohol-killed vaccines, *J. Bacteriol.* **97:**676–683.
31. Plant, J., and Glynn, A. A., 1974, Natural resistance to Salmonella infection, delayed hypersensitivity and Ir genes in different strains of mice, *Science* **248:**345–347.
32. Plant, J. E., Blackwell, J. M., O'Brien, A. D., Bradley, D. J., and Glynn, A. A., 1982, Are the Lsh and Ity disease resistance genes at one locus on mouse chromosome 1? *Nature* **297:**510–511.
33. Vidal, S. M., Malo, D., Vogan, K., Skamene, E., and Gros, P., 1993, Natural resistance to infection with intracellular parasites: Isolation of a candidate for *Bcg*, *Cell* **73:**469–485.
34. Lissner, C. R., Weinstein, D. L., and O'Brien, A. D., 1985, Mouse chromosome 1 *Ity* locus regulates microbicidal activity of isolated peritoneal macrophages against a diverse group of intracellular and extracellular bacteria, *J. Immunol.* **135:**544–547.
35. Benjamin, W. H., Jr., Hall, P., Roberts, S. J., and Briles, D. E., 1990, The primary effect of the *Ity* locus is on the rate of growth of *Salmonella typhimurium* that are relatively protected from killing, *J. Immunol.* **144:**3143–3151.
36. O'Brien, A. D., Scher, I., Campbell, G. H., MacDermott, R. P., and Formal, S. B., 1979, Susceptibility of CBA/N mice to infection with *Salmonella typhimurium*: Influence of the X-linked gene controlling B lymphocyte function, *J. Immunol.* **123:**720–724.
37. O'Brien, A. D., Rosenstreich, D. L., Scher, I., Campbell, G. H., MacDermott, R. P., and Formal, S. B., 1980, Genetic control of susceptibility to *Salmonella typhimurium* in mice: Role of the Lps gene, *J. Immunol.* **124:**20–24.
38. Eisenstein, T. K., Deakins, L. W., Killar, L., Saluk, P. H., and Sultzer, B. M., 1982, Dissociation

of innate susceptibility to Salmonella infection and endotoxin responsiveness in C3HeB/FeJ mice and other strains in the C3H lineage, *Infect. Immun.* **36:**696–703.
39. O'Brien, A. D., and Rosenstreich, D. L., 1977, Genetic control of the susceptibility of C3HeB/FeJ mice to *Salmonella typhimurium* is regulated by a locus distinct from known salmonella response genes, *J. Immunol.* **131:**2613–2619.
40. Hormaeche, C. E., 1979, Genetics of natural resistance to salmonellae in mice, *J. Immunol.* **37:**319–327.
41. Benjamin, W. H., Jr., Turnbough, C. L., Jr., Posey, B. S., and Briles, D. E., 1986, *Salmonella typhimurium* virulence genes necessary to exploit the *Ity*$^{s/s}$ genotype of the mouse, *Infect. Immun.* **51:**872–878.
42. Benbernou, N., and Nauciel, C., 1994, Influence of mouse genotype and bacterial virulence in the generation of interferon-gamma-producing cells during the early phase of *Salmonella typhimurium* infection, *Immunology* **83:**245–249.
43. Gowen, J. W., and Stadler, J., 1967, Genetic characteristics influencing vaccine-conferred immunity, *J. Infect. Dis.* **117:**129–150.
44. Robson, H. G., and Vas, S. I., 1972, Resistance of inbred mice to *Salmonella typhimurium*, *J. Infect. Dis.* **126:**378–386.
45. Hormaeche, C. E., 1979, Natural resistance to *Salmonella typhimurium* in different inbred mouse strains, *Immunology* **37:**311–318.
46. Nauciel, C., Vilde, F., and Ronco, E., 1985, Host response to infection with a temperature-sensitive mutant of *Salmonella typhimurium* in a susceptible and a resistant strain of mice, *Infect. Immun.* **49:**523–527.
47. Killar, L. M., and Eisenstein, T. K., 1984, Differences in delayed-type hypersensitivity responses in various mouse strains in the C3H lineage infected with *Salmonella typhimurium*, strain SL3235, *J. Immunol.* **133:**1190–1196.
48. Eisenstein, T. K., and Angerman, C. R., 1978, Immunity to experimental Salmonella infection: Studies on the protective capacity and immunogenicity of lipopolysaccharide, acetone-killed cells, and ribosome-rich extracts of *Salmonella typhimurium* in C3H/HeJ and CD-1 mice, *J. Immunol.* **121:**1010–1014.
49. Acharya, I. L., Lowe, C. U., Thapa, R., Gurubacharya, V. L., Shrestha, M. B., Bact, D., Cadoz, M., Schultz, D., Armand, J., Bryla, D. A., Trollfors, B., Cramton, T., Schneerson, R., and Robbins, J. B., 1987, Prevention of typhoid fever in Nepal with the Vi capsular polysaccharide of *Salmonella typhi* in mice, *N. Engl. J. Med.* **317:**1101–1104.
50. Klugman, K., Gilberson, I. T., Koornhof, H. J., Robbins, J. B., Schneerson, R., Schulz, D., Cadoz, M., and Armand, J., 1987, Vaccination advisory committee: Protective activity of Vi capsular polysaccharide vaccine against typhoid fever, *Lancet* **2:**1165–1169.
51. Ivanoff, B., Levine, M. M., and Lambert, P. H., 1994, Vaccination against typhoid fever: Present status [Review], *Bull. WHO* **72:**957–971.
52. Marmion, D. E., Naylor, G. R. E., and Stewart, I. P., 1953, Second attacks of typhoid fever, *J. Hyg.* **51:**260.
53. Islam, A., Butler, T., and Ward, L. R., 1987, Reinfection with a different Vi-phage type of *Salmonella typhi* in an endemic area, *J. Infect. Dis.* **155:**155–156.
54. Hochadel, J. F., and Keller, K. F., 1977, Protective effects of passively transferred immune T- or B-lymphocytes in mice infected with *Salmonella typhimurium*, *J. Infect. Dis.* **135:**813–823.
55. Morris, J. A., Wray, C., and Sojka, W. J., 1976, The effect of T and B lymphocyte depletion on the protection of mice vaccinated with a *Gal*E mutant of *Salmonella typhimurium*, *Br. J. Exp. Pathol.* **57:**354–360.
56. Davies, R., and Kotlarski, I., 1976, The role of thymus-derived cells in immunity to Salmonella infection, *Aust. J. Exp. Biol. Med. Sci.* **54:**221–236.

57. Nauciel, C., 1990, Role of CD4+ T cells and T-independent mechanisms in acquired resistance to *Salmonella typhimurium* infection, *J. Immunol.* **145:**1265–1269.
58. Mixter, P. F., Camerini, V., Stone, B. J., Miller, V. L., and Kronenberg, M., 1994, Mouse T lymphocytes that express a gamma delta T-cell antigen receptor contribute to resistance to Salmonella infection *in vivo*, *Infect. Immun.* **62:**4618–4621.
59. Chander, R., Sainis, K. B., and Lewis, N. F., 1995, Role of thymus-derived lymphocytes in acquired immunity to salmonellosis in mice, *Microbiol. Immunol.* **30:**1299–1306.
60. Guilloteau, L., Buzoni-Gatel, D., Bernard, F., Lantier, I., and Lantier, F., 1993, Salmonella abortusovis infection in susceptible BALB/cby mice: Importance of Lyt-2+ and L3T4+ T cells in acquired immunity and granuloma formation, *Microb. Pathog.* **14:**45–55.
61. Hougen, H. P., and Jensen, E. T., 1990, Experimental *Salmonella typhimurium* infections in rats. III. Transfer of immunity with primed lymphocyte subpopulations, *Acta Pathol. Microbiol. Immunol. Scand.* **98:**1015–1021.
62. Mastroeni, P., Villarreal-Ramos, B., and Hormaeche, C. E., 1993, Adoptive transfer of immunity to oral challenge with virulent salmonellae in innately susceptible BALB/c mice requires both immune serum and T cells, *Infect. Immun.* **61:**3981–3984.
63. Paul, C., Shalala, K., Warren, R., and Smith, R., 1985, Adoptive transfer of murine host protection to salmonellosis with T-cell growth factor-dependent, Salmonella-specific T-cell lines, *Infect. Immun.* **48:**40–43.
64. Paul, C. C., and Smith, R. A., 1988, Transfer of murine host protection by using interleukin-2-dependent T-lymphocyte lines, *Infect. Immun.* **56:**2189–2192.
65. Killar, L. M., and Eisenstein, T. K., 1985, Immunity to *Salmonella typhimurium* infection in C3H/HeJ and C3H/HeNCrlBR mice: Studies with an aromatic-dependent live *S. typhimurium* strain as a vaccine, *Infect. Immun.* **47:**605–612.
66. Maskell, D. J., Hormaeche, C. E., Harrington, K. A., Joysey, H. S., and Liew, F. Y., 1987, The initial suppression of bacterial growth in a salmonella infection is mediated by a localized rather than a systemic response, *Microb. Pathog.* **2:**295–305.
67. Hof, H., Emmerling, P., Hacker, J., and Hughes, C., 1982, The role of macrophages in primary and secondary infection of mice with *Salmonella typhimurium*, *Ann. Immunol.* (Inst. Pasteur) **133:**21–32.
68. O'Brien, A. D., Scher, I., and Formal, S. B., 1979, Effect of silica on the innate resistance of inbred mice to *Salmonella typhimurium* infection, *Infect. Immun.* **25:**513–520.
69. Killar, L. M., and Eisenstein, T. K., 1986, Delayed-type hypersensitivity and immunity to *Salmonella typhimurium*, *Infect. Immun.* **52:**504–508.
70. O'Brien, A. D., and Metcalf, E. S., 1982, Control of early *Salmonella typhimurium* growth in innately Salmonella-resistant mice does not require functional T lymphocytes, *J. Immunol.* **129:**1349–1351.
71. Hormaeche, C. E., Mastroeni, P., Arena, A., Uddin, J., and Joysey, H. S., 1990, T cells do not mediate the initial suppression of a Salmonella infection in the RES, *Immunology* **70:**247–250.
72. Schafer, R., and Eisenstein, T. K., 1992, Natural killer cells mediate protection induced by a Salmonella *aro*A mutant, *Infect. Immun.* **60:**791–797.
73. Nakoneczna, I., and Hsu, H. S., 1980, The comparative histopathology of primary and secondary lesions in murine salmonellosis, *Br. J. Exp. Pathol.* **61:**76–84.
74. Dunlap, N. E., Benjamin, W. H., Berry, K. A., Eldridge, J. H., and Briles, D. E., 1992, A "safe site" for *Salmonella typhimurium* is within the splenic polymorphonuclear cells, *Microb. Pathog.* **13:**181–190.
75. Matsumura, H., Onozuka, K., Terada, Y., Nakano, Y. and Nakano, M., 1990, Effect of murine recombinant interferon-γ in the protection of mice against Salmonella, *Int. J. Immunopharmacol.* **12:**49–56.
76. Nakano, Y., Onozuka, K., Terada, Y., Shinomiya, H., and Nakano, M., 1990, Protective effect

of recombinant tumor necrosis factor-α in murine salmonellosis, *J. Immunol.* **144:**1935–1941.
77. Muotiala, A., and Makela, P. H., 1990, The role of IFN-γ in murine *Salmonella typhimurium* infection, *Microb. Pathog.* **8:**135–141.
78. Nauciel, C., and Espinasse-Maes, F., 1992, Role of gamma interferon and tumor necrosis factor alpha in resistance to *Salmonella typhimurium* infection, *Infect. Immun.* **60:**450–454.
79. Muotiala, A., and Makela, P. H., 1993, Role of gamma interferon in late stages of murine salmonellosis, *Infect. Immun.* **61:**4248–4253.
80. Mastroeni, P., Villarreal-Ramos, B., and Hormaeche, C. E., 1993, Effect of late administration of anti-TNF alpha antibodies on a Salmonella infection in the mouse model, *Microb. Pathog.* **14:**473–480.
81. Mastroeni, P., Villarreal-Ramos, B., and Hormaeche, C. E., 1992, Role of T cells, TNF alpha and IFN gamma in recall of immunity to oral challenge with virulent salmonellae in mice vaccinated with live attenuated *aro*-Salmonella vaccines, *Microb. Pathog.* **13:**477–491.
82. Ramarathinam, L., Niesel, D. W., and Klimpel, G. R., 1993, *Ity* influences the production of IFN-γ by murine splenocytes stimulated *in vitro* with *Salmonella typhimurium*, *J. Immunol.* **150:** 3965–3972.
83. Morrissey, P. J., and Charrier, K., 1994, Treatment of mice with IL-1 before infection increases resistance to a lethal challenge with *Salmonella typhimurium.* The effect correlates with the resistance allele at the *Ity* locus, *J. Immunol.* **153:**212–219.
84. Nakano, Y., Kasahara, T., Mukaida, N., Ko, Y. C., Nakano, M., and Matsushima, K., 1994, Protection against lethal bacterial infection in mice by monocyte-chemotactic and -activating factor, *Infect. Immun.* **62:**377–383.
85. Carter, P. B., and Collins, F. M., 1974, The route of enteric infection in normal mice, *J. Exp. Med.* **139:**1189–1203.
86. Wahdan, M. H., Serie, C., Cerisier, Y., Sallam, S., and Germanier, R., 1982, A controlled field trial of live *Salmonella typhi* strain Ty 21a oral vaccine against typhoid: Three-year results, *J. Infect. Dis.* **145:**292–295.
87. Levine, M. M., and Hone, D. M., 1991, Typhoid fever, in: *Vaccines and Immunotherapy* (S. J. J. Cryz, ed.), Pergamon Press, New York, pp. 59–72.
88. Gilman, R. H., Hornick, R. B., Woodward, W. E., DuPont, H. L., Snyder, M. J., Levine, M. M., and Libonati, J. P., 1977, Evaluation of a UDP-glucose-4-epimeraseless mutant of *Salmonella typhi* as a live oral vaccine, *J. Infect. Dis.* **136:**717–723.
89. D'Amelio, R., Tagliabue, A., Nencioni, L., Di Addario, A., Villa, L., Manganaro, M., Boraschi, D., Le Moli, S., Nisini, R., and Matricardi, P. M., 1988, Comparative analysis of immunological responses to oral (Ty21a) and parenteral (TAB) typhoid vaccines, *Infect. Immun.* **56:**2731–2735.
90. Hoiseth, S. K., and Stocker, B. A. D., 1981, Aromatic-dependent *Salmonella typhimurium* are non-virulent and effective as live vaccines, *Nature* **291:**238–239.
91. Cardenas, L., and Clements, J. D., 1992, Oral immunization using live attenuated Salmonella spp. as carriers of foreign antigens [Review], *Clin. Microbiol. Rev.* **5:**328–342.
92. Flynn, J. L., Weiss, W. R., Norris, K. A., Seifert, H. S., Kumar, S., and So, M., 1990, Generation of a cytotoxic T-lymphocyte response using a Salmonella antigen delivery system, *Mol. Microbiol.* **4:**2111–2118.
93. Brown, A., Hormaeche, C. E., Demarco de Hormaeche, R., Winther, M., Dougan, G., and Stocker, B. A., 1987, An attenuated aroA *Salmonella typhimurium* vaccine elicits humoral and cellular immunity to cloned beta-galactosidase in mice, *J. Infect. Dis.* **155:**86–92.
94. Michetti, P., Mahan, M. J., Slauch, J. M., Mekalanos, J. J., and Neutra, M. R., 1992, Monoclonal secretory immunoglobulin A protects mice against oral challenge with the invasive pathogen *Salmonella typhimurium*, *Infect. Immun.* **60:**1786–1792.

95. Bao, J. X., and Clements, J. D., 1991, Prior immunologic experience potentiates the subsequent antibody response when Salmonella strains are used as vaccine carriers, *Infect. Immun.* **59:**3841–3845.
96. Lee, J. C., Gibson, C. W., and Eisenstein, T. K., 1985, Macrophage-mediated mitogenic suppression induced in mice of the C3H lineage by a vaccine strain of *Salmonella typhimurium*, *Cell. Immunol.* **91:**75–91.
97. Al-Ramadi, B. K., Meissler, J. J., Jr., Huang, D., and Eisenstein, T. K., 1992, Immunosuppression induced by nitric oxide and its inhibition by interleukin-4, *Eur. J. Immunol.* **22:**2249–2254.
98. Al-Ramadi, B. K., Greene, J. M., Meissler, J. J., Jr., and Eisenstein, T. K., 1992, Immunosuppression induced by attenuated Salmonella: Effect of LPS responsiveness on development of suppression, *Microb. Pathog.* **12:**267–278.
99. Eisenstein, T. K., Killar, L. M., Stocker, B. A., and Sultzer, B. M., 1984, Cellular immunity induced by avirulent Salmonella in LPS-defective C3H/HeJ mice, *J. Immunol.* **133:**958–961.

5

Shigella Infection

GERALD T. KEUSCH and DAVID W. K. ACHESON

1. INTRODUCTION

Shigella dysenteriae type 1, the first *Shigella* species isolated, was discovered by Kiyoshi Shiga in 1896[1] in the course of an epidemic of especially severe dysentery in Japan, with nearly 100,000 cases and 30,000 deaths. Shiga not only isolated the same organism from many patients, but his report in 1898 documented that patients developed agglutinating antibodies during convalescence. The same organism was soon reported elsewhere in the world,[2,3] and over the next 40 years, three additional groups of related organisms were defined and ultimately taxonomically placed in the genus *Shigella* and named *S. dysenteriae*, *S. flexneri*, *S. boydii*, and *S. sonnei* to honor the lead workers, Shiga, Flexner, Boyd, and Sonne.[4] In the current era of molecular microbiology, however, it is not so clear that *Shigella* should be a distinct genus, because the organisms are so similar to *E. coli* that no distinction can be made on the basis of DNA composition, DNA–DNA hybridization, or genetic organization, and *Shigella* can be thought of as differentiated, highly (human) host-adapted pathogenic *E. coli* species. In more practical clinical microbiological terms, *Shigella* are distinct from nearly all *E. coli* by their inability to ferment lactose, a property used to select for suspect lactose-negative colonies on differential agars. Clinicians also immediately recognize the genus name *Shigella* as a pathogen when reported in cultures.

Just 2 years after Shiga's complete description of *S. dysenteriae* type 1, Flexner found that parenteral injection of killed *Shigella* cultures in mice led

---

GERALD T. KEUSCH and DAVID W. K. ACHESON • Division of Geographic Medicine and Infectious Diseases, Tupper Research Institute, New England Medical Center, Boston, Massachusetts 02111.

Enteric Infections and Immunity, edited by Lois J. Paradise *et al.* Plenum Press, New York, 1996.

to their death and on this basis concluded that the disease was caused by "a toxic agent rather than infection per se."[2] Three years later, Conradi[5] found that autolysates of cultures of *S. dysenteriae* type 1 caused diarrhea, limb paralysis, and death within 48–72 hr of intravenous injection into young adult rabbits. Because of these findings, the factor was called Shiga neurotoxin, or just Shiga toxin. Todd[6] soon found that *S. flexneri* filtrates, which also caused diarrhea when injected, did not cause paralysis, indicating specific production of "neurotoxin" by *S. dysenteriae* type 1, which was ultimately confirmed when the gene was identified. In retrospect, these early investigators were no doubt seeing the combined effects of lipopolysaccharide (LPS) endotoxin and the protein Shiga toxin. Shiga toxin is now known to be the prototype of a family of structurally and functionally related toxins, including the *E. coli* Shiga-like toxins [SLTs; also known as Verotoxins (VTs)] implicated in the pathogenesis of enterohemorrhagic *E. coli* infection.[7] In the past few years, evidence of two additional nonlethal *Shigella* protein toxins related to toxins found in certain *E. coli.* has been presented.[8,9] Because they result in fluid secretion by the intestine, they have been named *Shigella* enterotoxins (ShET) 1 and 2.

Over the past four decades, the critical pathogenetic role of *Shigella* invasion into host cells has been delineated, and a start has been made to unravel the molecular mechanisms involved, leading to the selection of putative protective antigens for vaccines.[10] At the same time, it is also fair to say that we do not know the basis of immune protection in shigellosis or the nature of the protective or possibly harmful immune response in this complex infection. This chapter is devoted to an exploration of these issues.

2. EPIDEMIOLOGY

Shigellosis is a global problem. Infection occurs with a very small inoculum,[11] which permits the person-to-person spread of this highly human host-adapted organism, although food- and water-borne outbreaks do occur.[12] *S. flexneri* is the hyperendemic species in developing countries, with epidemic and endemic disease being caused by *S. dysenteriae* type 1, whereas *S. sonnei* is the predominant isolate in developed countries, involved in over 75% of the cases annually in the United States. The fourth species, *S. boydii*, was first found in India and to this day is uncommonly encountered except in the Indian subcontinent. Curiously, although *S. sonnei* is isolated three times more often than *S. flexneri* in the United States, the latter is the most common isolate in male homosexuals.[13] As a result, the incidence of *S. flexneri* in the United States has, for the first time in the past 25 years, increased significantly in the past decade, with the increase entirely among young adult males, as the rate in females and children has continued to decrease. The consequence of this is a rise in the average age of patients with *S. flexneri* from around 5 to 25 years.[14]

3. CLINICAL FEATURES

The manifestations of *Shigella* infection vary with the infecting species, the age of the host, the presence of risk factors, and, we presume, the specific immune status of the host. Clinical disease typically begins within 24–48 hr of ingestion of a few hundred to a few thousand organisms with constitutional symptoms such as fever, fatigue, malaise, and anorexia. Watery diarrhea develops soon thereafter in the majority, progressing in some within a few hours to a few days to bloody diarrhea and the dysentery syndrome. Progression is more likely with *S. flexneri* or *S. dysenteriae* type 1 infection, and least likely with *S. sonnei*. Large fluid losses and severe dehydration are rare in shigellosis[15]; however, profound hyponatremia can occur secondary to inappropriate secretion of antidiuretic hormone.[16] Dysentery is a triad of findings: (1) the classic dysenteric stool, which is of small volume and grossly bloody, with mucus and often detectable pus, (2) abdominal cramps, and (3) tenesmus, the sensation to urgently pass stool, with little result except painful spasm of the anal sphincter. Lesions are most prominent in the distal large bowel, with death and sloughing of epithelial cells leading to mucosal ulcerations, bleeding into the bowel lumen, and a marked inflammatory infiltrate in the lamina propria, with large numbers of neutrophils in the excreta. Stool frequency is increased, generally 10–20 movements per day, but on occasion much more frequently. This is rather distressing to the patient because of the pain and tenesmus associated with evacuations. Anorexia, which is a prominent finding initially, may persist into convalescence and contribute to the deterioration in the patient's nutritional status, which commonly occurs in shigellosis.

Shigellosis can be complicated by severe local and systemic events, including intestinal obstruction, toxic megacolon, or perforation of the dilated intestine.[17,18] The pathogenesis of toxic megacolon remains obscure but is highly associated with *S. dysenteriae* type 1 infection, suggesting that the intensity of mucosal inflammation, possibly related to Shiga toxin, may be an important factor, probably via the release of inflammatory mediators. Bacteremia is usually not looked for in the mild nondysenteric illness caused by *S. sonnei* commonly seen in the United States, but in poorly nourished young infants with shigellosis in Bangladesh, almost 10% are bacteremic with either the infecting species of *Shigella* or another gram-negative rod.[19] The mortality rate with *Shigella* sepsis is double that of nonbacteremic patients (21 versus 10%) and is fivefold increased in those with bacteremia caused by other gram-negative rods. Regardless of blood culture data, circulating endotoxin is readily detected in hospitalized patients with *Shigella* infection in developing countries such as Bangladesh, and the highest levels are recorded in *S. dysenteriae* type 1 infection.

Certain systemic complications accompany *S. dysenteriae* type 1 infection, including toxic megacolon, leukemoid reactions with leukocyte counts in excess of 50,000/dl, and hemolytic uremic syndrome (HUS), a microangiopathic hemolytic anemia resulting from damage to the small vessels in the kidney and elsewhere, associated with thrombocytopenia usually in the range of 30,000–100,000, and acute renal failure. HUS is often diagnosed as the patient begins to recover from the intestinal disease.[20] This suggests that the initiating events in HUS occur early, probably when both endotoxin (an important activator of endothelial cell synthesis of cytokines and other vasoactive mediators) and Shiga toxin (or SLTs), which are cytotoxic to endothelial cells, are circulating.

Neurologic complications, especially seizures, are common in young patients with shigellosis. Seizures are usually generalized, occur early in the course of disease, are often associated with a rapidly rising fever, usually do not recur, and are rarely associated with permanent sequelae. They resemble typical febrile seizures, except for the higher mean age in *Shigella* patients. The pathogenesis is unknown; suggestions that they may be caused by Shiga toxin, or by a 100- to 125-kDa cytotoxin distinct from Shiga toxin, remain unproven.[21] Encephalopathy, independent of hypoglycemia, can also develop, especially during infection with *S. flexneri*. Fatalities in HUS are often associated with CNS pathology and cerebral edema.

Other recognized complications include post-*Shigella* reactive arthritis or full-blown Reiter's syndrome, particularly with *S. flexneri* infection. An immunological basis for this is suggested by the strong correlation between reactive arthritis and expression of the HLA-B27 histocompatibility antigen. Monoclonal antibodies against B27 epitopes cross-react with 23- and 36-kDa proteins of *S. flexneri*,[22] and monoclonal antibodies raised against a 36-kDa *S. flexneri* outer membrane protein cross-react with lymphocytes expressing HLA-B27.[23] These findings suggest that molecular mimicry between organism and host may be the basis for the response. In addition, a 2-MDa plasmid containing genes with sequence homology with the α-1 domain of HLA-B27 has been found in arthritogenic strains of *S. flexneri*.[24] The role of molecular mimicry in pathogenesis of Reiter's syndrome remains speculative.[25]

4. PATHOGENESIS

4.1. Acid Resistance

The molecular basis of shigellosis is complex and virulence of *Shigella* is absolutely determined by the capacity of the organism to invade mammalian cells.[26] This property does not explain one of the unique characteristics of *Shigella* pathogenesis, the small inoculum of a few hundred to a few thousand organisms required to cause disease. Such a small inoculum would need to survive the gastric acid barrier, and it is known that gastric acid increases the infectious dose of acid-sensitive organisms such as *V. cholerae* and nontyphoidal salmonellae. However, *S. flexneri* is significantly more acid resistant *in vitro* than *Salmonella* or *E. coli.*[27] This resistance is growth phase dependent and acquired as bacteria reach late log and stationary phase. Acid resistance can be genetically deleted and transferred. Recombination experiments have demonstrated a global regulatory system in which a putative stationary-phase sigma factor encoded by *rpoˢ* (formerly designated *katF*) plays a key role, along with other unidentified genes.[28] *Shigella* excreted in stool are in stationary phase and thus are acid resistant, ready to survive the gastric acid barrier if ingested by the next host. Although acid-exposed bacteria have a reduced capacity to invade mammalian cells, at least *in vitro,* once past the stomach they can begin to grow, rapidly losing their acid resistance but acquiring the invasive phenotype.

4.2. Soluble Virulence Proteins

Shigella is an outstanding example of the polygenic nature of microbial virulence, and required virulence genes are found on both chromosomal and plasmid DNA[4] (Table I). Evidence exists that siderophores (chelating proteins used to acquire and transport essential iron) have some impact on virulence. While studies in tissue culture models do not show any difference between siderophore positive or negative organisms, *in vivo,* where iron is highly protein bound and free iron is present in limiting amounts, the kinetics of infection with *iuc* aerobactin mutants is delayed,[29] suggesting that a functional siderophore can facilitate the initial multiplication of the organism. The nature of the O-side chain of somatic antigens may also affect virulence. When genes encoding *E. coli* somatic antigens 8 or 25 are conjugally transferred into *S. flexneri* 2a, the O-8 transformant is no longer virulent while the O-25 construct, which resembles *S. flexneri* 2a somatic antigen in its high mannose content, retains virulence.[30] Rough *Shigella* lacking O-side chains invade and multiply intracellularly, but in contrast to smooth (O-

TABLE I
Chromosomal Loci Associated with Virulence in *S. flexneri*[a]

Locus	Virulence phenotype of mutants	Regulatory or effector function
T-locus	Y variant expressing only group 3,4 somatic antigen may exhibit decreased virulence	Integration site for incorporation of lysogenic phage 4 encoding type-specific somatic antigen
kcpA	Sereny[b] negative with limited intracellular and intercellular bacterial spread in tissue culture monolayers	Positive regulation of plasmid gene *virG* (*icsA*)
virR	Invasive in tissue culture when grown at 30°C	Repression of plasmid invasion loci, e.g., *ipaA–D* in response to temperature
stx	Decrease in vascular damage in the colonic epithelium of orally challenged monkeys[c]	Synthesis of Shiga toxin
rfb	Sereny negative, decreased intercellular spread in infected tissue culture monolayers	Synthesis of group-specific somatic antigen
ompR–envZ	Decreased invasion in tissue culture; Sereny negative	Induction of plasmid invasion loci
rfa	Delayed or negative Sereny test, decreased intercellular spread in infected tissue culture monolayers	Synthesis of somatic antigen basal core
iucA,B,C,D *iutA*	Delayed Sereny reaction, decreased mortality in orally challenged guinea pig; decreased histopathology and fluid response in ligated rabbit ileal loop; diminished virulence in orally challenged monkey model	Synthesis of aerobactin and 76-kDa aerobactin receptor protein
sodB	Sensitivity to oxygen-dependent killing by phagocytes; Sereny negative; greatly decreased histopathology in ligated rabbit ileal loop	Superoxide dismutase inactivates superoxide radicals produced by phagocytes during respiratory burst

[a]Modified from Hale,[4] which should be consulted for a complete list of references.
[b]Sereny test involves inoculation of *Shigella* into the conjunctival sac of a rabbit or guinea pig. Invasive virulent organisms cause a purulent keratoconjunctivitis, whereas noninvasive avirulent strains are negative.
[c]The *stx* gene is present only in the chromosome of *S. dysenteriae* type 1.

antigen complete) strains, they do not spread from cell to cell and are avirulent *in vivo*.(31) More recently, a Tn5 insertion into the *rfa* locus (involved in synthesis of the LPS basal core region) was shown to delay cell-to-cell spread *in vitro* and the extent of the inflammatory mucosal reactions *in vivo*.(32)

Superoxide dismutase (*sodB*), an enzyme dismutating the bactericidal superoxide radical to water, may be used by the organism to survive *in vivo* by blocking bactericidal effects of host-derived reactive oxygen radicals.[33] Thus, a *sodB*$^-$ *S. flexneri* mutant, obtained by allelic exchange, which is extremely sensitive to oxygen stress and is readily killed by mouse macrophages or human neutrophils, is avirulent in the ligated rabbit ileal loop model.

4.3. Toxins

Shiga toxin is not essential for virulence of *S. dysenteriae* type 1 in primates but contributes to severity of disease manifestations, especially bloody diarrhea/dysentery.[34] Because of the production of structurally and functional similar toxins by enterohemorrhagic *E. coli*, which cause HUS but lack all of the invasion mechanisms of *Shigella*, Shiga toxin is believed to play a role in pathogenesis of systemic microangiopathic lesions. These SLT-producing *E. coli* also cause the characteristic clinical disease, hemorrhagic colitis.[35]

Shiga toxin consists of two noncovalently liked peptides, a 32-kDa enzymatically active A subunit and five 7- to 8-kDa B subunits which are responsible for binding toxin to a cell surface glycolipid receptor, globotriaosylceramide (Gb3). Like cholera toxin, the Shiga toxin A subunit is activated by proteolytic processing and reduction of an internal disulfide linkage into an enzymatically active A1 fragment (27 kDa) and a carboxy-terminal A2 fragment (4 kDa). The A1 subunit possesses the identical *N*-glycosidase action as the toxic plant lectin, ricin, and by removing adenine from a single specific adenosine in the 28 S component of the 60 S ribosomal subunit it permanently inactivates that ribosome and ultimately stops protein synthesis completely in the affected cell, resulting in cell death.[36]

The B subunit of Shiga toxin and SLT-1 are identical. SLT-1 B subunit has been purified to homogeneity.[37,38] Its x-ray crystallographic structure has been solved,[39] revealing a pentameric planar molecule with a conserved carbohydrate binding site formed by β-sheet interactions between adjacent B monomers. Thus, there are five potential binding sites per pentamer. Site-directed mutagenesis of the B peptide suggests the importance of aspartate residues at positions 16 and 17 in the putative binding cleft of the three-dimensional structure.[40] The host cell receptor for Shiga toxin is a membrane blood group active neutral glycolipid, Gb3, containing a terminal galactose-α1→4-galactose disaccharide.[41] At least in rabbit gut, this glycolipid is developmentally regulated and is not expressed before day 16 of postnatal life.[42] Expression appears to be related to increasing activity of the biosynthetic enzyme, UDP-galactose:lactosylceramide galactosyltransferase, and decreasing activity of the degradative enzyme, α-galactosidase.[43] In addition, Shiga toxin receptors in the rabbit are expressed on mature villus, and not immature crypt, cells,[44] resulting in net fluid accumulation in the lumen secondary to reduced sodium absorption (a villus cell function) without a change in chloride secretion (a crypt cell function). Studies in cultured human intestinal cells confirm that Gb3 is preferentially expressed by villus and not crypt cells and can be upregulated by maturational influences.[45]

Shiga toxin is active on cultured endothelial cells, particularly when the cells are activated by LPS or cytokines.[46] Of interest, human renal endothelial cells appear to express constitutively high levels of Gb3 that are not further induced by LPS or cytokines,[47] perhaps explaining the targetting of toxin-mediated pathology to the kidney.

The distinctive *Shigella* enterotoxins, ShET 1 and 2, are, in contrast to Shiga toxin, produced by all *Shigella* species. These toxins alter electrolyte transport by rabbit small bowel tissue *in vitro* and cause fluid secretion *in vivo* in ligated rabbit ileal loops. Humans convalescent from shigellosis develop neutralizing antibody. However, the role of the ShET toxins, if any, in the watery diarrhea phase of shigellosis is a topic of investigation.

4.4. Cellular Invasion and Spread

Cellular invasion and spread of the infection is complex and a good example of multiple gene actions. The process can be arbitrarily divided into at least four stages: (1) cell invasion; (2) intracellular multiplication; (3) intra- and intercellular spread; and (4) host cell killing.[48] Most studies have used *S. flexneri* strains and it is presumed that these events in all other *Shigella* and the biologically similar enteroinvasive *E. coli* (EIEC) proceed by the same basic process.

4.4.1. Cell Invasion

S. flexneri invade epithelial cells by a process resembling neutrophil or macrophage phagocytosis.[49] There remains some controversy as to whether intestinal epithelial cells are directly invaded by *Shigella*[50] or if entry is via damaged brush borders of epithelial cells, via the M cells, or the nonapical membrane of epithelial cells accessed via intercellular junctions. Recent data obtained using monolayers of human colon-derived cryptlike T-84 cells grown on semipermeable filters favor the last route.[51] These data show that *S. flexneri* added to the apical surface of T-84 cells induces transmigration of neutrophils added to the basal surface of the cells. Neutrophil migration reaches a maximum after 1 hr of incubation, followed by bacterial invasion of the monolayer beginning at the intercellular junctions. The formation of the intracellular vacuole in which the organism is ingested involves a moving junction of host cell membrane that will surround the bacterial cell and ultimately pinch off to surround the organism.[52] This process is directed by at least three surface constituents of the microbe itself, known as invasion plasmid antigens (Ipa's) B, C, and D, that lead to the complex changes in cytoskeletal elements of the host cell required for phagocytosis to occur. Once inside the host cell, fully virulent *Shigella* rapidly dissolve the vacuolar membrane to enter the host cell cytoplasm, using the microbe's hemolysin protein for this purpose.[53]

Many of the microbial genes involved in invasion by *S. flexneri* are present on large 220-kb plasmids. Tn5 transposon mutagenesis has identified five independent contiguous regions, termed R1–R5, within a 30-kb segment of this plasmid. R1 contains a positive regulatory gene variously termed *virB* in *S. flexneri* 2a,[54] *ipaR* in *S. flexneri* 5a,[55] or *invE* in *S. sonnei.*[56] *virF*, another positive regulator, acts on *ipaR* (virB) to induce genes in other loci, most thoroughly documented in R2 where the *ipa* genes are located within a 6-kb region coordinately regulated by *ipaR* gene products via temperature-activated transcription. *mxi* genes encode proteins required for membrane export and insertion of the Ipa proteins, and are present in R3 and R4,[57] near the *spa* locus (surface presentation of invasion plasmid antigens), located in R5 downstream from *mxiA.*[58,59] Other loci, such as *mxiD* and *mxiJ*, located between *ipa* and *mxiA*, are required for secretion of Ipa proteins.[60] This operon also contains other genes of still uncertain role in virulence.

Regulation of virulence genes *in vivo* is critical to the success of pathogenic bacteria. Shiga toxin expression is iron regulated via the *fur* gene system and many of the invasion genes discussed above are regulated by temperature and/or osmotic pressure.[61] When grown at 30°C, virulent strains of *S. flexneri*, *S. sonnei*, and *S. dysenteriae* type 1 are unable to invade epithelial cells *in vitro* or *in vivo* and reexpression of virulence after shift to 37°C requires several generations of growth and depends on the synthesis of new proteins. Among the virulence factors regulated by temperature are Shiga toxin, the *ipaA–D* and *mxi* gene products, contact hemolytic activity, and the expression of the positive regulators *virF* and *ipaR*. *virR*, a chromosomal gene located just in front of the *stx* gene, is a negative regulator of the temperature-regulated *ipa* and *mxi/spa* gene products.[62]

4.4.2. *Intracellular Multiplication*

Shigella multiply within the target cell soon after invasion. This requires lysis of the vacuole, which requires the action of a hemolysin. Virulent strains grown at 30°C do not express this hemolysin nor do they multiply in host cells. Hemolysin activity increases 100- to 1000-fold as the pH drops from 7 to 5.5,[63] consistent with the drop in intravacuolar pH that occurs, further suggesting the relevance of the hemolysin activity. Intracellular bacterial growth is necessary, for if this is inhibited, the organism is less virulent. Blocking folic acid synthesis via the aromatic pathway or inducing mutations of the porins OmpC and OmpF or deleting the regulatory gene, *ompB*, impairs intracellular multiplication and reduces virulence.

4.4.3. *Intracellular Spread*

The third phase in the invasion process is intracellular migration of the nonmotile *Shigella*, leading to cell-to-cell spread by means of a process utilizing the host cell cytoskeleton[64] and the large plasmid-encoded IcsA (formerly called VirG).[65] As intracytoplasmic *Shigella* begin to divide, they move within the host cell to the inner surface of the plasma membrane, and IcsA is expressed at the back pole of the organism relative to its direction of movement.[66] This leads to actin polymerization in this location and the formation of an actin tail immediately behind the bacterium in the region of cell division. Plastin, an actin binding host protein with two actin binding sites per monomer, is deposited in the same region. This cross-links the actin, causing a sphincterlike contraction of actin filaments with propulsive force, the so-called "actin motor." IcsA is an ATPase which generates the energy needed for this process. IcsA is phosphorylated by cyclic nucleotide-dependent host cell protein kinases.[67] Site-directed mutagenesis of a sequence with a phosphorylation consensus motif almost completely abolishes the ability of IcsA to be phosphorylated by protein kinase A, resulting in an increased capacity to spread from cell to cell. These findings suggest that host cell phosphorylation of IcsA may be an important defense mechanism.

When the organism reaches the plasma membrane, it pushes out long fingerlike protrusions into the adjacent cell, leading to cell-to-cell invasion when these membranes are subsequently lysed by the organism. The host cell cadherin L-CAM appears to be required to complete this process. L-CAM-deficient cells form abortive protrusions, failing to support passage of organisms from one cell to the adjacent cell unless L-CAM is transfected into the deficient cell.[68] Homotypic binding of the cadherin extracellular domain appears to promote active internalization of the protrusion by the neighboring cell. The force for this is made possible by anchoring the actin tail through interactions with intermediate junction proteins. *Shigella* also can move within cells by binding to actin stress fibers in an organized manner, following the cytoskeletal architecture in what has been termed organellelike movement (Olm).[69] This is distinct from the IcsA-mediated movement just described. When examined in the intestinal cell line CaCo-2,

Olm movement was associated primarily with colonization of organisms in the actin filament ring of the perijunctional area, whereas IcsA movement was more related to cell-to-cell spread of infection. Together the two mechanisms provide a means by which a single organism can enter a single host cell, multiply, and spread in a contiguous fashion in a monolayer, without ever exiting from the intracellular milieu during the process.[70]

4.4.4. Cell Death

The mechanism whereby *Shigella* induces cell death is not clear. Macrophages are rapidly killed after phagocytosis of *S. flexneri* by apoptosis, independent of their capacity to produce the cytolethal protein, Shiga toxin.[71] However, binding of recombinant Shiga toxin B subunit to Gb3, expressed in high amounts on Burkitt lymphoma cells, also triggers apoptosis.[72] The clinical relevance of these findings for the pathogenesis of colitis or endothelial cell damage in shigellosis remains to be demonstrated, but may relate to the release of cytokines (see below).

5. IMMUNE RESPONSES

5.1. Evidence for Protective Immunity

The remarkable advances in the study of pathogenesis of *Shigella* infection in the past decade have served to highlight the continuing paucity of knowledge of host protective immune responses. Uptake of *Shigella* by M cells represents an important mechanism for host surveillance of, and response to, *Shigella* antigens, whether or not the M cell is an important route for the initial infection of the intestinal epithelium.[73] However, the nature and the role of the ensuing immune responses in controlling the infection or in preventing subsequent infections by the same organism are poorly understood.

General understanding of the secretory immune system suggests that mucosal antibody, presumably primarily secretory IgA (sIgA), should be protective in shigellosis, although it has been difficult to convincingly demonstrate this in humans. The only well-established principle is that protection in natural shigellosis, is serotype specific and, presumably, associated with immune responses to the oligosaccharide structure of LPS. Thus, an individual convalescent from *S. flexneri* 2a infection is protected against reinfection only with the homologous serotype. This has been the basis for a resurgence in interest in LPS determinants for immunization, even by the parenteral route.[74] An interesting speculation concerning immunity and the epidemiology of *Shigella* in developing countries, where the ratio of *S. flexneri* to *S. sonnei* is high compared to the reverse ratio of these organisms in industrialized countries, relates to the frequency with which surface water in developing countries is contaminated with *Pleisiomonas shigelloides*, which expresses the same O-antigen as *S. sonnei.*[75] The ability to protect water sources from microbial contamination in developed countries means that this "natural" immunogen to *S. sonnei* is not available for ongoing immunization and the population remains susceptible to *S. sonnei* into adulthood.

5.2. Antibody Responses in *Shigella* Infection

Antibody responses to the somatic antigens of *Shigella* develop early in infection and follow the typical course for anti-LPS antibodies, that is, an IgM response that peaks within weeks and wanes after 1–2 years. Epidemiologic studies can use anti-O-antigen antibodies to detect prior exposure to *Shigella.*[76] Structural analysis of the specific O-antigens of *Shigella*[77,78] plus methods to isolate pure LPS[79] and the availability of simple and sensitive enzyme immunoassays[80] have improved serological diagnosis of infection for epidemiologic, but not clinical, purposes. *S. dysenteriae* 1 LPS has a unique O-repeat unit,[81] shared only with *E. coli* expressing O-1 or O-120 antigens. Since neither *E. coli* O-1 nor O-120 is common among bacterial pathogens, testing for antibody to *S. dysenteriae* type 1 LPS is quite specific. *S. flexneri* types la–5b contain a tetrasaccharide repeat and extensive cross-reactions occur among them,[82] precluding specific serologic diagnosis, whereas LPS from *S. flexneri* type 6 is serologically distinctive. The *S. sonnei* antigen is an unusual disaccharide of 2-acetamido-2-deoxy-L-altruronic acid linked 1→4 to 2-acetamido-4-amino-2,4,6-trideoxy-D-galactose. Thus, specific assays for *S. dysenteriae* type 1, *S. flexneri* 6, and *S. sonnei* anti-LPS are available. Because of the uncommon isolation of *S. flexneri* in developed countries, even the nonspecific assay for *S. flexneri* la–5b can be useful to diagnose recent infections, since most individuals, regardless of age, are seronegative. In endemic countries this assay might be of use in young children under 3 years of age.

IgM antibody to Shiga toxin has been documented in *S. dysenteriae* type 1 infection, both in the field setting[83] and in experimental infection in humans.[84] This is consistent with the observation that toxin binds to CD77 (Gb3)-positive IgG- and IgA-producing B lymphocytes, and not T lymphocytes, and inhibits immunoglobulin synthesis in these cells.[85] CD77 is present predominantly in germinal center B cells[86] and is a differentiation marker. Another endogenous B-lymphocyte surface protein, CD19, an early marker of B-cell differentiation, disappears with terminal differentiation of B cells to plasma cells.[87] CD19 modulates signal transduction, proliferation, and maturation of B cells,[88] possibly by binding to CD77. CD19 also is reported to have sequence homology with Shiga toxin and the co-capping of CD19 and CD77 that occurs in the presence of anti-CD19 antibodies is prevented by the prior binding of SLT-1B to CD77.[89] The effect of this would be to inhibit B-cell activation. Although this may be one mechanism by which Shiga toxin blocks IgG and IgA antibody responses, holotoxin and the isolated binding B subunit also induce apoptotic death in the high-CD77-expressing Burkitt lymphoma cells.[72] These findings may be relevant to clinical disease. The presence of circulating IgM antibody to Shiga toxin does not preclude the clinical response of human volunteers to viable, virulent *Shigella* challenge.[84] However, circulating IgG antibody induced by immunization with formalinized Shiga toxoid[90] or with synthetic B subunit peptides[91] prevents the systemic effects of intact toxin.

Local antibodies bathing mucosal surfaces, in particular sIgA, are thought to provide local protection against infection, probably by preventing interactions of organisms with mucosal epithelial cells and reducing attachment and/or invasion. Local antibodies of any isotype could, in addition, bind to and neutralize the biological effect of microbial products involved in pathogenesis, such as Shiga toxin or other toxins. No evidence of the role of either mechanism has been obtained in human shigellosis, although IgA-secreting cells have been identified in the circulation of patients with natural *S. sonnei* infection.[92] Animal experiments have not provided encouraging data; thus, sIgA collected from the gut of rabbits immunized with an invasive *E. coli–Shigella* hybrid vaccine strain did not inhibit invasion of tissue culture cells by *S. flexneri* 2a.[93] In contrast, there is some, albeit limited, evidence of an antibody-dependent cellular cytotoxicity (ADCC) response in acute shigellosis that could contribute to the control of the infection.[94,95] The antibodies are IgG; however, there is no clue as to the antigens recognized. Classical cell-mediated immune responses have yet to be described in *Shigella* infection.

5.3. Cytokines in Shigellosis

As an intense inflammatory colitis, there has been increasing interest in the role of cytokines, produced either locally in the gut or systemically, in the pathogenesis of shigellosis and its complications. In Sri Lanka, serum IL-6 and TNF levels were increased in children with *S. dysenteriae* type 1 infections early in the course and especially in those with complications.[96] Stool IL-6 and TNF were also detectable in many, but not all, of the *Shigella* patients. Serum or stool cytokines were negative in the vast majority of the controls, including patients with *Salmonella*, *Cryptosporidium*, rotavirus, or healthy controls. A more recent study among adults in Bangladesh with acute shigellosis has shown that the concentrations of stool cytokines were more than 100-fold higher than plasma cytokines and were in proportion to the severity of the clinical illness.[97] This is consistent with the *in vitro* observation that invasion of LPS-stimulated murine macrophages by *S. flexneri* leads to release of IL-1 during apoptosis, but not of IL-6 or TNF-α.[98] In addition, in the Bangladesh study, more cytokines were elevated in stool (IL-1β, IL-6, IL-8, TNF-α, IFN-τ, GM-CSF, and IL-1Ra) than in plasma (IL-6, TNF-α, IFN-τ, and IL-1Ra), with no relation of plasma levels to disease severity and no time-dependent decline in levels except for IL-6. Of particular interest was the low level of IFN-τ in stool on admission, with a gradual rise in concentration during the course of the illness. The highest IFN-τ levels were observed in asymptomatic Bangladeshi controls and these levels were five times higher than among asymptomatic Swedish subjects.

Although there are some inconsistencies, these data suggest that local cytokine production in the gut may be of importance in shigellosis, reflecting the intense inflammatory response in the intestinal mucosa in this disease. The early low IFN-τ levels in patient stool compared to Bangladeshi controls suggest that shigellosis may turn off this component of the TH-1 response in acute disease and that a selective "defect" in IFN-τ production could, in turn, contribute to the intensity of infection. These results were in general confirmed by *in situ* immunohistochemical examination of rectal biopsy samples from these patients.[99] Thus, patients had significantly more cells producing IL-1α and β, IL-6, IL-4, IL-8, IL-10, TNF-α, INF-τ, IL-1Ra, and TGF β_{1-3} than healthy controls, with a particular increase in IL-1-, IL-6-, TNF-α, and IFN-τ-producing cells in those with severe inflammation. Curiously, no difference in the number of cytokine-producing cells was observed between acute and convalescent samples 30 days after illness began, although clinical disease abated by day 7 of therapy. However, this is consistent with the persistence of histological evidence of inflammation in the mucosa well into convalescence. These data also document that both TH-1 and TH-2 response cytokines are modulated by shigellosis and that both humoral and cytotoxic defense mechanisms could be activated during and after infection.

Systemic cytokines may also play a role in the pathogenesis of vascular lesions leading to HUS or to the pathogenetically similar syndrome, thrombotic thrombocytopenic purpura (TTP). Studies using human umbilical vein endothelial cells (HUVEC) *in vitro* have shown that LPS, IL-1β, or TNF-α induces toxin receptors[46] and markedly increases the sensitivity of the cells to toxin.[100] Thus, it is possible that local release of cytokines in the gut can effect entry to the circulation and modify distant endothelial bed responses to Shiga toxin gaining entry to the circulation. SLT infusion in a mouse transgenic for a CAT reporter for TNF leads to renal-specific induction of this cytokine.[101] Barrett[102] and Tesh[103] also observed that murine macrophages exposed to SLTs increase production of TNF, albeit slower and in less robust fashion than the response to LPS.

5.4. Induced Immune Response and Vaccine Protection

Because of the severity of clinical shigellosis, the frequency of antibiotic resistance, and the low ID_{50} dose needed to cause clinical disease, it has long been a goal to develop an effective *Shigella* vaccine. This goal has been elusive. To date, no vaccine has been both safe and effective, starting with parenteral heat- or acetone-killed whole-cell vaccines, which induced circulating antibody but little or no protection.[93] Contemporary vaccine development has therefore concentrated on use of live attenuated vaccine strains. These have been of three broad types: (1) *Shigella* strains attenuated by a chromosomal mutation, (2) *Shigella* strains attenuated by mutations in plasmid virulence genes, or (3) hybrid *Shigella* or *E. coli* strains expressing cloned *Shigella* virulence determinants. None of these vaccines has been clinically useful, either because of inherent instability of the mutation, failure of protection, or considerable reactogenicity, and it seems likely that an effective vaccine will need to induce immunity to more than one microbial antigen and still be highly attenuated. Current efforts have been focused on the development of aromatic amino acid auxotrophs, such as the *S. flexneri* vaccine strain Sfl-114, in which the *aroD* gene is insertionally inactivated by means of the transposon Tn10.[104] The insertion in *aroD* inhibits the biosynthesis of aromatic metabolites and, as a result, blocks the organism's ability to synthesize folic acid. Because of this, initial mucosal invasion is followed by rapid intracellular clearing. Although satisfactory immune responses have been observed, over 50% of volunteers given 10^{10} Sfl-114 by mouth have developed intestinal discomfort, albeit mild and not clinically important. Specific plasmid antigen mutations have also been tested, and are either ineffective or too reactogenic to be useful.[105] Overall, the data to date suggest that immune responses to individual *Shigella* virulence factors can be induced but that no vaccine thus far is sufficiently effective and clinically benign to be of clinical use.

Recently, the potential of killed parenteral vaccines has come back into vogue with improvements in chemical preparation of antigens and the ability to couple polysaccharides to protein carriers to induce T-cell-dependent responses. In shigellosis, this approach has been boosted by the lack of success with live vaccines to date and by the observation that resistance to shigellosis correlates closely with the presence of serum IgG anti-LPS antibodies.[106] Therefore, attempts to develop conjugate LPS vaccines for *Shigella* have been initiated.[74] This strategy depends on the ability of circulating IgG anti-LPS antibodies to leak into the bowel lumen, especially likely in the presence of infection-induced mucosal inflammatory responses. Then, if the concentration of antibody is high enough, organisms at and within the mucosa could be killed, while circulating antibody together with complement should be bactericidal for organisms that gain access to the circulation. Phase I study of *S. flexneri* and *S. sonnei* LPS-conjugate vaccines in humans has been completed, demonstrating both safety and immunogenicity,[(107)] and efficacy trials are now under way.

A model of Shiga toxin-induced disease has been developed using the rabbit enteropathogenic *E. coli* (EPEC) pathogen, RDEC-1, into which the virtually identical Shiga-like toxin 1 gene from *E. coli* has been inserted. This construct causes significant febrile responses and weight loss in experimental animals, believed to result from local and possibly systemic cytokine responses induced by the infecting organism. These systemic clinical manifestations are markedly attenuated by prior parenteral immunization with a purified recombinant Shiga-like toxin B subunit,[(108)] suggesting the potential of toxin-based immunization strategies for at least the systemic responses attributable to this toxin.

REFERENCES

1. Shiga, K., 1898, Ueber den Dysenteriebacillus (*Bacillus dysenteriae*), *Zentralbl. Bakteriol. Parasitenkd. Abt. I Orig.* **24:**817–824.
2. Flexner, S., 1900, On the etiology of tropical dysentery, *Bull. Johns Hopkins Hosp.* **11:**231–242.
3. Kruse, W., 1900, Ueber die Ruhr als Volkskrankheit und ihrer Erreger, *Dtsch. Med. Wochenschr.* **26:**637–639.
4. Hale, T. L., 1991, Genetic basis of virulence in *Shigella* species, *Microbiol. Rev.* **55:**206–224.
5. Conradi, H., 1903, Uber losliche, durch aseptische Autolyse erhaltene Giftstoffe von Ruhr- und Typhus bazillen, *Dtsch. Med. Wochenschr.* **20:**26–28.
6. Todd, C., 1904, On a dysentery toxin and antitoxin, *J. Hyg.* **4:**480–494.
7. Acheson, D. W. K., Donohue-Rolfe, A., and Keusch, G. T., 1991, The family of Shiga and Shiga-like toxins, in: *Sourcebook of Bacterial Protein Toxins* (J. E. Alouf and J. H. Freer, eds.), Academic Press, New York, pp. 415–433.
8. Fasano, A., Guandalini, S., Russell, R. G., Rae, B., Raimondi, F., and Levine, M. M., 1991, Enterotoxin production by *Shigella flexneri* 2a, *J. Pediatr. Gastroenterol. Nutr.* **13:**320.
9. Fasano, A., Kay, B. A., Russell, R. G., Maneval, D. R., Jr., and Levine, M. M., 1990, Enterotoxin and cytotoxin production by enteroinvasive *Escherichia coli, Infect. Immun.* **58:** 3717–3723.
10. Sansonetti, P. J. (ed.), 1992, *Pathogenesis of Shigellosis, Curr. Top. Microbiol. Immunol.* **180**, Springer-Verlag, Berlin.
11. DuPont, H. L., Levine, M. M., Hornick, R. B., and Formal, S. B., 1989, Inoculum size in shigellosis and implications for expected mode of transmission, *J. Infect. Dis.* **159:**1126–1128.
12. Acheson, D. W. K., and Keusch, G. T., 1995, Shigella and enteroinvasive *Escherichia coli*, in: *Infections of Gastrointestinal Tract* (M. J. Blaser, P. D. Smith, J. I. Raudin, H. B. Greenberg, and R. L. Guerrant, eds.), Raven Press, Ltd., New York, pp. 763–784.

13. Drusin, L. M., Genvert, G., Topf-Olstein, B., and Levy-Zombeck, E., 1976, Shigellosis: Another sexually transmitted disease? *Br. J. Vener. Dis.* **52:**348–350.
14. Tauxe, R. V., McDonald, R. C., Hargrett-Bean, N., and Blake, P. A., 1988, The persistence of *Shigella flexneri* in the United States: The increased role of the adult male, *Am. J. Public Health* **78:**1432–1435.
15. Butler, T., Speelman, P., Kabir, I., and Banwell, J., 1986, Colonic dysfunction during shigellosis, *J. Infect. Dis.* **154:**817–824.
16. Bennish, M. L., Azad, A. K., Rahman, O., and Phillips, R. E., 1990, Hypoglycemia during childhood diarrhea: Prevalence, pathophysiology and outcome, *N. Engl. J. Med.* **322:**1357–1363.
17. Azad, M. A. K., Islam, M., and Butler, T., 1986, Colonic perforation in *Shigella dysenteriae* type I infection, *Pediatr. Infect. Dis.* **5:**103–104.
18. Bennish, M. L., Azad, A. K., and Yousefzadeh, D., 1991, Intestinal obstruction during shigellosis: Incidence, clinical features, risk factors, and outcome, *Gastroenterology* **101:**626–634.
19. Struelens, M. J., Patte, D., Kabir, I., Salam, A., Nath, S. K., and Butler, T., 1985, *Shigella* septicemia: Prevalence, presentation, risk factors, and outcome, *J. Infect. Dis.* **152:**784–790.
20. Koster, F., Levin, J., Walker, L., Tung, K. S., Gilman, R. H., Rahaman, M. M., Majid, M. A., Islam, S., and Williams, R. C., Jr., 1978, Hemolytic-uremic syndrome after shigellosis: Relation to endotoxemia and circulating immune complexes, *N. Engl. J. Med.* **298:**927–933.
21. Ashkenazi, S., Cleary, K. R., Pickering, L. K., Murray, B. E., and Cleary, T. G., 1990, The association of Shiga toxin and other cytotoxins with the neurologic manifestations of shigellosis, *J. Infect. Dis.* **161:**961–965.
22. Raybourne, R. B., and Williams, K. M., 1990, Monoclonal antibodies against an HLA-B27-derived peptide react with an epitope present on bacterial proteins, *J. Immunol.* **145:**2539–2544.
23. Williams, K. M., and Raybourne, R. B., 1990, Demonstration of cross-reactivity between bacterial antigens and class I human leukocyte antigens by using monoclonal antibodies to *Shigella flexneri*, *Infect. Immun.* **58:**1774–1781.
24. Stieglitz, H., Formal, S. B., and Lipsky, P., 1989, Identification of a 2-Md plasmid from Shigella flexneri associated with reactive arthritis, *Arthritis Rheum.* **32:**937–946.
25. Kapasi, K., Chui, B., and Inman, R. D., 1992, HLA-B27/microbial mimicry: An in vivo analysis, *Immunology* **77:**456–461.
26. Falkow, S., Scrider, H., Baron, L. S., and Formal, S. B., 1963, Virulence of *Escherichia–Shigella* genetic hybrids in the guinea pig, *J. Bacteriol.* **86:**1251–1258.
27. Gorden, J., and Small, P. L. C., 1993, Acid resistance in enteric bacteria, *Infect. Immun.* **61:**364–367.
28. Small, P. L. C., and Falkow, S., 1992, A genetic analysis of acid resistance in *Shigella flexneri*; the requirement for a *katF* homologue, *Abstr. Gen. Meet. Am. Soc. Microbiol.* **B74:**38.
29. Nassif, X., Mazert, M. C., Mouniew, J., and Sansonetti, P. J., 1987, Evaluation with an *iuc*::Tn10 mutant of the role of aerobactin production in the virulence of *Shigella flexneri*, *Infect. Immun.* **55:**1963–1969.
30. Gemski, P., Sheahan, D. G., Washington, O., and Formal, S. B., 1972, Virulence of *Shigella flexneri* hybrids expressing *Escherichia coli* somatic antigens, *Infect. Immun.* **6:**104–111.
31. Okamura, N., Nagai, T., Nakaya, R., Kondo, S., Murakami, M., and Hisatsune, K., 1983, HeLa cell invasiveness and O antigen of *Shigella flexneri* as separate and prerequisite attributes of virulence to evoke keratoconjunctivitis in guinea pigs, *Infect. Immun.* **39:**505–513.
32. Okada, N., Sasakawa, C., Tobe, T., Yamada, M., Nagai, S., Talukder, K. A., Komatsu, K., Kanegasaki, S., and Yoshikawa, M., 1991, Virulence-associated chromosomal loci of *Shigella flexneri* identified by random Tn5 insertion mutagenesis, *Mol. Microbiol.* **5:**187–195.
33. Franzon, V. L., Arondel, J., and Sansonetti, P. J., 1990, Contribution of superoxide dismutase and catalase activities to *Shigella flexneri* pathogenesis, *Infect. Immun.* **58:**529–535.

34. Fontaine, A., Arondel, J., and Sansonetti, P. J., 1988, Role of Shiga toxin in the pathogenesis of bacillary dysentery studied using a tox$^-$ mutant of *Shigella dysenteriae* I, *Infect. Immun.* **56:**3099–3109.
35. O'Brien, A. D., Tesh, V. L., Donohue-Rolfe, A., Jackson, M. P., Olsnes, S., Sandvig, K., Lindberg, A. A., and Keusch, G. T., 1992, Shiga toxin: Biochemistry, genetics, mode of action and role in pathogenesis, in: *Pathogenesis of Shigellosis* (P. J. Sansonetti, ed.), Springer-Verlag, Berlin, pp. 65–94.
36. Endo, Y., Tsurugi, K., Yutsudo, T., Takeda, Y., Ogasawara, T., and Igarashi, K., 1988, Site of action of verotoxin (VT2) from *Escherichia coli* O157:H7 and of Shiga toxin on eukaryotic ribosomes. RNA N-glycosidase activity of the toxins, *Eur. J. Biochem.* **171:**45–50.
37. Ramotar, K., Boyd, B., Tyrrell, G., Gariepy, J., Lingwood, C., and Brunton, J., 1990, Characterization of Shiga-like toxin I B subunit purified from overproducing clones of the SLT-I B cistron, *Biochem. J.* **272:**805–811.
38. Acheson, D. W. K., Calderwood, S. B., Boyko, S. A., Lincicome, L., Kane, A. V., Donohue-Rolfe, A., and Keusch, G. T., 1993, Comparison of Shiga-like toxin I B-subunit expression and localization in *Escherichia coli* and *Vibrio cholerae* by using *trc* or iron-regulated promoter systems, *Infect. Immun.* **61:**1098–1104.
39. Stein, P. E., Boodhoo, A., Tyrrell, G. J., Brunton, J. L., and Read, R. J., 1992, Crystal structure of the cell-binding B oligomer of verotoxin-1 from *E. coli*, *Nature* **355:**748–750.
40. Jackson, M. P., Wadolkowski, E. A., Weinstein, D. L., Holmes, R. K., and O'Brien, A. D., 1990, Functional analysis of the Shiga toxin and Shiga-like toxin type II variant binding subunits by using site-directed mutagenesis, *J. Bacteriol.* **172:**653–658.
41. Jacewicz, M., Clausen, H., Nudelman, E., Donohue-Rolfe, A., and Keusch, G. T., 1986, Pathogenesis of *Shigella* diarrhea. XI. Isolation of shigella toxin-binding glycolipid from rabbit jejunum and HeLa cells and its identification as globotriaosylceramide, *J. Exp. Med.* **163:**1391–1404.
42. Mobassaleh, M., Gross, S. K., McCluer, R. H., Donohue-Rolfe, A., and Keusch, G. T., 1989, Quantitation of the rabbit intestinal glycolipid receptor for Shiga toxin. Further evidence for the developmental regulation of globotriaosylceramide in microvillus membranes, *Gastroenterology* **97:**384–391.
43. Mobassaleh, M., Koul, O., Mishra, K., McClure, R. H., and Keusch, G. T., 1994, Developmental regulation of intestinal Gb3 galactosyltransferase and α-galactosidase control Shiga toxin receptors, *Am. J. Physiol.* **267:**G618–G624.
44. Kandel, G., Donohue-Rolfe, A., Donowitz, M., and Keusch, G. T., 1989, Pathogenesis of shigella diarrhea. XVI. Selective targeting of Shiga toxin to villus cells or rabbit jejunum explains the effect of the toxin on intestinal electrolyte transport, *J. Clin. Invest.* **84:**1509–1517.
45. Jacewicz, M. S., Acheson, D. W. K., Mobassaleh, M., Donohue-Rolfe, A., Balasubramanian, K. A., and Keusch, G. T., 1995, Maturational regulation by butyrate of globotriaosylceramide, the Shiga-like toxin I receptor, in cultured human gut epithelial cells, *J. Clin. Invest.* **96:**1328–1335.
46. van de Kar, N. C., Monnens, L. A., Karmali, M. A., and van Hinsbergh, V. W., 1992, Tumor necrosis factor and interleukin-1 induce expression of the verocytotoxin receptor globotriaosylceramide on human endothelial cells: Implications for the pathogenesis of the hemolytic uremic syndrome, *Blood* **80:**2755–2764.
47. Obrig, T. G., Louise, C. B., Lingwood, C. A., Boyd, B., Barley-Maloney, L., and Daniel, T. O., 1993, Endothelial heterogeneity in Shiga toxin receptors and responses, *J. Biol. Chem.* **268:**15484–15488.
48. Sansonetti, P. J., 1992, Molecular and cellular biology of *Shigella flexneri* invasiveness: From cell assay systems to shigellosis, in: *Pathogenesis of Shigellosis* (P. J. Sansonetti, ed.), Springer-Verlag, Berlin, pp. 1–19.

49. Goldberg, M. B., and Sansonetti, P. J., 1993, *Shigella* subversion of the cellular cytoskeleton: A strategy for epithelial colonization, *Infect. Immun.* **61:**4941–4946.
50. Mounier, J., Vasselon, T., Hellio, R., Lesourd, M., and Sansonetti, P. J., 1992, *Shigella flexneri* enters human colonic Caco-2 epithelial cells through the basolateral pole, *Infect. Immun.* **60:**237–248.
51. Perdomo, J. J., Gounon, P., and Sansonetti, P. J., 1994, Polymorphonuclear leukocyte transmigration promotes invasion of colonic epithelial monolayer by *Shigella flexneri*, *J. Clin. Invest.* **93:**633–643.
52. Clerc, P., and Sansonetti, P. J., 1987, Entry of *Shigella flexneri* into HeLa cells: Evidence for directed phagocytosis involving actin polymerization and myosin accumulation, *Infect. Immun.* **55:**2681–2688.
53. Sansonetti, P. J., Ryter, A., Clerc, P., Maurelli, A. T., and Mounier, J., 1986, Multiplication of *Shigella flexneri* within HeLa cells: Lysis of the phagocytic vacuole and plasmid-mediated contact hemolysis, *Infect. Immun.* **51:**461–469.
54. Adler, B., Sasakawa, C., Tobe, T., Makino, S., Komatsu, K., and Yoshikawa, M., 1989, A dual transcriptional activation system for the 230kb plasmid genes coding for virulence associated antigens of *Shigella flexneri*, *Mol. Microbiol.* **3:**627–635.
55. Buysse, J. M., Venkatesan, J. A., Mills, J. A., and Oaks, E. V., 1990, Molecular characterization of a trans-acting positive effector (*ipa*R) of invasion plasmid antigen synthesis in *Shigella flexneri* serotype 5, *Microb. Pathog.* **8:**197–211.
56. Watanabe, H., Arakawa, E., Ito, K., Kato, J. I., and Nakamura, A., 1990, Genetic analysis of an invasion region by use of a Tn3-lac transposon and identification of a second positive regulator gene, *invE* for cell invasion of *Shigella sonnei*: Significant homology of *invE* with *parB* of plasmid Pl, *J. Bacteriol.* **172:**619–629.
57. Hromockyj, A. E., and Maurelli, A. T., 1989, Identification of Shigella invasion genes by construction of temperature-regulated *inv::lacZ* operon fusions, *Infect. Immun.* **57:**2963–2970.
58. Andrews, G. P., Hromockyj, A. E., Coker, C., and Maurelli, A. T., 1991, Two novel virulence loci, *mxiA* and *mxiB*, in *Shigella flexneri* 2a facilitate excretion of invasion plasmid antigens, *Infect. Immun.* **59:**1997–2005.
59. Venkatesan, M. M., Buysse, J. M., and Oaks, E. V., 1992, Surface presentation of *Shigella flexneri* invasion plasmid antigens requires the products of the *spa* locus, *J. Bacteriol.* **174:** 1990–2001.
60. Allaoui, A., Menard, R., Sansonetti, P. J., and Parsot, C., 1993, Characterization of the *Shigella flexneri ipgD* and *ipgF* genes, which are located in the proximal part of the *mxi* locus, *Infect. Immun.* **61:**1707–1714.
61. Maurelli, A. T., Hromockyj, A. E., and Bernardini, M. L., 1992, Environmental regulation of *Shigella* virulence, in: *Pathogenesis of Shigellosis* (P. J. Sansonetti, ed.), Springer-Verlag, Berlin, pp. 95–116.
62. Maurelli, A. T., and Sansonetti, P. J., 1988, Identity of a chromosomal gene controlling temperature regulated expression of *Shigella* virulence, *Proc. Natl. Acad. Sci. USA* **85:**2820–2824.
63. Clerc, P., Ryter, A., Mounier, J., and Sansonetti, P. J., 1987, Plasmid mediated early killing of eukaryotic cells by *Shigella flexneri* as studied by infection of J774 macrophages, *Infect. Immun.* **55:**521–527.
64. Goldberg, M. B., and Sansonetti, P. J., 1993, *Shigella* subversion of the cellular cytoskeleton: A strategy for epithelial colonization, *Infect. Immun.* **61:**4941–4946.
65. Makino, S., Sasakawa, C., Kamata, K., Kurata, T., and Yoshikawa, M., 1986, A genetic determinant required for continuous reinfection of adjacent cells on large plasmid in *Shigella flexneri* 2a, *Cell* **46:**551–555.
66. Goldberg, M. B., Theriot, J. A., and Sansonetti, P. J., 1994, Regulation of surface presentation of IcsA, a *Shigella* protein essential to intracellular movement and spread, is growth phase dependent, *Infect. Immun.* **62:**5664–5668.

67. d'Hauteville, H., and Sansonetti, P. J., 1992, Phosphorylation of IcsA by cAMP-dependent protein kinase and its effect on intercellular spread of *Shigella flexneri*, *Mol. Microbiol.* **6:** 833–841.
68. Sansonetti, P. J., Mounier, J., Prevost, M. C., and Mege, R.-M., 1994, Cadherin expression required for formation and internalization of *Shigella flexneri*-induced intercellular protrusions involved in spread between epithelial cells, *Cell* **76:**829–839.
69. Vasselon, T., Mounier, J., Prevost, M. C., Hellio, R., and Sansonetti, P. J., 1991, A stress fiber-based movement of *Shigella flexneri* within cells, *Infect. Immun.* **59:**1723–1732.
70. Vasselon, T., Mounier, J., Hellio, R., and Sansonetti, P. J., 1992, Movement along actin filaments of the perijunctional area and de novo polymerization of cellular actin are required for *Shigella flexneri* colonization of epithelial Caco-2 cell monolayers, *Infect. Immun.* **60:**1031–1040.
71. Zychlinsky, A., Prevost, M. C., and Sansonetti, P. J., 1992, *Shigella flexneri* induces apoptosis in infected macrophages, *Nature* **358:**167–169.
72. Mangeney, M., Lingwood, C. A., Taga, S., Caillou, B., Tursz, T., and Wiels, J., 1993, Apoptosis induced in Burkitt's lymphoma cells via Gb3/CD77, a glycolipid antigen, *Cancer Res.* **53:**5314–5319.
73. Wassef, J. S., Keren, D. F., and Mailloux, J. L., 1989, Role of M cells in initial antigen uptake and in ulcer formation in the rabbit intestinal loop model of shigellosis, *Infect. Immun.* **57:**858–863.
74. Robbins, J. B., Chu, C., and Schneerson, R., 1992, Hypothesis for vaccine development: Protective immunity to enteric diseases caused by non-typhoidal *Salmonellae* and *Shigellae* may be conferred by serum IgG antibodies to the O-specific polysaccharide of their lipopolysaccharides, *Clin. Infect. Dis.* **15:**346–361.
75. Sack, D. A., Hoque, A. T., Hug, A., and Etheridge, M., 1994, Is protection against shigellosis induced by natural infection with *Pleisiomonas shigelloides*? *Lancet* **343:**1413–1415.
76. Lindberg, A. A., Cam, P. D., Chan, N., Phu, L. K., Trach, D. D., Lindberg, C., Karlsson, K., Karnell, A., and Ekwall, E., 1991, Shigellosis in Vietnam: Seroepidemiologic studies with use of lipopolysaccharide antigens in enzyme immunoassays, *Rev. Infect. Dis.* **13:**S231–S237.
77. Ewing, W. H., and Lindberg, A. A., 1984, Serology of *Shigella*, in: *Methods in Microbiology*, Volume 14 (T. Bergan, ed.), Academic Press, New York, pp. 113–142.
78. Lindberg, A. A., Karnell, A., and Weintraub, A., 1991, The lipopolysaccharide of *Shigella* bacteria as a virulence factor, *Rev. Infect. Dis.* **13**(Suppl. 4)**:**S279–S284.
79. Lindberg, A. A., and Holme, T., 1972, Evaluation of some extraction methods for the preparation of bacterial lipopolysaccharides for structural analysis, *Acta Pathol. Microbiol. Immunol. Scand. Sect. B* **80:**751–759.
80. Carlsson, H. E., and Lindberg, A. A., 1978, Application of enzyme immunoassay for diagnosis of bacterial and mycotic infections, *Scand. J. Immunol.* **8**(Suppl. 7)**:**97–110.
81. Dmitriev, B. A., Knirel, Y. A., Kochetkov, N. K., and Hofman, I. L., 1976, Somatic antigens of *Shigella*. Structural investigations of the O-specific polysaccharide chain of *Shigella dysenteriae* type I lipopolysaccharide, *Eur. J. Biochem.* **66:**559–566.
82. Carlin, N. I. A., Lindberg, A. A., Bock, K., and Bundle, D. R., 1984, The *Shigella flexneri* O-antigenic polysaccharide chain. Nature of the biological repeating unit, *Eur. J. Biochem.* **139:**189–194.
83. Keusch, G. T., and Jacewicz, M., 1973, Serum enterotoxin-neutralizing antibody in human shigellosis, *Nature New Biol.* **241:**31–32.
84. Keusch, Jacewicz, M., Levine, M. M., Hornick, R. B., and Kochwa, S., 1976, Pathogenesis of Shigella diarrhea. Serum anti-cytotoxin antibody response produced by toxigenic and "non-toxigenic" *S. dysenteriae* 1, *J. Clin. Invest.* **57:**194–202.
85. Cohen, A., Madrid-Marina, V., Estrov, Z., Freedman, M. H., Lingwood, C. A., and Dosch, H. M., 1990, Expression of glycolipid receptors to Shiga-like toxin on human B lympho-

cytes: A mechanism for the failure of long-lived antibody response to dysenteric disease, *Int. Immunol.* **2:**1–8.

86. Mangeney, M., Richard, Y., Coulaud, D., Tursz, T., and Wiels, J., 1991, CD77: An antigen of germinal center B cells entering apoptosis, *Eur. J. Immunol.* **21:**1131–1140.
87. Nadler, L. M., Anderson, K. C., Marti, G., Bates, M., Park, E., Daley, J. F., and Schlossman, S. F., 1983, B4, a human B lymphocyte associated antigen expressed on normal, mitogen activated and malignant B lymphocytes, *J. Immunol.* **131:**244–250.
88. Callard, R. E., Rigley, K. P., Smith, S. H., Thurstan, S., and Shields, J. G., 1992, CD19 regulation of human B cell responses. B cell proliferation and antibody secretion are inhibited or enhanced by ligation of the CD19 surface glycoprotein depending on the stimulating signal used, *J. Immunol.* **148:**2983–2987.
89. Maloney, M. D., and Lingwood, C. A., 1994, CD19 has a potential CD77 (globotriaosylceramide)-binding site with sequence similarity to verotoxin B-subunits: Implications of molecular mimicry for B cell adhesion and enterohemorrhagic *Escherichia coli* pathogenesis, *J. Exp. Med.* **180:**191–201.
90. McIver, J., Grady, G. F., and Formal, S. B., 1977, Immunization with *Shigella dysenteriae* type 1: Evaluation of antitoxic immunity in prevention of experimental disease in rhesus monkeys (*Macaca mulatta*), *J. Infect. Dis.* **136:**416–421.
91. Harari, I., Donohue-Rolfe, A., Keusch, G. T., and Arnon, R., 1988, Synthetic peptides of Shiga toxin B-subunit induce antibodies which neutralize the biological activities of the toxin, *Infect. Immun.* **56:**1618–1624.
92. Orr, N., Robin, G., Lowell, G., and Cohen, D., 1992, Presence of specific immunoglobulin A secreting cells in peripheral blood following natural infection with *Shigella sonnei*, *J. Clin. Microbiol.* **30:**2165–2168.
93. Hale, T. L., and Kern, D. F., 1992, Pathogenesis and immunology in shigellosis: Applications for vaccine development, in: *Pathogenesis of Shigellosis* (P. J. Sansonetti, ed.), Springer-Verlag, Berlin, pp. 117–137.
94. Lowell, G. H., McDermott, R. P., Sommers, P. L., Reeder, A. A., Bertovich, M. J., and Formal, S. B., 1980, Antibody-dependent cell-mediated antibacterial activity: K lymphocytes, monocytes, and granulocytes are effective against *Shigella*, *J. Immunol.* **125:**2778–2784.
95. Tagliabue, A., Nencioni, L., Villa, L., Keren, D. F., Lowell, G. H., and Boraschi, D., 1983, Antibody-dependent cell-mediated antibacterial activity of intestinal lymphocytes with secretory IgA, *Nature* **306:**184–186.
96. Harendra de Silva, G. D., Mendis, L. N., Sheron, N., Alexander, G. J., Candy, D. C., Chart, H., and Rowe, B., 1993, Concentrations of interleukin 6 and tumour necrosis factor in serum and stools of children with *Shigella dysenteriae* 1 infection, *Gut* **34:**194–198.
97. Raqib, R., Wretlind, B., Andersson, J., and Lindberg, A. A., 1995, Cytokine secretion in acute shigellosis is correlated to disease activity and directed more to stool than to plasma, *J. Infect. Dis.* **171:**376–384.
98. Zychlinsky, A., Fitting, C., Cavaillon, J.-M., and Sansonetti, P. J., 1994, Interleukin 1 is released by murine macrophages during apoptosis induced by *Shigella flexneri*, *J. Clin. Invest.* **94:**1328–1332.
99. Raqib, R., Lindberg, A. A., Wretlind, B., Bardhan, P. K., Andersson, V., and Andersson, J., 1995, Persistence of local cytokine production in shigellosis in acute and convalescent stages, *Infect. Immun.* **63:**289–296.
100. Obrig, T. G., Del Vecchio, P. J., Brown, J. E., Moran, T. P., Rowland, B. M., Judge, T. K., and Rothman, S. W., 1988, Direct cytotoxic action of Shiga toxin on human vascular endothelial cells, *Infect. Immun.* **56:**2373–2378.
101. Harel, Y., Silva, M., Giroir, B., Weinberg, A., Cleary, T. B., and Beutler, B., 1993, A reporter transgene indicates renal-specific induction of tumor necrosis factor (TNF) by Shiga-like toxin, *J. Clin. Invest.* **92:**2110–2116.

102. Barrett, T. J., Potter, M. E., and Strockbine, N. A., 1990, Evidence for participation of the macrophage in Shiga toxin-II induced lethality in mice, *Microb. Pathog.* **9**:95–99.
103. Tesh, V. L., Ramegowda, B., and Samuel, J. E., 1994, Purified Shiga-like toxins induce expression of proinflammatory cytokines from murine peritoneal macrophages, *Infect. Immun.* **62**:5085–5094.
104. Lindberg, A. A., Karnell, A., Pal, T., Sweiha, H., Hultenby, K., and Stocker, B. A., 1990, Construction of an auxotrophic *Shigella flexneri* strain for use as a live vaccine, *Microb. Pathog.* **8**:433–440.
105. Sansonetti, P. J., and Arondel, J., 1989, Construction and evaluation of a double mutant of *Shigella flexneri* as a candidate for oral vaccination against shigellosis, *Vaccine* **7**:443–450.
106. Cohen, D., Green, M. S., Block, C., Slepon, R., and Ofek, I., 1991, A prospective study on the association between serum antibodies to lipopolysaccharide and attack rate of shigellosis, *J. Clin. Microbiol.* **29**:386–389.
107. Taylor, D. N., Trofa, A. C., Sadoff, J. C., Chu, C., Bryla, D., Shiloach, J., Cohen, D., Ashkenazi, S., Lerman, Y., and Egan, W., 1993, Synthesis, characterization, and clinical evaluation of conjugate vaccines composed of the O-specific polysaccharides of *Shigella dysenteriae* type 1, *Shigella flexneri* type 2 and *Shigella sonnei* (*Pleisiomonas shigelloides*) bound to bacterial toxoids, *Infect. Immun.* **61**:3678–3687.
108. Mullet, C., Acheson, D. W. K, Tseng, L.-Y., and Boedeker, E., 1994, High titer serum antibody against the B subunit of Shiga-like toxin protects in a model of enterohemorrhagic Escherichia coli (EHEC) colitis, *Gastroenterology* **106**:A740.

6

Escherichia coli

RONALD P. RABINOWITZ
and MICHAEL S. DONNENBERG

1. INTRODUCTION

There is probably more known about *Escherichia coli* than any other organism. Much of our knowledge of intermediary metabolism, DNA structure and synthesis, transcription and translation has been based on studies of *E. coli*. Although most *E. coli* strains exist as symbionts in the human gut, several distinct groups are armed with specific virulence factors and capable of causing serious illness. These pathogenic strains include several categories that cause diarrhea by different mechanisms, affect specific age groups in diverse geographic locations, and lead to a variety of clinical presentations. These enterovirulent strains will be the focus of this chapter.

The first widely accepted evidence that *E. coli* cause diarrhea was provided by John Bray in the early 1940s.(1) Shortly thereafter a system for serologic typing of *E. coli* and other organisms was developed.(2) This system is based on antigenic properties of lipopolysaccharide (O), flagellar (H), and capsular (K) determinants of the organism. The combination of O and H antigens defines scores of different *E. coli* serotypes. Each category of pathogenic *E. coli* is associated with a different set of characteristic O:H serotypes. This serotyping system was critical for early studies that helped define pathogenic categories of diarrheagenic *E. coli*. However, based on antigenic characteristics that may not be relevant to pathogenesis, serotyping lacks the sensitivity and specificity required of a diagnostic test. Thus,

RONALD P. RABINOWITZ and MICHAEL S. DONNENBERG • Division of Infectious Diseases, University of Maryland School of Medicine, R. A. Cowley Shock Trauma Center, Baltimore, Maryland 21201.

Enteric Infections and Immunity, edited by Lois J. Paradise *et al.* Plenum Press, New York, 1996.

definitive assignment of *E. coli* to a particular pathogenic category requires testing for putative virulence traits. Unfortunately, such tests are not commonly available in clinical microbiology laboratories.

There are currently six distinct categories of *E. coli* that cause diarrhea[(3–6)] (see Table I): enterotoxigenic *E. coli* (ETEC), enteroinvasive *E. coli* (EIEC), enteropathogenic *E. coli* (EPEC), enterohemorrhagic *E. coli* (EHEC), enteroaggregative *E. coli* (EAggEC), and diffusely adherent *E. coli* (DAEC). Of these, the last two have only recently been shown to be epidemiologically linked to diarrhea.[(3–5)] Despite many differences in the epidemiology, pathogenesis, and clinical syndromes associated with these categories, certain common themes are apparent. Strains belonging to most of these categories produce specific fimbriae associated with attachment to enterocytes. Strains belonging to several of these categories produce specific toxins that can result in net intestinal fluid loss. And finally, virulence in most of the diarrheagenic *E. coli* is associated with the presence of specific plasmids encoding toxin, invasion, and/or adherence genes.

2. ENTEROTOXIGENIC *E. COLI*

Enterotoxigenic *E. coli* (ETEC) were first described in 1967,[(7)] shortly after the discovery of cholera toxin. ETEC are an important cause of childhood diarrhea in developing countries. ETEC are also notorious for their association with travelers' diarrhea and may be responsible for up to half of all such cases.[(6,8,9)] In addition to ETEC strains that cause disease in humans, specific strains cause disease in a variety of domesticated animals. This host specificity appears to be related to different adhesins used by each strain.

2.1. Pathogenesis

ETEC pathogenesis can be divided into two facets, colonization and toxin production. Specific adherence factors are necessary for ETEC to colonize the small intestine and resist being washed away with the normal flow of fecal contents.[(10–12)] Once resident in the bowel, ETEC releases one or more enterotoxins that stimulate second messengers to decrease absorption and cause fluid secretion.[(13,14)]

2.1.1. Colonization Factor Antigens

ETEC may possess multiple fimbrial and nonfimbrial adhesins that aid in intestinal colonization. There are currently at least eight colonization factor antigens (CFAs) identified in strains of ETEC that infect humans and

Table I
***E. coli* That Cause Diarrhea**

Category	Epidemiology	Clinical syndrome	Putative virulence characteristics
Enterotoxigenic *E. coli*	Travelers; children in developing countries	Watery diarrhea	Colonization factor antigens, heat-stable toxin, heat-labile toxin
Enteroinvasive *E. coli*	Occasional epidemics, children	Dysentery	Cellular invasion, cell-to-cell spread
Enteropathogenic *E. coli*	Infants in developing countries	Watery diarrhea	Bundle-forming pilus, intimin, attaching and effacing
Enterohemorrhagic *E. coli*	All ages, outbreaks	Bloody diarrhea, hemolytic uremic syndrome	Shiga-like toxin, intimin, attaching and effacing
Enteroaggregative *E. coli*	Children	Watery diarrhea, persistent disease	Fimbriae, toxins?
Diffusely adherent *E. coli*	Children	Watery diarrhea, persistent disease	Adhesins

several CFAs from ETEC strains that cause disease in animals.[15] These fimbriae can be distinguished morphologically, serologically, and by their receptor specificity. In contrast to extraintestinal strains of *E. coli*, for which fimbrial genes are usually chromosomally encoded, most adhesin genes of ETEC have been located on plasmids. However, the existence of chromosomal loci has been suggested by recent work.[17] Regulatory genes are often carried on plasmids.[16,17] In addition, toxin genes are also coded for on the same plasmids and appear to be tightly coregulated. Individual organisms can carry the genetic determinants for several CFAs but in general only one adhesin is expressed at a time.[12] This phase variation (or switching) of CFAs may provide a mechanism for escape from the body's immune system as antibodies to one CFA do not cross-react with different CFAs.[15,18] There has been a great deal of interest regarding the exact nature of the receptors for CFAs. The receptors identified to date are all glycoconjugates including mannosides, *N*-acetylgalactosamine, and *N*-acetyl-D-glucosamine.[19] Many are still unknown.

CFAs are virulence factors as demonstrated by testing mutants deficient in CFA production in animal models. Studies have revealed that both CFA and toxin production are needed for disease.[20–22] Adherence to epithelial cells in the gut via these factors is essential for efficient delivery of toxins to their targets and may assist the organism in acquiring nutrients to compete with commensal flora.[12,23]

2.1.2. *Enterotoxins*

The first enterotoxin of ETEC was discovered in 1967 when a cell-free heat-stable solution from a porcine *E. coli* strain was demonstrated to cause fluid accumulation in a ligated pig intestine. Shortly thereafter a distinct heat-labile toxin antigenically similar to cholera toxin (CT) was recognized.[24] These two toxins are known as heat-stable toxin (ST) and heat-labile toxin (LT).[25] Subsequently, several forms of each toxin have been characterized.[26,27] Two types of ST (STI and STII) have been described based on solubility in methanol and activity in infant mice.[28] STI has been further subdivided into STIa and STIb, which differ slightly in amino acid sequence and length.[29] There are two types of LT (designated LTI and LTII). ETEC strains are capable of producing either toxin class alone or in combination. Human ETEC may produce LT and STI but not STII, which is produced in animal ETEC strains.

a. Heat-Stable Enterotoxins. STI has a molecular weight of approximately 2000.[28] The toxin can withstand temperatures up to 100°C for 15 min, is resistant to several proteases, and is resistant to acid but not basic pH. It can be inactivated by reduction of disulfide bonds. It is soluble in water and organic solvents.[25] STIa is an 18-amino-acid protein produced by animal and human isolates.[26] STIb is a 19-amino-acid protein made only by human isolates. Both are synthesized as larger precursor proteins that are subsequently cleaved to the active toxins. The two toxins share antigenic determinants and a central biologically active core of 13 amino acids at the carboxyl terminus. This 13-amino-acid sequence has six cysteine residues which form three disulfide bonds. All of these bonds are critical for activity of the toxin.[30]

STs are homologous to the mammalian hormone guanylin. Guanylin appears to play a role in the regulation of fluid and electrolyte absorption in the human gut.[31] ST and guanylin bind to the same receptor on intestinal epithelial cells.[31,32] ST activates guanylate cyclase in the intestinal epithelial cell,[33,34] which leads to increased cGMP and cGMP-dependent protein kinases.[13] Increased chloride secretion and decreased sodium absorption ensue via an unknown mechanism.[13]

b. Heat-Labile Toxins. LT has been crystallized and its three-dimensional structure has been solved to a resolution of 2.3 Å.[35] LT is composed of one A and five B subunits with a quaternary structure similar to those of CT and Shiga-like toxin of EHEC.[35–37] CT and LT have considerable sequence homology.[28,38] The A subunit is split into two subunit sections linked via a disulfide bond. The enzymatically active A1 subunit is released via reduction of this bond. The B subunit is necessary for binding and eventual internalization of the active A1 subunit. The B subunit binds to a glycolipid toxin receptor, Gm ganglioside.[39] Although the genes for LTI reside on plasmids, chromosomal genes may be involved in regulation.[40] The G+C content of the genes encoding LT suggests that this toxin may have originated in another species.[40,41]

LTII is antigenically distinct from LTI but is structurally similar. LTII appears to be much more potent than LTI, but is produced at lower levels.[26] LTII binds to another glycolipid, GD1b. The genes for LTII production are chromosomally encoded.[42,43] The A subunits of LTI and LTII share 57% nucleotide sequence identity. However, the B subunits have little homology.[44]

Despite their differences, the mechanism of action of the two LTs and of CT is identical. The A subunit catalyzes the transfer of ADP-ribose from NAD to an amino acid in a stimulatory guanine nucleotide binding regulatory protein (Gs).[45] This locks the G protein in the stimulatory mode and causes constitutive activation of adenylate cyclase, which in turn leads to elevated cAMP. This cyclic nucleotide decreases sodium absorption in villus tip cells and activates chloride secretion in crypt cells. There is no apparent intestinal epithelial cell damage; disease is the result of the massive fluid shifts that follow the change in ion fluxes.

2.2. Clinical Illness

The predominant mode of transmission of ETEC infection is through ingestion of contaminated food or water. A large inoculum (~ 10^8 organisms) is required to cause disease. The incubation period is 12–72 hr. Disease is characterized by nausea, abdominal pain, and watery, nonbloody diarrhea. On average there are five to seven loose bowel movements per day. Fever is seen in only a third of the patients. Leukocytosis is rare as are WBCs in the stool. The average duration of illness in the otherwise healthy individual is 2–4 days. The most serious consequence of ETEC infection is dehydration secondary to voluminous watery diarrhea. Occasionally, severe fluid loss reminiscent of cholera can occur. The diagnosis of ETEC is usually clinical. DNA probes and ELISAs are useful research tools to identify ETEC on the basis of virulence genes and toxins. Treatment consists of fluid replacement, usually via oral rehydration solution. The duration of illness can be shortened by antimicrobial therapy.

2.3. Immunity

Infection with ETEC consists of initial colonization and the subsequent production of toxin. Any method of preventing or decreasing these effects would be the logical targets for protection against diarrhea. Because ETEC are noninvasive, mucosal immunity likely plays a predominant role in the body's response to challenge with ETEC.

The most widely studied mechanism of mucosal immunity in the gut is the secretory immunoglobulin A (SIgA) system.[46] SIgA may act to prevent binding of pathogens and to neutralize toxins.[47,48] Experimentally, SIgA can also activate antibody-dependent T-cell-mediated cytotoxicity.[48,49] SIgA may also interfere with acquisition of necessary growth factors by the pathogen (e.g., iron).[50] Prospective surveillance for diarrheal disease in developing countries suggests that multiple ETEC infections occur early in life and eventually induce immunity that carries into adulthood.[51] Serum antibodies to CFAs and somatic O antigens are present after infection and correlate with protection against subsequent infection from ETEC strains with the same factors.[52] Antibodies to LT alone do not seem to confer immunity in volunteer rechallenge experiments.[53] Tacket *et al.*[54] conducted a randomized double-blind study of the efficacy of hyperimmune bovine colostrum in the prevention of ETEC infection. Volunteers were fed colostrum either high in ETEC antibodies or devoid of ETEC antibodies and then were challenged orally with ETEC. Diarrhea developed in none of ten volunteers who received high-titer colostrum compared to nine of ten volunteers who received control colostrum. They concluded that milk immunoglobulin could be effective prophylaxis against ETEC infection. Their study demonstates the potential for SIga to produce complete protection against disease.

2.4. Vaccine Development

There is currently no highly effective vaccine against ETEC.[55] Based on what we understand of ETEC pathogenesis it would appear that the most practical targets of vaccine development would be the CFAs and the toxins.[50] Despite the large number of CFAs found in human ETEC, three main CFAs are known (CFAI, CFAII, and CFAIV). It would make sense to include these in any practical vaccine preparation. Because of the potential for long-lasting immunity after a single dosage, the best approach probably involves development of live oral vaccines. A limited number of volunteer studies have shed light on the prospects of ETEC vaccines. Volunteers fed a fimbriated strain devoid of toxins were rechallenged 1 month later with a strain producing toxin as well as the same fimbriae. All of six naive control volunteers, but only 3 of 12 vaccinees, developed diarrhea. The mechanism of protection seemed to be prevention of colonization, as shown by duodenal cultures.[6] Another approach involves the construction of a vaccine composed of killed ETEC and a toxoid component. A prototype ETEC with the above CFAs fixed with formalin and combined with CT B subunit (CTB) (which cross-reacts with LTB) has been formulated.[55] It is stable in gastric juices, is immunogenic in volunteers and is well tolerated.[55–57] This vaccine may actually confer immunity for both LT and ST producing ETEC as ST may be found in strains producing the CFAs included in the formulation. This vaccine has been tested in Swedish volunteers.[56,58] It is safe and produces significant levels of intestinal anti-CFA and anti-LT antibodies in a majority of volunteers. The magnitude of immune response was comparable to that seen in convalescent patients from Bangladesh, suggesting that naive volunteers are able to mount adequate responses.

3. ENTEROINVASIVE *E. COLI*

EIEC infection causes an illness similar to that caused by shigella species. Since the initial description of the pathogen, research has shown that EIEC are genetically nearly identical to *Shigella* spp.[59] The pathogenesis of EIEC infection is also similar to shigellosis, consisting of colonization, invasion, cell-to-cell spread, and a brisk polymorphonuclear inflammatory response.[60] Invasion requires several genes on a large (140 MDa) virulence plasmid and chromosomal loci all working in concert for full molecular and clinical expression of this disease.[60]

Despite their remarkable similarities, there are several biochemical differences between these two species that allow laboratories to distinguish between EIEC and Shigella. These differences do not seem to be related to their pathogenic pathways. EIEC can use acetate and mucate as biochemical substrates, whereas Shigella rarely can use either substrate.[61,62]

Because of the great similarity between EIEC and Shigella, it is surprising that EIEC are less common. EIEC cause only 2–4% of acute episodes of bloody diarrhea in the developing world versus 20–25% for Shigella.[63] One potential explanation for the relative rarity of EIEC infection is that the infectious dose is four orders of magnitude higher than that of Shigella. Outbreaks of EIEC have been linked to infected cheese and other foods.[64,65] There has also been a report of a large outbreak in Mexico that subsequently spread to Texas.[65] Recent work with Tn*pho*A mutagenesis has confirmed the genetic similarities between these two organisms and has shown that these minor differences may be related to a cumulative number of small changes in many genes.[59]

Because of the similarity of these organisms, we refer the reader to the discussion of Shigella elsewhere in this text (Chapter 5).

4. ENTEROPATHOGENIC *E. COLI*

EPEC were the first *E. coli* to be associated with diarrhea. The epidemiological link between what we now call EPEC and disease became clear in the 1940s.[1] Since that time, multiple studies have shown EPEC to be an important cause of childhood and infantile diarrhea worldwide, occasionally eclipsing rotavirus in prevalence.[66–71] Because we are just beginning to understand the pathogenesis of EPEC infection, controversy exists over exactly which *E. coli* can be called EPEC. We favor applying the description to strains of diarrheagenic *E. coli* pathogenic for humans that attach intimately to and efface microvilli of enterocytes but do not produce high levels of Shiga-like toxins.[72] The attaching and effacing effect is characterized by intimate adherence of the bacteria to epithelial cells with destruction of microvilli and disruption of the cytoskeleton beneath the adherent organisms[73] (Fig. 1).

Several O:H serotypes are commonly recognized as EPEC.[74] Multilocus enzyme electrophoresis, a technique for measuring genetic relatedness, has been used to show that EPEC are comprised of two clonal heritages each comprising a number of classic serotypes.[74,75] However, as molecular

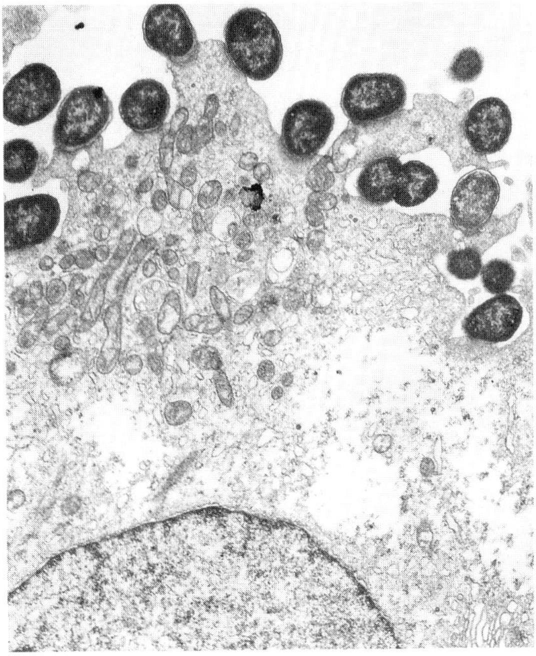

FIGURE 1. Attaching and effacing adherence of enteropathogenic *E. coli* to epithelial cells. Such lesions are found in biopsy specimens from patients infected with enteropathogenic *E. coli* as well as in animal or tissue culture models of enteropathogenic *E. coli* and enterohemorrhagic *E. coli*.

biology techniques have allowed better identification of EPEC, serotyping as a means of diagnosis has become outmoded.

The histopathology of severe EPEC infections often reveals severe villous atrophy throughout the small and large intestine. A mononuclear infiltrate in the lamina propria is often found but there can also be polymorphonuclear cells and macrophages.[(49,76,77)] Ultrastructural examination shows effaced microvilli and bacteria intimately attached to epithelial cells. The loss of microvilli is correlated with diminished activity of enzymes associated with the brush border.[(49,78)]

4.1. Pathogenesis

Application of the tools of molecular biology has greatly enhanced our understanding of EPEC pathogenesis. The study of mutants deficient in facets of EPEC interactions with epithelial cells has allowed a composite picture to be proposed comprising three distinct stages.[79] Although the temporal sequence of these stages is uncertain, they will be discussed sequentially for simplicity.

4.1.1. *Localized Adherence*

Cravioto *et al.* first noted that, unlike most *E. coli*, EPEC adhere avidly to tissue culture cells.[80] Because EPEC form tight microcolonies rather than covering the surface of the cell, the phenomenon has been termed "localized adherence"[81] (Fig. 2A). Localized adherence is plasmid mediated.[82–84] Plasmid cured strains lose the capacity for localized adherence and are less virulent in volunteer studies.[85] Girón *et al.* identified a fimbria called the bundle-forming pilus (BFP), a leading candidate for the factor responsible for localized adherence.[86,87] The BFP is a member of the type IV fimbria family found in numerous pathogenic gram-negative bacteria

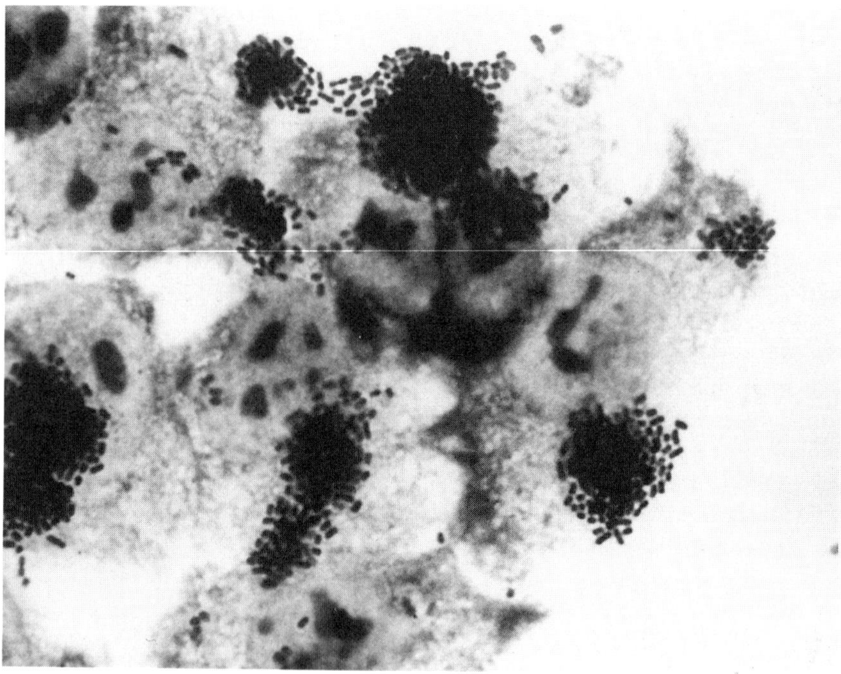

A

FIGURE 2. (A) Localized adherence to HEp-2 cells characteristic of enteropathogenic *E. coli.* (B) Diffuse adherence pattern of HEp-2 cells associated with diffusely adherent *E. coli.* (C) Pattern of aggregative adherence to HEp-2 cells characteristic of enteroaggregative *E. coli.* (Figure courtesy of Dr. James Nataro.)

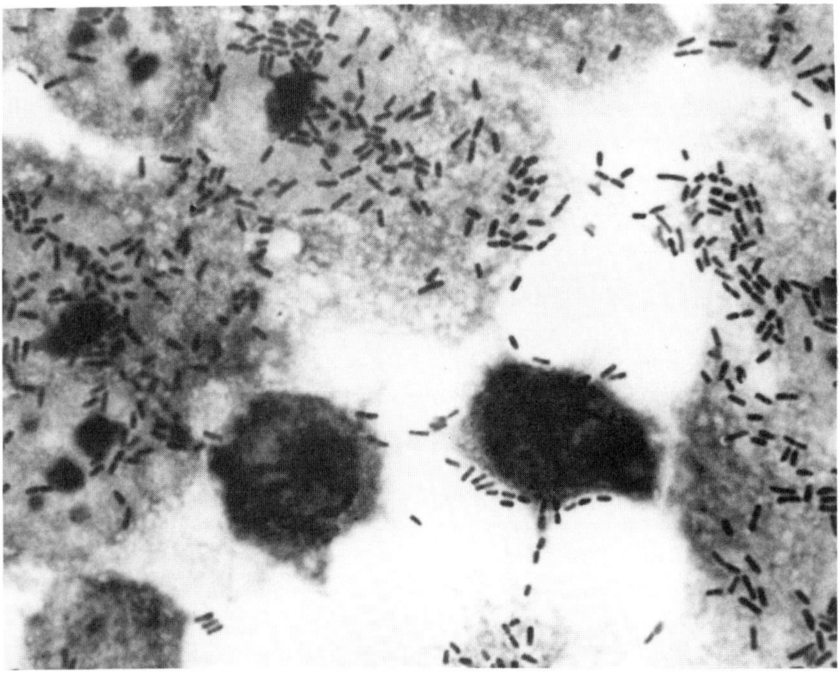

B

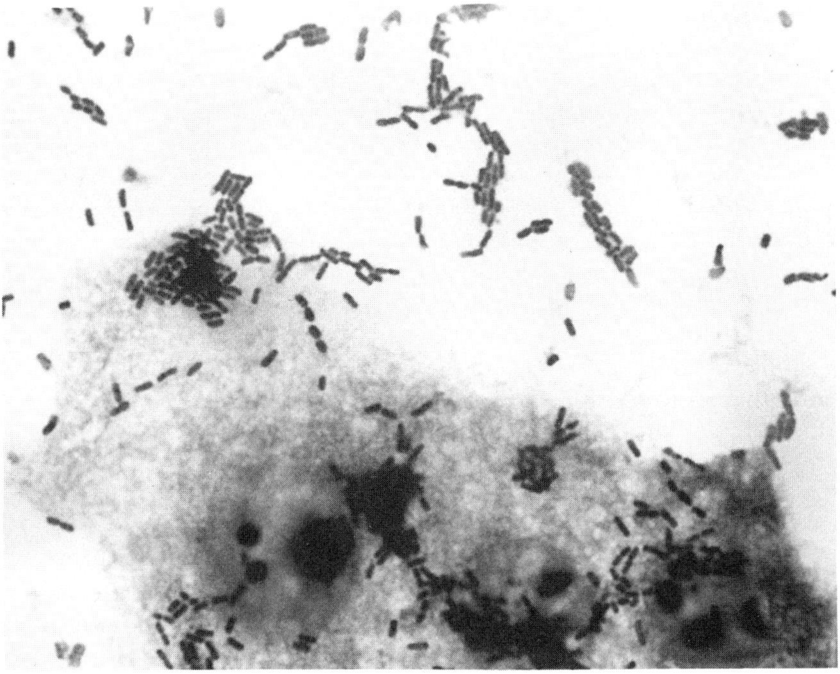

C

FIGURE 2. (*Continued*)

including *Vibrio cholerae, Pseudomonas aeruginosa,* and pathogenic *Neisseria* species. Additional studies have shown that many genes are necessary for localized adherence, probably involved in export, processing, assembly, and regulation of BFP.(83,88)

4.1.2. *Intimate Attachment*

The hallmark of EPEC histopathology is the ability of the organism to attach intimately to enterocytes and efface microvilli. Originally described in 1969 by Staley *et al.*,(89) the term *attaching and effacing* (A/E) was coined by Moon *et al.* in 1983.(90) The A/E lesion consists of bacteria in close proximity to enterocytes. The epithelial cell forms a cuplike pedestal on which the bacteria sit.(91,92) Beneath the attached organism, reorganization of cytoskeletal elements takes place with accumulation of actin, a light chain of myosin, α-actinin, talin, and ezrin.(93,94) Such A/E lesions have been described in biopsies from patients infected with several serogroups of EPEC as well as in animal models of EHEC, *Citrobacter freundii,* and *Hafnia alvei* infections.(95–98)

Jerse *et al.* were the first to identify the *eaeA* locus (*E. coli* *a*ttaching and *e*ffacing gene *A*) by screening transposon mutants of EPEC.(99) Mutants deficient in this gene cannot form A/E lesions but are capable of signal transduction (see below).(99–101) Sequences of *eaeA* have been found in the other organisms capable of A/E as well as in animal pathogens that can produce these lesions.(102–105) This gene is located as part of a large chromosomal insertion common to A/E *E. coli.*(106) The product of *eaeA* is a 94-kDa outer membrane protein called intimin that has sequence similarity to the invasin protein of *Yersinia enterocolitica.*(107) The precise function of intimin is unknown but it may be the intimate adhesin that binds the bacteria to the epithelial cell.(108) Intimin is the first virulence factor to be unambiguously identified in EPEC. In a double-blind volunteer study comparing wild-type EPEC to an isogenic *eaeA* deletion mutant, diarrhea developed in all recipients of the wild-type strain but in only 4 of 11 volunteers who ingested the mutant.(109) Recently, additional genes involved in intimate attachment have been identified(110) and genes on the EPEC adherence plasmid have been found that regulate expression of intimin.(107,111)

4.1.3. *Signal Transduction*

The cytoskeletal effects induced by EPEC result from interference by the bacteria with host cell signal transduction. Rosenshine *et al.*(101) have detected a 90-kDa host cell tyrosine kinase substrate in cells infected by EPEC. The identity of this protein is not yet known. Fluxes of both calcium and inositol phosphate occur in cells infected with EPEC and these second messengers may be involved in actin accumulation.[92,112,113] The following model of signal transduction has been proposed: EPEC activate a host cell tyrosine kinase which in turn activates phospholipase C. The activation of phospholipase C increases the levels of inositol phosphates and diacylglycerol. Inositol phosphates may cause intracellular calcium stores to be released.[114] Diacylglycerol may activate protein kinase C and lead to actin nucleation.[115] Mutants deficient in *eaeA* are still able to initiate these signal transduction events but additional mutants have been isolated that are deficient in tyrosine kinase stimulation and cytoskeletal disruption.[88,101]

Despite our increased knowledge of the molecular pathogenesis of EPEC infection, the actual mechanism of diarrhea is unknown. The loss of microvilli seen makes it tempting to postulate that malabsorption plays a key role in the disease process,[116] but the fact that diarrhea may persist in patients on total parenteral nutrition indicates that secretory factors also play a role. The short incubation time in volunteers who ingest EPEC (between 3 and 12 hr) also suggests a secretory effect. The observed calcium fluxes may activate calcium-sensitive ion channels resulting in secretory diarrhea. It is probably safe to assume that both malabsorption and secretion play a role in EPEC diarrhea.[76]

4.2. Clinical Illness

There is very little clinical information regarding EPEC disease. The original reports from the 1940s and 1950s indicate that it was a devastating illness with profuse watery diarrhea.[117–119] Fever and vomiting were also common. Mortality rates were as high as 25–70%! Several studies in developing countries have shown fever, vomiting, and dehydration to be important features of EPEC disease. Diarrhea lasts a mean of 8 days but is not distinctive compared to other pathogens.[69,120] EPEC can cause persistent disease in some patients.[49,76,77] Volunteer studies in adults have shown that diarrhea ensues 3–12 hr after ingestion and lasts for approximately 2 days, but disease in infants is probably quite different.[121] The diarrhea can be voluminous and can be accompanied by cramps, nausea, vomiting, malaise, and fever. Fecal leukocytes are sometimes seen. Therapy is mainly supportive; the role of antibiotics is not well established.[122,123] The diagnosis of EPEC infection was traditionally made by serogrouping and serotyping. These tests are neither specific nor sensitive. Testing for localized adherence is accurate but subjective and user-dependent and requires tissue culture facilities and light microscopy. DNA probes that recognize the genes for intimin and BFP are the most objective and widely tested molecular diagnostic assays for EPEC. Positive results correlate well with localized adherence. However, despite their use in multiple epidemiological studies, these remain research tools.

4.3. Immunity

EPEC, a disease of the very young, is very rare in adults. This finding may indicate either the development of immunity with age or an inherent resistance in older populations. Serological surveys show increases with age in strain-specific antibodies directed against EPEC O antigens.[124] Volunteers convalescing from experimental EPEC infection develop serum IgG responses to LPS, intimin, and type I fimbriae.[85,122,125,126] Intestinal secretory IgA against EPEC has also been detected in volunteers.[85,122,125,126] There is evidence for natural immunity against EPEC infection. In a community-based longitudinal study from Mexico, 34 of 53 (64%) of infants developed diarrhea after initial colonization with EPEC compared to only 2 of 8 colonized a second time. However, volunteers rechallenged with EPEC showed no decrease in attack rate or illness[127] and protective immunity has not been definitively shown in experimental infection. Although volunteer studies have shown that adults are susceptible to EPEC, the inoculum required is much larger than that probably required to cause outbreaks in children.[121,123,128] Taken together, these findings suggest that EPEC may induce protective immunity but may also indicate an inherently increased susceptibility in infants as compared to older children and adults.

4.4. Vaccines

There are currently no vaccines to prevent EPEC infection.[7] Ongoing research defining virulence genes could lead to development of an attenuated strain. In addition, subunit vaccines composed of intimin and/or the BFP are considerations for future research.

5. ENTEROHEMORRHAGIC *E. COLI*

Konowalchuk[129] first identified a new cytotoxin from certain strains of *E. coli* active on Vero cells. Related toxins have become known as Shiga-like toxins (SLT) or verocytotoxins and strains that produce them as enterohemorrhagic *E. coli* (EHEC) or verocytotoxigenic *E. coli*. EHEC have been identified as the most important cause of hemorrhagic colitis in outbreaks and surveillance studies across North America and elsewhere.[130–134] In addition to hemorrhagic colitis, EHEC have been associated with hemolytic uremic syndrome (HUS) and thrombotic thrombocytopenic purpura (TTP).[135–138] Although as many as 57 different *E. coli* serotypes have been associated with hemorrhagic colitis, by far the most important serotype is O157:H7, accounting for 70–95% of the cases.[63] Prior to the 1980s, this serotype had been very unusual. EHEC infection is most common in summer months with 60% of cases occurring between June and September. Children less than 5 years of age and the elderly seem to be most susceptible to serious EHEC infection and its complications.[139]

There is a clear association of EHEC infection with undercooked ground beef; several of the major outbreaks that have been reported involve fast-food restaurants that serve hamburgers.[132,133] In addition to beef, unpasteurized milk, turkey, apple cider, and water-borne outbreaks have also been reported.[140,141] EHEC have been isolated from cows, pigs, and poultry.[142,143] EHEC can also be spread person to person as shown in a nursing home outbreak[142,144] and in day-care facilities.[143,145] Pathological examination of biopsy specimens from patients with EHEC infection shows hemorrhage and edema in the colon with focal necrosis.[146]

5.1. Pathogenesis

As with other diarrheal pathogens, the ability to adhere to the gastrointestinal mucosa is an important aspect of pathogenesis. The other key factor involved in EHEC infection is the elaboration of cytotoxins.

5.1.1. Attaching and Effacing

EHEC and EPEC share a common ancestry.[75] Their divergence may have started with the acquisition by EHEC of its toxins. EHEC are able to form A/E lesions in animals and tissue culture that are indistinguishable from those of EPEC.[92] Indeed, the genetic determinants for A/E appear to be interchangeable between the two species.[147] Despite this fact, the initial adherence of EHEC and EPEC are quite distinct. EHEC do not react with an EPEC adherence plasmid probe and do not produce the BFP of EPEC.[148] O157:H7 strains do possess a 60-MDa plasmid, but this plasmid is distinct from that of EPEC.[149] Although the plasmid is conserved among EHEC strains, its role in adherence and pathogenesis is uncertain.[140,150]

5.1.2. Toxins

EHEC may produce one or both of two types of toxins, SLT1 and SLT2, that are closely related to Shiga toxin of *Shigella dysenteriae.* These toxins are composed of two subunits: a single A polypeptide, which is the active subunit, surrounded by five B (binding) subunits.[151] SLT1 is biologically identical to Shiga toxin, differing in only one amino acid, and is neutralized by anti-Shiga-toxin antisera.[152,153] SLT2 toxin is 50–60% homologous to Shiga toxin and is not inactivated by anti-Shiga-toxin antisera. All of these toxins work by the same mechanism.[151] The B subunit binds to the galactose-α1,4-galactose-containing glycolipid globotriosyl ceramide.[152–154] The A subunit then enters the cell and cleaves an *N*-glycoside bond at a specific adenine base at position 4324 of 28 S ribosomal RNA.[151] This cleavage prevents protein synthesis, which eventually leads to cell death. Shiga-like toxins are active on a variety of cell types including intestinal epithelial cells and endothelial cells. The systemic absorption of Shiga-like toxins is presumed to be a critical factor in the development of HUS and TTP.

5.2. Clinical Illness

EHEC infection is characterized by bloody diarrhea and severe abdominal cramps. A nonbloody diarrheal prodrome may be present, but within several days many patients develop bloody stools. Fever, present in no more than a third of patients, is typically low-grade.[155] Vomiting is present in about half of afflicted persons. Depending on the age of the patient, EHEC infection may be confused with appendicitis, inflammatory bowel disease, or ischemic colitis. The incubation period is 3–4 days and the average duration of illness is 1 week. It is estimated that 3–4% of patients can develop HUS or TTP.[155] Treatment of EHEC is supportive. Antibiotic therapy has been associated with a higher risk of serious sequelae.[142,144] However, a recent prospective, randomized study by Proulx *et al.*[156] showed no difference in progression to HUS between children treated with trimethoprim–sulfamethoxazole and untreated controls. Unfortunately, therapy was begun on day 7 of illness, which may have affected the results. A presumptive diagnosis of EHEC infection can be made by culturing stool on sorbitol–MacConkey agar. EHEC O157:H7 do not ferment sorbitol whereas normal commensal *E. coli* can. This assay is readily available in most clinical microbiology laboratories. The diagnosis can be confirmed at specialized laboratories using DNA probes against the EHEC plasmid, toxin, and attachment factors, by serotyping, or assaying for toxin production.[140]

5.3. Immunity/Vaccine Development

Because of the severe nature of EHEC infection, volunteer studies to assess immune response and vaccine challenge studies are not possible. However, serological surveys indicate that up to 20% of healthy control subjects have antibodies to SLTs. It can be postulated that these individuals have had previous asymptomatic exposure to EHEC. Such exposure may be protective during an outbreak.[140]

6. ENTEROAGGREGATIVE *E. COLI*

Since the initial description of HEp-2-adherent *E. coli*, it has become apparent that there are several different categories of adherence patterns (see Fig. 2). EAggEC is defined by its aggregative adherence phenotype, which is characterized by a "stacked brick" appearance and autoagglutination of bacteria to each other and often to the glass slide (Fig. 2C).(70) The serotypes that comprise EAggEC strains have not been well described. Volunteer and epidemiological studies have helped clarify the pathogenic potential of EAggEC strains. Of four such strains, one caused diarrhea in three of five recipients and one volunteer had enteric symptoms.(157) None of the other strains caused disease. Mathewson *et al.*(158) fed two strains of EAggEC to volunteers. Three of eight developed diarrhea after a challenge of 10^{10} organisms and two others had "enteric" symptoms of borborygmi or cramps. These studies have provided the most compelling evidence of the pathogenicity of at least some EAggEC strains. Epidemiological evidence supporting the pathogenicity of these organisms comes from numerous studies conducted throughout the world.(5,66,70,159–162) These studies involved children typically less than 5 and usually less than 3 years of age. In at least four studies that used appropriate matched controls, EAggEC was statistically associated with acute diarrhea, and in five studies, isolation of EAggEC was associated with persistent diarrhea (defined as greater than 14 days). All of these studies were conducted in developing countries. Conversely, a number of studies have found very high excretion rates of EAggEC in the absence of clinical disease.(3,4,159) The weight of the evidence, however, including the epidemiological and volunteer data, indicates that at least some strains of EAggEC are pathogenic.

6.1. Pathogenesis

As with other diarrheagenic *E. coli*, there is evidence to support the role of fimbriae and toxins in the pathogenesis of EAggEC infection.

6.1.1. Adherence

Initial work with EAggEC has shown the presence of a 65-MDa plasmid necessary for aggregative adherence.(163,164) This plasmid confers the aggregative phenotype on transfer to the nonadherent strain HB101.(164,165) The plasmid does not cross-react with DNA probes used to identify the large plasmid of EPEC. Nataro *et al.* have identified a fimbria—termed the *a*ggregative *a*dherence *f*imbria *I*(AAF/I)(164,165)—that mediates adherence and agglutination of RBCs. The genes necessary for this fimbria are encoded on the plasmid in two distinct regions. The first region encodes the structural units of the fimbria and the second cluster encodes a transcriptional activator of expression.

6.1.2. *Toxins*

During study of the EAggEC adherence plasmid, an open reading frame (ORF) predicted to encode a polypeptide with homology to ST of ETEC was identified. This putative toxin has been designated EAST 1.(166,167) The predicted protein sequence has four cysteine residues, which is the same number as guanylin but two less than ST of ETEC. Culture supernatants from clones carrying the EAST 1 gene cause an increase in short-circuit current in Ussing chamber models, as do other enterotoxins. An additional toxin has been described in supernatants of EAggEC that increases intracellular calcium levels and induces phosphorylation of several HEp-2 cytoplasmic proteins.(168) These changes were neutralized by antiserum to *E. coli* hemolysin, suggesting antigenic similarity. Another toxin has been described that lyses sheep erythrocytes under specific conditions of temperature and pH, when provided in close association with the cells. This toxin has been called *contact hemolysin* and may play a role in pathogenesis.(169)

The limited information available makes it tempting to draw parallels between ETEC and EAggEC pathogenesis. The AAF/I is a putative factor in the colonization of the bowel. Once colonization takes place, the production of EAST 1 or other toxins may lead to fluid secretion as in ETEC infection. However, the roles of these putative virulence factors in disease remain to be tested in animal models and volunteer studies.

6.2. Clinical Illness

There is very little information about the clinical presentation of EAggEC. What is known comes from the above-mentioned volunteer and epidemiological studies.(66,160) The illness appears to be similar to that of EPEC. It is usually of short duration characterized by relatively small volume of stool. Fever may be present and bloody diarrhea is seen in up to a third of the cases. EAggEC is one of a small number of pathogens that has been linked with persistent diarrhea. Persistent diarrhea is a major contributor to infant mortality in developing countries. The cycle of diarrheal illness, weight loss, malnutrition, depressed immunity, and further episodes of diarrhea is exacerbated by prolonged episodes. Thus, EAggEC is probably a very dangerous pathogen. The diagnosis of EAggEC can be established by observing the classic aggregative pattern in an adherence assay. Again, this is user-dependent, subjective, and requires tissue culture facilities. The use of DNA probes against the 65-MDa plasmid has had varying success and remains a research tool.(170) A simple laboratory assay that takes advantage of the ability of EAggEC to form clumps at the surface of Mueller–Hinton broth cultures may allow fast and presumptive diagnosis of EAggEC.(171)

6.3. Immunity/Vaccine Development

The immune response to EAggEC remains to be described. There have been no vaccine trials with EAggEC, but empiric observations suggest an approach similar to that of other *E. coli* with emphasis on preparations against fimbriae and toxins. The ability to cause chronic infection implies that EAggEC may be able to evade the host immune system for prolonged periods.

7. DIFFUSELY ADHERENT *E. COLI*

DAEC remain the least well characterized and probably the most heterogeneous group of *E. coli* proposed to cause diarrhea. Contributing to the heterogeneity and confusion is the frequent grouping of DAEC with EAggEC and often EPEC into a wastebasket group called enteroadherent *E. coli*, a practice that should be abandoned.(172,173) DAEC are defined by their ability to adhere uniformly over the entire surface of epithelial cells (Fig. 2B).

Early epidemiological studies on DAEC failed to demonstrate a significant association with diarrhea.(5,59,66,159,160) However, recent studies have linked DAEC to illness. Jallat *et al.*(174) found that 38% of 262 specimens from patients hospitalized in France with diarrhea displayed diffuse adherence to epithelial cells compared with 8% of controls. Gunzburg *et al.*(175) found a statistically significant association of DAEC with diarrhea in aboriginal Australian children $\geq$18 months old compared with age-matched controls. Levine *et al.*(176) studied children up to 59 months of age and found a significant association of DAEC with diarrhea only in children greater than 4 years old. And lastly, Baqui *et al.*(162) studied children less than 5 years old in Bangladesh and found significantly more DAEC in children with diarrhea compared with asymptomatic children. In some studies, DAEC have been associated with persistent diarrhea. In a volunteer study,(177) none of 13 volunteers challenged with two different DAEC strains developed diarrhea as defined in the study, although two in each group had stools that narrowly missed the case definition of diarrhea.

7.1. Pathogenesis

The pathogenesis of this infection is poorly understood. There have been two adherence factors described in two different strains. Bilge *et al.*(178,179) characterized the F1845 fimbria from a strain of DAEC. The gene encoding this fimbria can be found on the chromosome or on a plasmid. The gene product is predicted to be a 14-kDa protein similar to structural subunits of mannose-resistant fimbria. The *a*dhesin *i*nvolved in *d*iffuse *a*dherence (AIDA-I) described by Benz *et al.*(180,181) in a different strain is a 100-kDa plasmid-encoded protein. Antisera against both of these adherence factors inhibit bacterial attachment in *in vitro* assays. These data suggest a possible pathogenic role for adhesins in this infection as in other pathogens. DNA probes for F1845 are positive in 30–75% of DAEC strains whereas strains probe-positive for AIDA-I are much less common.(170)

7.2. Clinical Illness

Although the clinical illness associated with DAEC is not well characterized, it does not seem to be distinctive. Diarrhea may be watery and contain mucus. Fever and vomiting are seen in approximately a third of the patients. The median duration of symptoms is 8 days. Diagnosis is by classic DA phenotype in adherence assays(3) (Fig. 2c). DNA probes described above may be useful for confirming diagnosis in the future.

7.3. Immunity

There is very little information regarding the immune response to DAEC infection. A minority of volunteers developed significant serum or secretory responses to whole-cell antigens after ingesting DAEC strains.(177) Colonization of the jejunum had no relation to antibody response. The early status of research on these strains suggests that development of a DAEC vaccine will not be forthcoming in the near future.

8. SUMMARY

Enterovirulent *E. coli* cause considerable morbidity and mortality worldwide. These *E. coli* may be divided into at least six categories that cause disease by distinct mechanisms often involving plasmid-mediated factors. ETEC produce CFAs necessary for adherence and enterotoxins that elicit fluid secretion. EIEC relies on its ability to invade enterocytes and spread from cell to cell causing cell damage and an intense inflammatory response. EPEC adhere to intestinal cells, disrupting the cytoskeleton and possibly causing both malabsorption and secretory diarrhea. EHEC cause both bloody diarrhea and systemic disease by attaching to cells and secreting a potent cytotoxin. EAggEC also attach to enterocytes and produce toxins that may lead insidiously to persistent diarrhea. Lastly, DAEC use adhesins to initiate a disease process that is poorly understood. Although much has been learned about diarrheagenic *E. coli*, little is understood about the immune response and the potential for prevention of infection caused by these organisms.

REFERENCES

1. Bray, J., 1945, Isolation of antigenically homogeneous strains of *Bact. coli* neapolitanum from summer diarrhoea of infants. *J. Pathol. Bacteriol.* **57:**239–247.
2. Kauffmann, F., 1947, The serology of the coli group, *J. Immunol.* **57:**71–100.
3. Girón, J. A., Jones, T., Millán-Velasco, F., Castro-Muñoz, E., Zárate, L., Fry, J., Frankel, G., Moseley, S. L., Baudry, B., Kaper, J. B., Schoolnik, G. K., and Riley, L. W., 1991, Diffuse-adhering *Escherichia coli* (DAEC) as a putative cause of diarrhea in Mayan children in Mexico, *J. Infect. Dis.* **163:**507–513.
4. Vial, P. A., Robins-Browne, R., Lior, H., Prado, V., Kaper, J. B., Nataro, J. P., Maneval, D., Elsayed, A., and Levine, M. M., 1988, Characterization of enteroadherent-aggregative *Escherichia coli*, a putative agent of diarrheal disease, *J. Infect. Dis.* **158:**70–79.
5. Wanke, C. A., Schorling, J. B., Barrett, L. J., Desouza, M. A., and Guerrant, R. L., 1991, Potential role of adherence traits of *Escherichia coli* in persistent diarrhea in an urban Brazilian slum, *Pediatr. Infect. Dis. J.* **10:**746–751.
6. Levine, M. M., 1987, *Escherichia coli* that cause diarrhea: Enterotoxigenic, enteropathogenic, enteroinvasive, enterohemorrhagic, and enteroadherent, *J. Infect. Dis.* **155:** 377–389.
7. Levine, M. M., and Noriega, F., 1993, Vaccines to prevent enteric infections, *Baillieres Clin. Gastroenterol.* **7:**501–517.
8. Hyams, K. C., Bourgeois, A. L., Merrell, B. R., Rozmajzl, P. Escamilla, J., Thornton, S. A., Wasserman, G. M., Burke, A., Echeverria, P., Green, K. Y., Kapikian, A. Z., and Woody, J. N., 1991, Diarrheal disease during operation Desert Shield, *N. Engl. J. Med.* **325:**1423–1428.
9. Merson, M. H., Morris, G. K., Sack, D. A., Wells, J. G., Feeley, J. C., Sack, R. B., Creech, W. B., Kapikian, A. Z., and Gangarosa, E. J., 1976, Travelers' diarrhea in Mexico. A prospective study of physicians and family members attending a congress, *N. Engl. J. Med.* **294:**1299–1305.
10. Levine, M. M., Kaper, J. B., Black, R. E., and Clements, M. L., 1983, New knowledge on pathogenesis of bacterial enteric infections as applied to vaccine development, *Microbiol. Rev.* **47:**510–550.
11. Ørskov, I., and Ørskov, F., 1985, *Escherichia coli* in extra-intestinal infections, *J. Hyg.* **95:**551–575.
12. de Graaf, F. K., 1990, Genetics of adhesive fimbriae of intestinal *Escherichia coli*, *Curr. Top. Microbiol. Immunol.* **151:**29–53.
13. O'Loughlin, E. V., Scott, R. B., and Gall, D. G., 1991, Pathophysiology of infectious diarrhea: Changes in intestinal structure and function, *J. Pediatr. Gastroenterol. Nutr.* **12:**5–20.
14. Robertson, D. C., McDonel, J. L., and Dorner, F., 1985, *E. coli* heat-labile enterotoxin, *Pharmacol. Ther.* **28:**303–339.
15. Hacker, J., 1992, Role of fimbrial adhesins in the pathogenesis of *Escherichia coli* infections, *Can. J. Microbiol.* **38:**720–727.

16. Klemm, P., 1985, Fimbrial adhesions of *Escherichia coli, Rev. Infect. Dis.* **7:**321–340.
17. Perez-Casal, J., Swartley, J. S., and Scott, J. R., 1990, Gene encoding the major subunit of CSl pili of human enterotoxigenic *Escherichia coli, Infect. Immun.* **58:**3594–3600.
18. Krogfelt, K. A., 1991, Bacterial adhesion: Genetics, biogenesis, and role in pathogenesis of fimbrial adhesins of *Escherichia coli, Rev. Infect. Dis.* **13:**721–735.
19. Moseley, S. L., Dougan, G., Schneider, R. A., and Moon, H. W., 1986, Cloning of chromosomal DNA encoding the F41 adhesin of enterotoxigenic *Escherichia coli* and genetic homology between adhesins F41 and K88, *J. Bacteriol.* **167:**799–804.
20. Gaastra, W., and de Graaf, F. K., 1982, Host-specific fimbrial adhesins of noninvasive enterotoxigenic *Escherichia coli* strains, *Microbiol. Rev.* **46:**129–161.
21. Smith, H. W., and Linggood, M. A., 1971, Observations on the pathogenic properties of the K88, HLY and ENT plasmids of *Escherichia coli* with particular reference to porcine diarrhoea, *J. Med. Microbiol.* **4:**467–485.
22. Smith, H. W., and Linggood, M. A., 1972, Further observation on *Escherichia coli* enterotoxins with particular regard to those produced by atypical piglet strains and by calf and lamb strains: The transmissible nature of these enterotoxins and of a K antigen possessed by calf and lamb strains, *J. Med. Microbiol.* **5:**243–250.
23. Zafriri, D., Oron, Y., Eisenstein, B. I., and Ofek, I., 1987, Growth advantage and enhanced toxicity of *Escherichia coli* adherent to tissue culture cells due to restricted diffusion of products secreted by cells, *J. Clin. Invest.* **79:**1210–1216.
24. Smith, H. W., and Halls, S., 1967, Studies on *Escherichia coli* enterotoxin, *J. Pathol.* **93:**531–543.
25. Smith, H. W., and Gyles, C. L., 1970, The relationship between two apparently different enterotoxins produced by enteropathogenic strains of *Escherichia coli* of porcine origin, *J. Med. Microbiol.* **3:**387–401.
26. Guth, B. E., Twiddy, E. M., Trabulsi, L. R., and Holmes, R. K., 1986, Variation in chemical properties and antigenic determinants among type II heat-labile enterotoxins of *Escherichia coli, Infect. Immun.* **54:**529–536.
27. Burgess, M. N., Bywater, R. J., Cowley, C. M., Mullan, N. A., and Newsome, P. M., 1978, Biological evaluation of a methanol-soluble, heat-stable *Escherichia coli* enterotoxin in infant mice, pigs, rabbits, and calves, *Infect. Immun.* **21:**526–531.
28. Moseley, S. L., Hardy, J. W., Huq, M. I., Echeverria, P., and Falkow, S., 1983, Isolation and nucleotide sequence determination of a gene encoding a heat-stable enterotoxin of *Escherichia coli, Infect. Immun.* **39:**1167–1174.
29. Weikel, C. S., Tiemens, K. M., Moseley, S. L., Huq, I. M., and Guerrant, R. L., 1986, Species specificity and lack of production of STb enterotoxin by *Escherichia coli* strains isolated from humans with diarrheal illness, *Infect. Immun.* **52:**323–325.
30. Okamoto, K., and Takahara, M., 1990, Synthesis of *Escherichia coli* heat-stable enterotoxin STa as a pre-pro form and role of the pro sequence in secretion, *J. Bacteriol.* **172:**5260–5265.
31. Wiegand, R. C., Kato, J., Huang, M. D., Fok, K. F., Kachur, J. F., and Currie, M. G., 1992, Human guanylin: cDNA isolation, structure, and activity, *FEBS Lett.* **311:**150–154.
32. Cantey, J. R., 1993, *Escherichia coli* diarrhea, *Gastroenterol. Clin. North Am.* **22:**609–622.
33. Field, M., Graf, L. H., Jr., Laird, W. J., and Smith, P. L., 1978, Heat-stable enterotoxin of *Escherichia coli*: In vitro effects on guanylate cyclase activity, cyclic GMP concentration, and ion transport in small intestine, *Proc. Natl. Acad. Sci. USA* **75:**2800–2804.
34. Hughes, J. M., Murad, F., Chang, B., and Guerrant, R. L., 1978, Role of cyclic GMP in the action of heat-stable enterotoxin of *Escherichia coli, Nature* **271:**755–756.
35. Sixma, T. K., Pronk, S. E., Kalk, K. H., Wartna, E. S., van Zanten, B. A., Witholt, B., and Hol, W. G., 1991, Crystal structure of a cholera toxin-related heat-labile enterotoxin from *E. coli, Nature* **351:**371–377.

36. Richards, K. L., and Douglas, S. D., 1978, Pathophysiological effects of *Vibrio cholerae* and enterotoxigenic *Escherichia coli* and their exotoxins on eucaryotic cells, *Microbiol. Rev.* **42:** 592–613.
37. Pronk, S. E., Hofstra, H., Groendijk, H., Kingma, J., Swarte, M. B., Dorner, F., Drenth, J., Hol, W. G., and Witholt, B., 1985, Heat-labile enterotoxin of *Escherichia coli*, Characterization of different crystal forms, *J. Biol. Chem.* **260:**13580–13584.
38. Dallas, W. S., and Falkow, S., 1980, Amino acid sequence homology between cholera toxin and *Escherichia coli* heat-labile toxin. *Nature* **288:**499–501.
39. Moss, J., Garrison, S., Fishman, P. H., and Richardson, S. H., 1979, Gangliosides sensitize unresponsive fibroblasts to *Escherichia coli* heat-labile enterotoxin, *J. Clin. Invest.* **64:**381–384.
40. Katayama, S., Ninomiya, M., Minami, J., Okabe, A., and Hayashi, H., 1990, Transcriptional control plays an important role for the production of heat-labile enterotoxin in enterotoxigenic *Escherichia coli* of human origin, *Microbiol. Immunol.* **34:**11–24.
41. Neill, R. J., Twiddy, E. M., and Holmes, R. K., 1983, Synthesis of plasmid-coded heat-labile enterotoxin in wild-type and hypertoxinogenic strains of *Escherichia coli* and in other genera of Enterobacteriaceae, *Infect. Immun.* **41:**1056–1061.
42. Holmes, R. K., Twiddy, E. M., and Pickett, C. L., 1986, Purification and characterization of type II heat-labile enterotoxin of *Escherichia coli*, *Infect. Immun.* **53:**464–473.
43. Fukuta, S., Magnani, J. L., Twiddy, E. M., Holmes, R. K., and Ginsburg, V., 1988, Comparison of the carbohydrate-binding specificities of cholera toxin and *Escherichia coli* heat-labile enterotoxins LTh-I, LT-IIa, and LT-IIb, *Infect. Immun.* **56:**1748–1753.
44. Pickett, C. L., Twiddy, E. M., Coker, C., and Holmes, R. K., 1989, Cloning, nucleotide sequence, and hybridization studies of the type IIb heat-labile enterotoxin gene of *Escherichia coli*, *J. Bacteriol.* **171:**4945–4952.
45. Gyles, C. L., 1992, *E. coli* cytotoxins and enterotoxins, *Can. J. Microbiol.* **38:**738–746.
46. Mestecky, J., and McGhee, J. R., 1987, Immunoglobulin A (IgA): Molecular and cellular interactions involved in IgA biosynthesis and immune response, *Adv. Immunol.* **40:**153–245.
47. Holmgren, J., and Svennerholm, A. M., 1983, Cholera and the immune response, *Prog. Allergy* **33:**106–119.
48. Quiding, M., Nordstrom, I., Kilander, A., Andersson, G., Hanson, L. A., Holmgren, J., and Czerkinsky, C., 1991, Intestinal immune responses in humans. Oral cholera vaccination induces strong intestinal antibody responses and interferon-gamma production and evokes local immunological memory, *J. Clin. Invest.* **88:**143–148.
49. Ulshen, M. H., and Rollo, J. L., 1980, Pathogenesis of *Escherichia coli* gastroenteritis in man—Another mechanism, *N. Engl. J. Med.* **302:**99–101.
50. Holmgren, J., and Svennerholm, A. M., 1992, Bacterial enteric infections and vaccine development, *Gastroenterol. Clin. North Am.* **21:**283–302.
51. Cravioto, A., Reyes, R. E., Ortega, R., Fernández, G., Hernandez, R., and Lopez, D., 1988, Prospective study of diarrhoeal disease in a cohort of rural Mexican children: Incidence and isolated pathogens during the first two years of life, *Epidemiol. Infect.* **101:**123–128.
52. Svennerholm, A., Ahren, C., Wenneras, C., and Homgren, J., 1991, Vaccines against enterotoxigenic *E. coli*: Part I, in: *Molecular Pathogenesis of Gastrointestinal Infections* (A. Svennerholm, T. Wadstrom, P. Makela, and H. Wolf-Wapz, eds.), Plenum Press, New York, pp. 287–294.
53. Levine, M. M., Nalin, D. R., Hoover, D. L., Berquist, E. J., Hornick, R. B., and Young, C. R., 1979, Immunity to enterotoxigenic *Escherichia coli*, *Infect. Immun.* **23:**729–736.
54. Tacket, C. O., Losonsky, G., Link, H., Hoang, Y., Guesry, P., Hilpert, H., and Levine, M. M., 1988, Protection by milk immunoglobulin concentrate against oral challenge with enterotoxigenic *Escherichia coli*, *N. Engl. J. Med.* **318:**1240–1243.

55. Holmgren, J., and Svennerholm, A. M., 1990, Development of oral vaccines against cholera and enterotoxinogenic *Escherichia coli* diarrhea, *Scand. J. Infect. Dis. Suppl.* **76:**47–53.
56. Holmgren, J., and Svennerholm, A. M., 1990, New vaccines against bacterial enteric infections, *Scand. J. Infect. Dis. Suppl.* **76:**149–156.
57. Clemens, J. D., Sack, D. A., Harris, J. R., Chakraborty, J., Khan, M. R., Stanton, B. F., Kay, B. A., Khan, M. U., Yunus, M., and Atkinson, W. 1986, Field trial of oral cholera vaccines in Bangladesh, *Lancet* **2:**124–127.
58. Clemens, J. D., Harris, J. R., Sack, D. A., Chakraborty, J., Ahmed, F., Stanton, B. F., Khan, M. U., Kay, B. A., Huda, N., and Khan, M. R., 1988, Field trial of oral cholera vaccines in Bangladesh: Results of one year of follow-up, *J. Infect. Dis.* **158:**60–69.
59. Hsia, R.-C., Small, P. L. C., and Bavoil, P. M., 1993, Characterization of virulence genes of enteroinvasive *Escherichia coli* by TnphoA mutagenesis: Identification of *inv*X, a gene required for entry into HEp-2 cells, *J. Bacteriol.* **175:**4817–4823.
60. Hale, T. L., 1991, Genetic basis of virulence in *Shigella* species, *Microbiol. Rev.* **55:**206–224.
61. Farmer, J. J., Davis, B. R., Hickman-Brenner, F. W., McWhorter, A., Huntley-Carter, G. P., Asbury, M. A., Riddle, C., Wathen-Grady, H. G., Elias, C., and Fanning, G. R., 1985, Biochemical identification of new species and biogroups of Enterobacteriaceae isolated from clinical specimens, *J. Clin. Microbiol.* **21:**46–76.
62. Echeverria, P., Sethabutr, O., and Pitarangsi, C., 1991, Microbiology and diagnosis of infections with Shigella and enteroinvasive *Escherichia coli, Rev. Infect. Dis.* **13**(Suppl. 4): S220–S225.
63. Brook, M. G., and Bannister, B. A., 1993, Diarrhoea-causing *Escherichia coli, Dig. Dis.* **11:**288–297.
64. Doyle, M. P., 1990, Pathogenic *Escherichia coli, Yersinia enterocolitica,* and *Vibrio parahaemolyticus, Lancet* **336:**1111–1115.
65. Gordillo, M. E., Reeve, G., Pappas, J., Mathewson, J. J., DuPont, H. L., and Murray, B., 1992, Molecular characterization of strains of enteroinvasive *Escherichia coli* O143, including isolates from a large outbreak in Houston, Texas, *J. Clin. Microbiol.* **30:**889–893.
66. Cravioto, A., Tello, A., Navarro, A., Ruiz, J., Villafán, H., Uribe, F., and Eslava, C., 1991, Association of *Escherichia coli* HEp-2 adherence patterns with type and duration of diarrhoea, *Lancet* **337:**262–264.
67. Nataro, J. P., Baldini, M. M., Kaper, J. B., Black, R. E., Bravo, N., and Levine, M. M., 1985, Detection of an adherence factor of enteropathogenic *Escherichia coli* with a DNA probe, *J. Infect. Dis.* **152:**560–565.
68. Gunzburg, S. T., Chang, B. J., Burke, V., and Gracey, M., 1992, Virulence factors of enteric *Escherichia coli* in young Aboriginal children in north-west Australia, *Epidemiol. Infect.* **109:** 283–289.
69. Gomes, T. A. T., Rassi, V., Macdonald, K. L., Ramos, S. R. T. S., Trabulsi, L. R., Vieira, M. A. M., Guth, B. E. C., Candeias, J. A. N., Ivey, C., Toledo, M. R. F., and Blake, P. A., 1991, Enteropathogens associated with acute diarrheal disease in urban infants in São Paulo, Brazil, *J. Infect. Dis.* **164:**331–337.
70. Nataro, J. P., Kaper, J. B., Robins-Browne, R., Prado, V., Vial, P., and Levine, M. M., 1987, Patterns of adherence of diarrheagenic *Escherichia coli* to HEp-2 cells, *Pediatr. Infect. Dis. J.* **6:**829–831.
71. Robins-Browne, R., Still, C. S., Miliotis, M. D., Richardson, N. J., Koornhof, H. J., Frieman, I., Schoub, B. D., Lecatsas, G., and Hartman, E., 1980, Summer diarrhoea in African infants and children, *Arch. Dis. Child.* **55:**923–928.
72. Donnenberg, M. S., 1994, Enteropathogenic *Escherichia coli,* in: *Infections in the Gastrointestinal Tract* (M. J. Blaser, P. D. Smith, J. I. Ravdin, H. B. Greenberg, and R. L. Guerrant, eds.), Raven Press, New York, in press.

73. Dabiri, G. A., Sanger, J. M., Portnoy, D. A., and Southwick, F. S., 1990, *Listeria monocytogenes* moves rapidly through the host-cell cytoplasm by inducing directional actin assembly, *Proc. Natl. Acad. Sci. USA* **87**:6068–6072.
74. Ørskov, F., Whittam, T. S., Cravioto, A., and Ørskov, I., 1990, Clonal relationships among classic enteropathogenic *Escherichia coli* (EPEC) belonging to different O groups, *J. Infect. Dis.* **162**:76–81.
75. Whittam, T. S., Wolfe, M. L., Wachsmuth, I. K., Ørskov, F., Ørskov, I., and Wilson, R. A., 1993, Clonal relationships among *Escherichia coli* strains that cause hemorrhagic colitis and infantile diarrhea, *Infect. Immun.* **61**:1619–1629.
76. Hill, S. M., Phillips, A. D., and Walker-Smith, J. A., 1991, Enteropathogenic *Escherichia coli* and life threatening chronic diarrhoea, *Gut* **32**:154–158.
77. Rothbaum, R., McAdams, A. J., Giannella, R., and Partin, J. C., 1982, A clinicopathological study of enterocyte-adherent *Escherichia coli*: A cause of protracted diarrhea in infants, *Gastroenterology* **83**:441–454.
78. Taylor, C. J., Hart, A., Batt, R. M., McDougall, C., and McLean, L., 1986, Ultrastructural and biochemical changes in human jejunal mucosa associated with enteropathogenic *Escherichia coli* (O111) infection, *J. Pediatr. Gastroenterol. Nutr.* **5**:70–73.
79. Donnenberg, M. S., and Kaper, J. B., 1992, Minireview: Enteropathogenic *Escherichia coli*, *Infect. Immun.* **60**:3953–3961.
80. Cravioto, A., Gross, R. J., Scotland, S. M., and Rowe, B., 1979, An adhesive factor found in strains of *Escherichia coli* belonging to the traditional infantile enteropathogenic serotypes, *Curr. Microbiol.* **3**:95–99.
81. Scaletsky, I. C. A., Silva, M. L. M., and Trabulsi, L. R., 1984, Distinctive patterns of adherence of enteropathogenic *Escherichia coli* to HeLa cells, *Infect. Immun.* **45**:534–536.
82. Baldini, M. M., Kaper, J. B., Levine, M. M., Candy, D. C., and Moon, H. W., 1983, Plasmid-mediated adhesion in enteropathogenic *Escherichia coli*, *J. Pediatr. Gastroenterol. Nutr.* **2**:534–538.
83. Nataro, J. P., Maher, K. O., Mackie, P., and Kaper, J. B., 1987, Characterization of plasmids encoding the adherence factor of enteropathogenic *Escherichia coli*, *Infect. Immun.* **55**: 2370–2377.
84. McConnell, M. M., Chart, H., Scotland, S. M., Smith, H. R., Willshaw, G. A., and Rowe, B., 1989, Properties of adherence factor plasmids of enteropathogenic *Escherichia coli* and the effect of host strain on expression of adherence to HEp-2 cells, *J. Gen. Microbiol.* **135**:1123–1134.
85. Levine, M. M., Nataro, J. P., Karch, H., Baldini, M. M., Kaper, J. B., Black, R. E., Clements, M. L., and O'Brien, A. D., 1985, The diarrheal response of humans to some classic serotypes of enteropathogenic *Escherichia coli* is dependent on a plasmid encoding an enteroadhesiveness factor, *J. Infect. Dis.* **152**:550–559.
86. Girón, J. A., Ho, A. S. Y., and Schoolnik, G. K., 1991, An inducible bundle-forming pilus of enteropathogenic *Escherichia coli*, *Science* **254**:710–713.
87. Donnenberg, M. S., Girón, J. A., Nataro, J. P., and Kaper, J. B., 1992, A plasmid-encoded type IV fimbrial gene of enteropathogenic *Escherichia coli* associated with localized adherence, *Mol. Microbiol.* **6**:3427–3437.
88. Donnenberg, M. S., Calderwood, S. B., Donohue-Rolfe, A., Keusch, G. T., and Kaper, J. B., 1990, Construction and analysis of TnphoA mutants of enteropathogenic *Escherichia coli* unable to invade HEp-2 cells, *Infect. Immun.* **58**:1565–1571.
89. Staley, T. E., Jones, E. W., and Corley, L. D., 1969, Attachment and penetration of *Escherichia coli* into intestinal epithelium of the ileum in newborn pigs, *Am. J. Pathol.* **56**: 371–392.
90. Moon, H. W., Whipp, S. C., Argenzio, R. A., Levine, M. M., and Giannella, R. A., 1983,

Attaching and effacing activities of rabbit and human enteropathogenic *Escherichia coli* in pig and rabbit intestines, *Infect. Immun.* **41:**1340–1351.
91. Knutton, S., Lloyd, D. R., and McNeish, A. S., 1987, Adhesion of enteropathogenic *Escherichia coli* to human intestinal enterocytes and cultured human intestinal mucosa. *Infect. Immun.* **55:**69–77.
92. Knutton, S., Baldwin, T., Williams, P. H., and McNeish, A. S., 1989, Actin accumulation at sites of bacterial adhesion to tissue culture cells: Basis of a new diagnostic test for enteropathogenic and enterohemorrhagic *Escherichia coli, Infect. Immun.* **57:**1290–1298.
93. Finlay, B. B., Rosenshine, I., Donnenberg, M. S., and Kaper, J. B., 1992, Cytoskeletal composition of attaching and effacing lesions associated with enteropathogenic *Escherichia coli* adherence to HeLa cells, *Infect. Immun.* **60:**2541–2543.
94. Manjarrez-Hernandez, H. A., Amess, B., Sellers, L., Baldwin, T. J., Knutton, S., Williams, P. H., and Aitken, A., 1991, Purification of a 20 kDa phosphoprotein from epithelial cells and identification as a myosin light chain: Phosphorylation induced by enteropathogenic *Escherichia coli* and phorbol ester, *FEBS Lett.* **292:**121–127.
95. Francis, D. H., Collins, J. E., and Duimstra, J. R., 1986, Infection of gnotobiotic pigs with an *Escherichia coli* O157:H7 strain associated with an outbreak of hemorrhagic colitis, *Infect. Immun.* **51:**953–956.
96. Tzipori, S., Wachsmuth, I. K., Chapman, C., Birner, R., Brittingham, J., Jackson, C., and Hogg, J., 1986, The pathogenesis of hemorrhagic colitis caused by *Escherichia coli* O157:H7 in gnotobiotic pigs, *J. Infect. Dis.* **154:**712–716.
97. Albert, M. J., Alam, K., Islam, M., Montanaro, J., Rahman, A. S. M. H., Haider, K., Hossain, M. A., Kibriya, A. K. M. G., and Tzipori, S., 1991, *Hafnia alvei*, a probable cause of diarrhea in humans, *Infect. Immun.* **59:**1507–1513.
98. Johnson, E., and Barthold, S. W., 1979, The ultrastructure of transmissible murine colonic hyperplasia, *Am. J. Pathol.* **97:**291–314.
99. Jerse, A. E., Yu, J., Tall, B. D., and Kaper, J. B., 1990, A genetic locus of enteropathogenic *Escherichia coli* necessary for the production of attaching and effacing lesions on tissue culture cells, *Proc. Natl. Acad. Sci. USA* **87:**7839–7843.
100. Donnenberg, M. S., and Kaper, J. B., 1991, Construction of an *eae* deletion mutant of enteropathogenic *Escherichia coli* by using a positive-selection suicide vector, *Infect. Immun.* **59:**4310–4317.
101. Rosenshine, I., Donnenberg, M. S., Kaper, J. B., and Finlay, B. B., 1992, Signal exchange between enteropathogenic *Escherichia coli* (EPEC) and epithelial cells: EPEC induce tyrosine phosphorylation of host cell protein to initiate cytoskeletal rearrangement and bacterial uptake, *EMBO J.* **11:**3551–3560.
102. Jerse, A. E., Gicquelais, K. G., and Kaper, J. B., 1991, Plasmid and chromosomal elements involved in the pathogenesis of attaching and effacing *Escherichia coli, Infect. Immun.* **59:** 3869–3875.
103. Albert, M. J., Faruque, S. M., Ansaruzzaman, M., Islam, M. M., Haider, K., Alam, K., Kabir, I., and Robins-Browne, R., 1992, Sharing of virulence-associated properties at the phenotypic and genetic levels between enteropathogenic *Escherichia coli* and *Hafnia alvei, J. Med. Microbiol.* **37:**310–314.
104. Schauer, D. B., and Falkow, S., 1993, Attaching and effacing locus of a *Citrobacter freundii* biotype that causes transmissible murine colonic hyperplasia, *Infect. Immun.* **61:**2486–2492.
105. Barrett, T. J., Kaper, J. B., Jerse, A. E., and Wachsmuth, I. K., 1992, Virulence factors in Shiga-like toxin-producing *Escherichia coli* isolated from humans and cattle, *J. Infect. Dis.* **165:**979–980.

106. McDaniel, T. K., Jarvis, K. G., Donnenberg, M. S., and Kaper, J. B., 1994, A large chromosomal gene cluster is present in enteropathogenic *Escherichia coli* (EPEC) but absent from non-pathogenic *E. coli*, ASM General Meeting B-286 (Abstract).
107. Jerse, A. E., and Kaper, J. B., 1991, The *eae* gene of enteropathogenic *Escherichia coli* encodes a 94-kilodalton membrane protein, the expression of which is influenced by the EAF plasmid, *Infect. Immun.* **59:**4302–4309.
108. Frankel, G., Candy, D. C. A., Everest, P., and Dougan, G., 1994, Characterization of the C-terminal domains of intimin-like proteins of enteropathogenic and enterohemorrhagic *Escherichia coli, Citrobacter freundii,* and *Hafnia alvei, Infect. Immun.* **62:**1835–1842.
109. Donnenberg, M. S., Tacket, C. O., James, S. P., Losonsky, G., Nataro, J. P., Wasserman, S. S., Kaper, J. B., and Levine, M. M., 1993, The role of the *eaeA* gene in experimental enteropathogenic *Escherichia coli* infection, *J. Clin. Invest.* **92:**1412–1417.
110. Donnenberg, M. S., Yu, J., and Kaper, J. B., 1993, A second chromosomal gene necessary for intimate attachment of enteropathogenic *Escherichia coli* to epithelial cells, *J. Bacteriol.* **175:**4670–4680.
111. Gómez, O., and Kaper, J. B., 1992, Plasmid encoded factors regulate the expression of *eae* gene of enteropathogenic *E. coli, Abstr. 92nd Gen. Meet. Am. Soc. Microbiol.* B-179, p. 55.
112. Baldwin, T. J., Ward, W., Aitken, A., Knutton, S., and Williams, P. H., 1991, Elevation of intracellular free calcium levels in HEp-2 cells infected with enteropathogenic *Escherichia coli, Infect. Immun.* **59:**1599–1604.
113. Foubister, V., Rosenshine, I., and Finlay, B. B., 1994, A diarrheal pathogen, enteropathogenic *Escherichia coli* (EPEC), triggers a flux of inositol phosphates in infected epithelial cells, *J. Exp. Med.* **179:**993–998.
114. Berridge, M. J., 1993, Inositol trisphosphate and calcium signalling, *Nature* **361:**315–325.
115. Shariff, A., and Luna, E. J., 1992, Diacylglycerol-stimulated formation of actin nucleation sites at plasma membranes, *Science* **256:**245–247.
116. Rothbaum, R. J., Partin, J. C., Saalfield, K., and McAdams, A. J., 1983, An ultrastructural study of enteropathogenic *Escherichia coli* infection in human infants, *Ultrastruct. Pathol.* **4:**291–304.
117. Taylor, J., Powell, B. W., and Wright, J., 1949, Infantile diarrhea and vomiting: A clinical and bacteriological investigation, *Br. Med. J.* **2:**117–141.
118. Smith, J., 1949, The association of certain types (α and β) of *Bact. coli* with infantile gastroenteritis, *J. Hyg.* **47:**221–226.
119. Giles, C., Sangster, G., and Smith, J., 1949, Epidemic gastroenteritis of infants in Aberdeen during 1947, *Arch. Dis. Child.* **24:**45–53.
120. Thorén, A., Stintzing, G., Tufvesson, B., Walder, M., and Habte, D., 1982, Aetiology and clinical features of severe infantile diarrhoea in Addis Ababa, Ethiopia, *J. Trop. Pediatr.* **28:**127–131.
121. Bower, J. R., Congeni, B. L., Cleary, T. G., Stone, R. T., Wanger, A., Murray, B. E., Mathewson, J. J., and Pickering, L. K., 1989, *Escherichia coli* O114:nonmotile as a pathogen in an outbreak of severe diarrhea associated with a day care center, *J. Infect. Dis.* **160:**243–247.
122. Ferguson, W. W., and June, R. C., 1952, Experiments of feeding adult volunteers with *Escherichia coli* 111, B4, a coliform organism associated with infant diarrhea, *Am. J. Hyg.* **55:**155–169.
123. Levine, M. M., Bergquist, E. J., Nalin, D. R., Waterman, D. H., Hornick, R. B., Young, C. R., Sotman, S., and Rowe, B., 1978, *Escherichia coli* strains that cause diarrhoea but do not produce heat-labile or heat-stable enterotoxins and are non-invasive, *Lancet* **1:**1119–1122.
124. Neter, E., Westphal, O., Lüderitz, O., Gino, R. M., and Gorzynski, E. A., 1955, Demonstra-

tion of antibodies against enteropathogenic *Escherichia coli* in sera of children of various ages, *Pediatrics* **16:**801–808.

125. Donnenberg, M. S., Tzipori, S., McKee, M., O'Brien, A. D., Alroy, J., and Kaper, J. B., 1993, The role of the *eae* locus of enterohemorrhagic *Escherichia coli* in intimate attachment in vitro and in a porcine model, *J. Clin. Invest.* **92:**1418–1424.
126. Karch, H., Heesemann, J., Laufs, R., Kroll, H. P., Kaper, J. B., and Levine, M. M., 1987, Serological response to type 1-like somatic fimbriae in diarrheal infection due to classical enteropathogenic *Escherichia coli, Microb. Pathog.* **2:**425–434.
127. Donnenberg, M. S., Tacket, C. O., Losonsky, G., Nataro, J. P., Kaper, J. B., and Levine, M. M., 1992, The role of the *eae* gene in experimental human enteropathogenic *Escherichia coli* (EPEC) infection, *Clin. Res.* **40:**214A (Abstract).
128. June, R. C., Ferguson, W. W., and Worfel, M. T., 1953, Experiments in feeding adult volunteers with *Escherichia coli* 55, B5, a coliform organism associated with infant diarrhea, *Am. J. Hyg.* **57:**222–236.
129. Konowalchuk, J., Speirs, J. I., and Stavric, S., 1977, Vero response to a cytotoxin of *Escherichia coli, Infect. Immun.* **18:**775–779.
130. Wells, J. G., Davis, B. R., Wachsmuth, I. K., Riley, L. W., Remis, R. S., Sokolow, R., and Morris, G. K., 1983, Laboratory investigation of hemorrhagic colitis outbreaks associated with a rare *Escherichia coli* serotype, *J. Clin. Microbiol.* **18:**512–520.
131. Anonymous, 1982, Isolation of *E. coli* O157:H7 from sporadic cases of hemorrhagic colitis—United States, *Morbid. Mortal. Weekly Rep.* **31:**580, 585.
132. Riley, L. W., Remis, R. S., Helgerson, S. D., McGee, H. B., Wells, J. G., Davis, B. R., Hebert, R. J., Olcott, E. S., Johnson, L. M., Hargrett, N. T., Blake, P. A., and Cohen, M. L., 1983, Hemorrhagic colitis associated with a rare *Escherichia coli* serotype, *N. Engl. J. Med.* **308:**681–685.
133. CDC, 1993, Update: Multistate outbreak of *Escherichia coli* O157:H7 infections from hamburgers—Western United States, 1992–1993, *Morbid, Mortal. Weekly Rep.* **42:**258–263.
134. Macdonald, K. L., and Osterholm, M. T., 1993, The emergence of *Escherichia coli* O157:H7 infection in the United States: The changing epidemiology of foodborne disease, *J. Am. Med. Assoc.* **269:**2264–2266.
135. Martin, D. L., Macdonald, K. L., White, K. E., Soler, J. T., and Osterholm, M. T., 1990, The epidemiology and clinical aspects of the hemolytic uremic syndrome in Minnesota, *N. Engl. J. Med.* **323:**1161–1167.
136. Griffin, P. M., and Tauxe, R. V., 1991, The epidemiology of infections caused by *Escherichia coli* O157: H7, other enterohemorrhagic *E. coli*, and the associated hemolytic uremic syndrome, *Epidemiol. Rev.* **13:**60–98.
137. Wachsmuth, K., Barrett, T. J., Griffin, P. M., Toth, I., Strockbine, N. A., and Wells, J. G., 1991, *Escherichia coli* O157:H7: A mini-review, in: *Application of Molecular Biology in Diagnosis of Infectious Diseases* (O. Olsvik and G. Bukholm, eds.), Norwegian College of Veterinary Medicine, Oslo, pp. 5–15.
138. Cordovéz, A., Prado, V., Maggi, L., Cordero, J., Martinez, J., Misraji, A., Rios, R., Soza, G., Ojeda, A., and Levine, M. M., 1992, Enterohemorrhagic *Escherichia coli* associated with hemolytic-uremic syndrome in Chilean children, *J. Clin. Microbiol.* **30:**2153–2157.
139. Ostroff, S. M., Hopkins, D. P., Tauxe, R. V., Sowers, E. G., and Strockbine, N. K., 1991, Surveillance of *Escherichia-coli* O157 isolation and confirmation, United States, 1988, *Morbid. Mortal. Weekly Rep.* **40**(No. SS-1)**:**1–5.
140. Cohen, M. B., and Giannella, R. A., 1992, Hemorrhagic colitis associated with *Escherichia coli* O157:H7, *Adv. Intern. Med.* **37:**173–195.
141. Besser, R. E., Lett, S. M., Weber, T., Doyle, M. P., Barrett, T. J., Wells, J. G., and Griffin, P. M., 1993, An outbreak of diarrhea and hemolytic uremic syndrome from *Escherichia coli* O157:H7 in fresh-pressed apple cider, *J. Am. Med. Assoc.* **269:**2217–2220.

142. Griffin, P. M., Ostroff, S. M., Tauxe, R. V., Greene, K. D., Wells, J. G., Lewis, J. H., and Blake, P. A., 1988, Illnesses associated with *Escherichia coli* O157:H7 infections: A broad clinical spectrum, *Ann. Intern. Med.* **109:**705–712.
143. Lerman, Y., Cohen, D., Gluck, A., Ohad, E., and Sechter, I., 1992, A cluster of cases of *Escherichia coli* O157 infection in a day-care center in a communal settlement (kibbutz) in Israel, *J. Clin. Microbiol.* **30:**520–521.
144. Carter, A. O., Borczyk, A. A., Carlson, J. A. K., Harvey, B., Hockin, J. C., Karmali, M. A., Krishnan, C., Korn, D. A., and Lior, H., 1987, A severe outbreak of *Escherichia coli* O157:H7-associated hemorrhagic colitis in a nursing home, *N. Engl. J. Med.* **317:**1496–1500.
145. Belongia, E. A., Osterholm, M. T., Soler, J. T., Ammend, D. A., Braun, J. E., and MacDonald, K. L., 1993, Transmission of *Escherichia coli* O157:H7 infection in Minnesota child day-care facilities, *J. Am. Med. Assoc.* **269:**883–888.
146. Griffin, P. M., Olmstead, L. C., and Petras, R. E., 1990, *Escherichia coli* O157:H7-associated colitis. A clinical and histological study of 11 cases, *Gastroenterology* **99:**142–149.
147. Donnenberg, M. S., Tzipori, S., McKee, M., O'Brien, A. D., Alroy, J., and Kaper, J. B., 1993, The role of the *eae* locus of enterohemorrhagic *Escherichia coli* in intimate attachment in vitro and in a porcine model, *J. Clin. Invest.* **92:**1418–1424.
148. Girón, J. A., Donnenberg, M. S., Martin, W. C., Jarvis, K. G., and Kaper, J. B., 1993, Distribution of the bundle-forming pilus structural gene (*bfp*A) among enteropathogenic *Escherichia coli*, *J. Infect. Dis.* **168:**1037–1041.
149. Levine, M. M., Xu, J.-G., Kaper, J. B., Lior, H., Prado, V., Tall, B., Nataro, J., Karch, H., and Wachsmuth, K., 1987, A DNA probe to identify enterohemorrhagic *Escherichia coli* of O157:H7 and other serotypes that cause hemorrhagic colitis and hemolytic uremic syndrome, *J. Infect. Dis.* **156:**175–182.
150. Wells, J. G., Shipman, L. D., Greene, K. D., Sowers, E. G., Green, J. H., Cameron, D. N., Downes, F. P., Martin, M. L., Griffin, P. M., Ostroff, S. M., Potter, M. E., Tauxe, R. V., and Wachsmuth, I. K., 1991, Isolation of *Escherichia-coli* serotype O157:H7 and other Shiga-like-toxin-producing *E. coli* from dairy cattle, *J. Clin. Microbiol.* **29:**985–989.
151. O'Brien, A. D., and Holmes, R. K., 1987, Shiga and Shiga-like toxins, *Microbiol. Rev.* **51:**206–220.
152. Jacewicz, M., Clausen, H., Nudelman, E., Donohue-Rolfe, A., and Keusch, G. T., 1986, Pathogenesis of shigella diarrhea. XI. Isolation of a shigella toxin-binding glycolipid from rabbit jejunum and HeLa cells and its identification as globotriaosylceramide, *J. Exp. Med.* **163:**1391–1404.
153. Lingwood, C. A., Law, H., Richardson, S., Petric, M., Brunton, J. L., De Grandis, S., and Karmali, M., 1987, Glycolipid binding of purified and recombinant *Escherichia coli* produced verotoxin in vitro, *J. Biol. Chem.* **262:**8834–8839.
154. Waddell, T., Head, S., Petric, M., Cohen, A., and Lingwood, C., 1988, Globotriosyl ceramide is specifically recognized by the *Escherichia coli* verocytotoxin 2, *Biochem. Biophys. Res. Commun.* **152:**674–679.
155. Ostroff, S. M., Kobayashi, J. M., and Lewis, J. H., 1989, Infections with *Escherichia coli* O157:H7 in Washington State. The first year of statewide disease surveillance [see comments], *J. Am. Med. Assoc.* **262:**355–359.
156. Proulx, F., Turgeon, J. P., Delage, G., Lafleur, L., and Chicoine, L., 1992, Randomized, controlled trial of antibiotic therapy for *Escherichia coli* O157:H7 enteritis, *J. Pediatr.* **121:** 299–303.
157. Nataro, J. P., Levine, M. M., Savarino, S., Losonsky, G., Guers, L., and Tackett, C. O., 1993, Volunteer studies of enteroaggregative *Escherichia coli*, *Abstr. Gen. Meet. Am. Soc. Microbiol.* B315, p. 82.
158. Mathewson, J. J., Johnson, P. C., and Dupont, H. L., 1986, Pathogenicity of enteroadherent *Escherichia coli* in adult volunteer, *J. Infect. Dis.* **154:**524–527.

159. Bhan, M. K., Khoshoo, V., Sommerfelt, H., Raj, P., Sazawal, S., and Srivastava, R., 1989, Enteroaggregative *Escherichia coli* and *Salmonella* associated with nondysenteric persistent diarrhea, *Pediatr. Infect. Dis. J.* **8:**499–502.
160. Bhan, M. K., Raj, P., Levine, M. M., Kaper, J. B., Bhandari, N., Srivastava, R., Kumar, R., and Sazawal, S., 1989, Enteroaggregative *Escherichia coli* associated with persistent diarrhea in a cohort of rural children in India, *J. Infect. Dis.* **159:**1061–1064.
161. Henry, F. J., Udoy, A. S., Wanke, C. A., and Aziz, K. M. A., 1992, Epidemiology of persistent diarrhea and etiologic agents in Mirzapur, Bangladesh, *Acta Paediatr. Scand. Suppl.* **381:**27–31.
162. Baqui, A. H., Sack, R. B., Black, R. E., Haider, K., Hossain, A., Alim, A. R. M. A., Yunus, M., Chowdhury, H. R., and Siddique, A. K., 1992, Enteropathogens associated with acute and persistent diarrhea in Bangladeshi children < 5 years of age, *J. Infect. Dis.* **166:**792–796.
163. Nataro, J. P., Scaletsky, I. C. A., Kaper, J. B., Levine, M. M., and Trabulsi, L. R., 1985, Plasmid-mediated factors conferring diffuse and localized adherence of enteropathogenic *Escherichia coli, Infect. Immun.* **48:**378–383.
164. Nataro, J. P., Deng, Y., Maneval, D. R., German, A. L., Martin, W. C., and Levine, M. M., 1992, Aggregative adherence fimbriae I of enteroaggregative *Escherichia coli* mediate adherence to HEp-2 cells and hemagglutination of human erythrocytes, *Infect. Immun.* **60:** 2297–2304.
165. Nataro, J. P., Yikang, D., Giron, J. A., Savarino, S. J., Kothary, M. H., and Hall, R., 1993, Aggregative adherence fimbria I expression in enteroaggregative *Escherichia coli* requires two unlinked plasmid regions, *Infect. Immun.* **61:**1126–1131.
166. Savarino, S. J., Fasano, A., Robertson, D. C., and Levine, M. M., 1991, Enteroaggregative *Escherichia coli* elaborate a heat-stable enterotoxin demonstrable in an in vitro rabbit intestinal model, *J. Clin. Invest.* **87:** 1450–1455.
167. Savarino, S. J., Fasano, A., Watson, J., Martin, B. M., Levine, M. M., Guandalini, S., and Guerry, P., 1993, Enteroaggregative *Escherichia coli* heat-stable enterotoxin 1 represents another subfamily of *E. coli* heat-stable toxin, *Proc. Natl. Acad. Sci. USA* **90:**3093–3097.
168. Baldwin, T. J., Knutton, S., Sellers, L., Manjarrez Hernandez, H. A., Aitken, A., and Williams, P. H., 1992, Enteroaggregative *Escherichia coli* strains secrete a heat-labile toxin antigenically related to *E. coli* hemolysin, *Infect. Immun.* **60:**2092–2095.
169. Haque, M. A., Ohki, K., Kikuchi, M., and Kohashi, O., 1994, Contact hemolysin production by strains of enteroaggregative *Escherichia coli* isolated from children with diarrhea, *J. Clin. Microbiol.* **32:**1109–1111.
170. Nataro, J. P., 1995, Enteroaggregative and diffusely adherent *E. coli*, in: *Infections of the Gastrointestinal Tract* (M. J. Blaser, P. D. Smith, J. I. Ravdin, H. B. Greenberg, and R. L. Guerrant, eds.), Raven Press, New York, 1995.
171. Albert, M. J., Qadri, F., Haque, A., and Bhuiyan, N. A. 1992, Bacterial clump formation at the surface of liquid culture as a rapid test for identification of enteroaggregative *Escherichia coli, J. Clin. Microbiol.* **31:**1397–1399.
172. Mathewson, J. J., Johnson, P. C., DuPont, H. L., Morgan, D. R., Thornton, S. A., Wood, L. V., and Ericsson, C. D., 1985, A newly recognized cause of travelers' diarrhea: Enteroadherent *Escherichia coli, J. Infect. Dis.* **151:**471–475.
173. Vial, P. A., Mathewson, J. J., DuPont, H. L., Guers, L., and Levine, M. M., 1990, Comparison of two assay methods for patterns of adherence to HEp-2 cells of *Escherichia-coli* from patients with diarrhea, *J. Clin. Microbiol.* **28:**882–885.
174. Jallat, C., Livrelli, V., Darfeuille-Michaud, A., Rich, C., and Joly, B., 1993, *Escherichia coli* strains involved in diarrhea in France: High prevalence and heterogeneity of diffusely adhering strains, *J. Clin. Microbiol.* **31:**2031–2037.
175. Gunzburg, S. T., Chang, B. J., Elliott, S. J., Burke, V., and Gracey, M., 1993, Diffuse and enteroaggregative patterns of adherence of enteric *Escherichia coli* isolated from aboriginal children from the Kimberley region of Western Australia, *J. Infect. Dis.* **167:**755–758.

176. Levine, M. M., Ferreccio, C., Prado, V., Cayazzo, M., Abrego, P., Martinez, J., Maggi, L., Baldini, M. M., Martin, W., and Maneval, D., 1993, Epidemiologic studies of *Escherichia coli* diarrheal infections in a low socioeconomic level peri-urban community in Santiago, Chile, *Am. J. Epidemiol.* **138:**849–869.
177. Tacket, C. O., Moseley, S. L., Kay, B., Losonsky, G., and Levine, M. M., 1990, Challenge studies in volunteers using *Escherichia coli* strains with diffuse adherence to HEp-2 cells, *J. Infect. Dis.* **162:**550–552.
178. Bilge, S. S., Apostol, J. M., Jr., Fullner, K. J., and Moseley, S. L., 1993, Transcriptional organization of the F1845 fimbrial adhesin determinant of *Escherichia coli, Mol. Microbiol.* **7:**993–1006.
179. Bilge, S. S., Clausen, C. R., Lau, W., and Moseley, S. L., 1989, Molecular characterization of a fimbrial adhesin, F1845, mediating diffuse adherence of diarrhea-associated *Escherichia coli* to HEp-2 cells, *J. Bacteriol.* **171:**4281–4289.
180. Benz, I., and Schmidt, M. A., 1989, Cloning and expression of an adhesin (AIDA-I) involved in diffuse adherence of enteropathogenic *Escherichia coli, Infect. Immun.* **57:**1506–1511.
181. Benz, I., and Schmidt, M. A., 1992, Isolation and serologic characterization of AIDA-I, the adhesin mediating the diffuse adherence phenotype of the diarrhea-associated *Escherichia coli* strain 2787 (O126:H27), *Infect. Immun.* **60:**13–18.

7

Molecular Pathogenesis of *Campylobacter jejuni* Enteritis

MICHAEL E. KONKEL and WITOLD CIEPLAK, JR.

1. INTRODUCTION

Campylobacter jejuni and *Campylobacter coli* are now recognized as among the most prevalent causes of bacterial enteritis in both developed and developing countries. Infection rates with *C. jejuni* in the United States and Great Britain have been estimated at 10 per 1000 population, thus equaling and likely surpassing the rates of *Salmonella* infection.[1] In addition to causing acute enterocolitis, *C. jejuni* infection has recently been implicated as a frequent antecedent to the development of Guillain–Barré-type polyneuropathy.[2,3] Despite the recognition of *Campylobacter* spp. as an important etiologic agent of enteritis, the biologic and molecular pathogenic aspects of *Campylobacter* enteritis remain ill-defined. There is considerable disagreement whether *C. jejuni*, like more well-recognized enteropathogens (e.g., *Shigella*, *Escherichia coli*, *Vibrio cholerae*, and *Yersinia*), produce any readily identifiable pathogenic/virulence determinants such as cytotoxins, enterotoxins, and invasion-related molecules during cultivation *in vitro*. Enteritis caused by *C. jejuni* and related species may represent an example of the interactions that occur between pathogens and the host environment and result in the expression of virulence determinants during the pathogenesis of disease.[4] This review summarizes the microbiology and immunology of

MICHAEL E. KONKEL • Department of Microbiology, Washington State University, Pullman, Washington 99164-4233. WITOLD CIEPLAK, JR. • Laboratory of Intracellular Parasites, Rocky Mountain Laboratories, National Institute of Allergy and Infectious Diseases, National Institutes of Health, Hamilton, Montana 59840.

Enteric Infections and Immunity, edited by Lois J. Paradise *et al.* Plenum Press, New York, 1996.

C. jejuni and concentrates on the available knowledge concerning the molecular bases of disease causation by this organism.

2. MICROBIOLOGY OF *C. JEJUNI*

The genus *Campylobacter* contains 14 species that can be differentiated on the basis of antibiotic susceptibility and metabolic tests, such as the hydrolysis of hippurate which can distinguish *C. jejuni* from *C. coli.*[5,6] Of these species, only three, *C. jejuni, C. coli,* and *C. fetus,* are frequently associated with human disease. *C. fetus* is a relatively common veterinary pathogen but is also known to produce both extraintestinal disease in immunocompromised humans and diarrheal illness in normal individuals.[7,8] Although many of the issues to be discussed are applicable to *C. coli* also, this review will refer to *C. jejuni* exclusively as it causes a greater number of human infections.

C. jejuni are small (1–1.5 μm in length), nonsporulating, gram-negative rods that possess a characteristic spiral or vibrio-like appearance. The bacteria may possess uni- or bipolar flagella and exhibit rapid, darting motility when examined by light microscopy. The synthesis of flagella exhibits both phase and antigenic variation.[9,10] Optimal growth of *C. jejuni* occurs in a microaerophilic atmosphere (5–10% O_2) at 42°C. Several selective media for the primary isolation of *C. jejuni* have been developed.[6,11–13] Typically, *C. jejuni* and *C. coli* are propagated in the laboratory in CO_2-enriched atmospheres produced by commercially available gas-generating packs. Growth generally appears within 24 hr, but several days may be necessary for the appearance of some isolates or strains. Colonies are typically small and nonmucoid in appearance.

3. PATHOLOGIC AND EPIDEMIOLOGIC ASPECTS OF *C. JEJUNI* ENTERITIS

The dominating presentation of *Campylobacter* enteritis is acute enterocolitis.[8,14] Grossly bloody diarrhea containing leukocytes and mucus frequently occurs after an incubation period of 2–7 days. Vomiting and other systemic presentations may occur. Fever, myalgia, and abdominal pain may precede the onset of diarrhea. In developing countries, where *C. jejuni* infection is often endemic, many patients exhibit watery diarrhea similar to that associated with enterotoxigenic *E. coli.* However, visitors from developed nations typically do not develop watery diarrheal disease when infected with apparent diarrheagenic strains indigenous to developing nations.[1] This variation in symptomology may reflect differences in host immune status or response. *Campylobacter* enteritis is often self-limiting and patients are recovering by the time a culture diagnosis is made. In severe cases, erythromycin is the chemotherapeutic agent of choice. *C. jejuni* enteritis is not easily distinguishable from enteritis caused by *Salmonella* or *Shigella* on the basis of clinical presentation alone. Complications of *C. jejuni* infection are only infrequently reported but include meningitis[15] and hemolytic uremic syndrome.[16] Deleterious sequelae associated with *Campylobacter* infection include reactive arthritis[17] and peripheral polyneuropathy (Guillain–Barré syndrome).[2]

Animals provide the principal reservoir of infection for humans. Personal and public hygiene appear to play only minor roles in the transmission of infection. *C. jejuni* enteritis, like *Salmonella* disease, is largely a food-borne illness. *C. jejuni* may be found in a wide variety of domestic and feral animals and is commonly associated with birds. This latter association may be facilitated by the temperature optimum of *Campylobacter* growth, which approximates the core temperature of birds. Domestic poultry are commonly colonized by *Campylobacter* early in life and represent the most common source of infection in the United States and other developed countries. Infection of poultry is generally asymptomatic, as it is in various other animals. It has been estimated that 50–90% of domestic chicken carcasses in the United States and other developed countries are contaminated by the time of sale.[18,19] The frequency of *Campylobacter* infection is greater during the summer months when less care is taken in the preparation of chicken. Mass outbreaks of *Campylobacter* enteritis are rare because of the relatively fastidious nature of *C. jejuni* and *C. coli*. However, consumption of untreated water and raw (unpasteurized) milk have led to several documented outbreaks.[18–21a] Serologic evidence indicates that *C. jejuni* infection is hyperendemic in developing countries. Overt disease in these nations is almost exclusively limited to young children.[1,22] This epidemiologic pattern suggests that repeated infection during the first few years of life leads to subsequent immunity from disease.

4. IMMUNE RESPONSE TO *C. JEJUNI* INFECTION

Serologic typing of *C. jejuni* indicates the presence of considerable antigenic diversity among isolates from various geographic areas. A serotyping scheme based on extractable heat-stable antigens (lipopolysaccharide) identifies approximately 55 serotypes of *C. jejuni* while another scheme based on heat-labile antigens (proteins) discriminates in excess of 120 serologic variants of *C. jejuni* and *C. coli*.[22a,23] Several studies have demonstrated the presence of antibodies to several surface proteins and lipopolysaccharide in both naturally and experimentally infected humans.[24–26] Human convalescent sera routinely react with 44-kDa major outer membrane protein, 65-kDa flagellin, and 25- to 29-kDa surface proteins. Studies with experimentally infected macaques and human volunteer studies indicate that serum antibodies appear 6–7 days after the onset of illness and increase rapidly during the early stages of disease.[27,28] The hypothesis that humoral immunity plays an essential role in the resolution of disease is supported by the observation that clinical symptoms of enteritis subside in a fashion coincident with the appearance of circulating anti-*C. jejuni* antibodies, particularly those against surface proteins.[14,29] Human volunteers experimentally infected with virulent *C. jejuni* develop serum antibodies that are protective against subsequent challenges with the homologous strain of *C. jejuni*.[28] These serum antibodies are serotype specific and opsonize *C. jejuni* in a complement-dependent fashion.[30] Pennie *et al.*[31] examined the ability of immune sera from *C. jejuni*-infected individuals to kill *C. jejuni in vitro.* Sera from individuals infected with heterologous *C. jejuni* strains were found to have little bactericidal activity. Convalescent sera showed greater than 99.9% killing of homologous bacteria. The bactericidal effects of both the acute and convalescent sera were abrogated by heat and EDTA treatment, indicating a direct role for complement in the bactericidal effect.

The role of local or mucosal immune responses in protection against *Campylobacter* enteritis requires additional investigation. Both fecal and urinary antibodies to *C. jejuni* have been detected during natural infection of humans[32] and specific IgA has been identified in jejunal aspirates of experimentally infected humans.[30] Rabbit models of infection have been used to demonstrate the rapid clearance of *C. jejuni* from the intestines of orally immunized animals, suggesting a role for mucosal antibodies in clearance.[33] McSweegan *et al.*[34] reported that the intestinal mucus collected from rabbits colonized with *C. jejuni* inhibited adherence of the bacteria to cultured cells and that the inhibitory effect was related to the presence of anti-*Campylobacter* secretory IgA. The IgA-containing immune mucus caused the clumping/aggregation of the bacteria with entrapment in the mucus, thereby preventing bacterial adherence. More recently, investigators have attempted to exploit the ability of the heat-labile enterotoxin (LT) from enterotoxigenic *E. coli* to stimulate mucosal immune responses to unrelated antigens in the development of strategies to elicit intestinal IgA responses to killed *C. coli.*[35] These studies showed that oral administration of killed *C. coli* with LT could induce the appearance of specific IgA in intestinal lavage fluids in mice and rabbits, and promoted rapid clearance of homologous, but not heterologous, heat-labile serotypes of *C. coli.*

Although affording protection from disease, the antibody response to

C. jejuni may have deleterious consequences particularly with respect to the development of Guillain–Barré-type polyneuropathy. Patients with Guillain–Barré syndrome associated with *C. jejuni* infection possess antibodies to G_{M1} gangliosides. Such antibodies may play an important role in the production of this syndrome. Structural studies have shown that the core oligosaccharide portion of the lipopolysaccharide derived from certain heat-stable serotypes of *C. jejuni* closely resembles human gangliosides.(3,36) Thus, infection with appropriate heat-stable serotypes of *C. jejuni* may induce antibodies that cross-react with nerve tissue gangliosides.

Systemic and local antibody responses are important in the resolution of *C. jejuni* enteritis. Humoral immunity has been determined to be strain or serotype specific by both *in vitro* antibody activity and protection from subsequent infection and disease. Development of a simple monovalent vaccine for *Campylobacter* enteritis is unlikely because of the antigenic variation and the existence of several related *Campylobacter* species that are capable of causing enteritis. Further work is necessary to identify antigens that are important in eliciting protective immunity and to determine the extent to which such antigens vary among strains.

5. PATHOGENIC MECHANISMS OF *C. JEJUNI*

The identification of the molecular pathogenic mechanisms associated with *C. jejuni* enteritis has been hampered by the lack of suitable and convenient small animal models. Mink, ferrets, dogs, cats, and hamsters have been described.(37) The utility of many of these models is mitigated by the presence of widespread natural *Campylobacter* infection among the various species, the lack of specific immunologic reagents, and the variable occurrence of overt disease. Mice and rabbits are most commonly used for *in vivo* studies of *Campylobacter* infection. Mice do not naturally harbor *Campylobacter* but can be colonized with *C. jejuni* after oral infection. However, diarrheal disease seldom occurs.(38) The removable intestinal tie adult rabbit diarrhea (RITARD) model developed for the study of *V. cholerae* and *E. coli* disease has been used to study *C. jejuni* enteritis. In this model, the terminal ileum is ligated and organisms are injected into the mid-small bowel. Several hours after instillation of the bacteria, the ligation is removed. This procedure effectively inhibits the normal initial peristaltic clearing mechanism and results in enhanced colonization and disease production. In contrast to orally infected rabbits, the RITARD model results in a high attack rate with mucoid diarrhea and death in a significant proportion of animals.(39) Colostrum-deprived newborn piglets orally challenged with *C. jejuni* have been shown to develop clinical symptoms and histopathologic lesions similar to those of humans with *C. jejuni* infections.[40] Laboratory monkeys exhibit naturally acquired *C. jejuni* infection with inflammatory diarhhea.[41] Infant (3½–4½ month old) *Macaca mulatta*[42] and *Macaca nemestrina*[27] have been used as models for *C. jejuni* infection which resulted in acute diarrhea and inflammation of the colon.

5.1. Interactions of *C. jejuni* with Cultured Epithelial Cells

In vitro assays examining the interactions of *C. jejuni* with cultured epithelial cells have been used to study the pathogenic behavior of this organism. These cell culture models have been used extensively to assess the ability of a number of different pathogenic bacteria, including *E. coli*, *Salmonella* spp., *Shigella flexneri*, and *Yersinia* spp., to bind to, enter, and replicate in eukaryotic cells.[43] Inoculation of a cell culture with a bacterial suspension allows infection of a uniform population of cells under defined conditions. The assays are relatively simple, reproducible, and allow for the quantitation of both adherent and internalized bacteria.

The ability of *C. jejuni* to bind to and colonize the cells lining the gastrointestinal tract is important for infection and disease production. Initial *in vitro* studies revealed that *C. jejuni* isolated from individuals with fever and diarrhea were able to bind to epithelial cells with greater efficiency than strains isolated from patients without diarrhea or fever.[44] *C. jejuni* outer membrane proteins, flagellin, and lipopolysaccharide have been proposed to mediate adherence to epithelial cells. De Melo and Pechere[45] identified four proteins in *C. jejuni* outer membrane preparations and acid extracts that bound epithelial cell membranes. These proteins possess apparent molecular masses of 28, 32, 36, and 42 kDa. Fauchere *et al.*[46] also detected a subset of outer membrane proteins, ranging in mass from 26 to 30 kDa, that bound to HeLa cells. Pei and Blaser[47] cloned a *C. jejuni* gene encoding a 28-kDa protein, termed PEB1, that promotes attachment to epithelial cells. This protein appears to be the homologue of membrane proteins from other gram-negative bacteria and functions in amino acid transport. Flagella have also been suggested to promote the binding of *C. jejuni* to epithelial cells. Examination of *C. jejuni* incubated with epithelial cells by scanning electron microscopy has provided evidence that the tips of the flagella initiate the contact with the target cell.[48] Also, McSweegan and Walker[49] reported that purified flagella bound to INT 407 cells. However, no differences have been observed in the binding of recombinant, nonflagellated *C. jejuni* mutants and their isogenic counterparts to cultured cells.[50] Flagella and associated motility are likely important virulence determinants since infection of humans and rabbits with aflagellate variants of *C. jejuni* results in selective recovery of flagellated forms.[9,28] These findings indicate that flagella, or their constituent protein flagellin, are not primary adhesins but may play a role in allowing the bacteria to reach target cells and facilitate postbinding events such as internalization (see below). It is likely that other surface proteins, such as PEB1, function in mediating the primary attachment of *C. jejuni* to eukaryotic cells *in vitro.*

The ability of *C. jejuni* to enter and survive in eukaryotic cells is proposed as an important virulence determinant associated with *C. jejuni* enteritis. This proposal is based on the observation that clinical enteritis caused by *C. jejuni* is often characterized by the presence of fecal leukocytes, suggestive of an inflammatory process resulting from bacterial invasion of the intestinal epithelium.[14] The observation of intracellular bacteria associated with colonic damage and diarrheal disease in experimentally infected monkeys again suggests that *C. jejuni* produce enteritis by an invasive mechanism.[42] The internalization of *C. jejuni* by cultured mammalian cells has been demonstrated using both immunofluorescence and transmission electron microscopy.[51–54] The ability of cell-associated organisms to survive in the presence of the aminoglycoside antibiotic gentamicin has been used to quantify the ability of bacterial pathogens to enter and survive within eukaryotic cells *in vitro.*[43] Eukaryotic cell membranes are impermeable to gentamicin, which allows the intracellular bacteria to remain viable. Studies have shown that *C. jejuni* is capable of entering epithelial cells at levels that exceed those of noninvasive bacteria (e.g., laboratory *E. coli* strains).[53–55] *C. jejuni* is internalized more efficiently when cells of human and nonhuman primate origin are employed.[53] The precise mechanism by which *C. jejuni* enters epithelial cells has not been defined. The internalization of *C. jejuni* has been shown to be inhibited at low temperature (4°C) and by cytochalasin D and dansylcadaverine, chemicals that inhibit cellular microfilament formation and receptor cycling, respectively.[53,54] Transmission electron microscopy of *C. jejuni*-infected cells revealed that the bacteria are localized in membrane-bound vacuoles.[53,54] These findings have led to the conclusion that *C. jejuni* enters cells through a microfilament-dependent process that resembles the endocytotic mechanism employed by invasive enteric pathogens including *Salmonella* and *Yersinia.*[43] Colchicine and nocodazole, inhibitors of microtubule formation, have been reported to block the internalization of *C. jejuni.*[55] This finding may indicate the involvement of a microtubule-dependent pathway in internalization, although it has also been reported that the effects of colchicine on *C. jejuni* internalization are variable and not dose dependent.[53]

The internalization/intracellular survival of *C. jejuni* in epithelial cell cultures has been shown to be significantly reduced by chloramphenicol, a specific inhibitor of bacterial protein synthesis, suggesting a direct role for newly synthesized proteins in the internalization process.[56] Two-dimensional gel electrophoretic analyses of metabolically labeled *C. jejuni* in the presence and absence of cultured mammalian cells have revealed that a number of proteins are synthesized preferentially or exclusively in the presence of the cells.[57] The new or enhanced proteins are detected within 60 min after addition of the bacteria to the epithelial cell monolayers, and their synthesis is temporally associated with an increase in *C. jejuni* internalization. A rabbit polyclonal serum raised against *C. jejuni* cultured in the presence of the epithelial cells recognized at least nine proteins not recognized by an antiserum prepared against conventionally cultured *C. jejuni*. The former polyclonal antiserum inhibited the internalization of *C. jejuni* in a dose-dependent fashion.[57] Quantitative and qualitative alterations in the protein profile of *C. jejuni* have also been observed during incubation in rabbit ileal loops.[58] These findings suggest that one or more of the proteins synthesized by *C. jejuni* in response to the eukaryotic cells may play a direct role in facilitating internalization. Flagella and/or associated motility are also required for the maximum internalization of epithelial cells in both polarized and nonpolarized cell culture models.[50,59]

Some cell lines have been shown to form polarized monolayers, with defined apical and basolateral cell surfaces separated by tight junctions, when grown on microporous membranes. The monolayers form an effective barrier against passive diffusion of noninvasive bacteria and small molecules. *C. jejuni* has been shown to translocate across human polarized colonic carcinoma (Caco-2) cell monolayers.[60,61] Because polarized Caco-2 cells resemble the architecture of enterocytes in the colon, this cell model is believed to more accurately reflect bacterial–target cell interactions that may occur *in vivo* as compared to models using conventionally grown epithelial cells. Everest *et al.*[61] reported that 86% of the *C. jejuni* strains isolated from individuals with colitis translocated across polarized Caco-2 cell monolayers, compared to only 48% of the strains from individuals with noninflammatory disease. Transmission electron microscopic examination of infected cell monolayers has revealed that *C. jejuni* may translocate across the cell monolayer by passing both through and between the cells.[60] The data obtained using *C. jejuni* infection of polarized Caco-2 cells indicate a pathogenic mechanism involving translocation either through or between the cells lining the gastrointestinal tract and subsequent proliferation or residence in the submucosal tissues. The translocation of *C. jejuni* across the mucosal epithelium may be an important mechanism associated with intestinal damage and inflammation.

The interaction of *C. jejuni* with mononuclear phagocytes has also been examined. Kiehlbauch *et al.*[62] examined the ability of *C. jejuni* to survive within several types of mononuclear phagocytes, including a murine macrophagelike cell line, murine resident peritoneal macrophages, and human peripheral blood monocytes. *C. jejuni* were found to be internalized by all three mononuclear phagocytes in the absence of specific opsonins. Although the bacteria rapidly converted from a spiral to a coccoid form following uptake, they remained viable for at least 6 days. The investigators suggested that the phagocytes provide *C. jejuni* with an environment that prolongs their viability and could play a role in the persistence of *C. jejuni* shedding *in vivo*. *C. jejuni* has also been found to possess a superoxide dismutase (SOD) activity that may play a role in intracellular invasion/survival by protecting the bacteria from oxidative damage by host-cell defense mechanisms.[63,64] Using the gentamicin-protection assay, the internalization and/or intracellular survival of a *C. jejuni* mutant lacking SOD activity was determined to be reduced by a factor of at least 10 compared to the wild-type parental strain.[64]

5.2. Role of Cytotoxins and Enterotoxins in *C. jejuni* Enteritis

A number of studies indicate that *C. jejuni* may produce cytotoxins and/or enterotoxins.[65–73] Cytotoxic activity is determined by trypan blue exclusion dye studies and observing morphologic alterations of the cells following incubation of *C. jejuni* extracts or culture filtrates with various cultured cell lines (e.g., HeLa, Vero, CHO). Guerrant *et al.*[65] reported that extracts from 5 of 12 *C. jejuni* isolates had a cytolethal effect on HeLa cells at a dilution of 1:4 or greater. Trypsin or heat treatment of culture filtrates reduced cytotoxic activity by more than 50%. Antibody inhibition experiments indicated that the *C. jejuni* cytotoxin is immunologically distinct from Shiga and *Clostridium difficile* cytotoxins. Infection of epithelial cells with *C. jejuni* for more than 48 hr results in cytolytic effects.[53] Mahajan and Rodgers[66] identified a protein, with an apparent molecular mass of 68 kDa, that produced rounding or overt cytolysis of several cell lines. Other studies have failed to confirm cytotoxin production and have found only low levels of activity in culture filtrates derived from clinical isolates or fecal filtrates of patients with confirmed *C. jejuni* enteritis.[74,75] Further, such patients failed to mount any immune response to the low-level cytotoxic activities observed, as judged by antibody inhibition studies, thus casting some doubt on the clinical relevance of cytotoxins in inflammatory *C. jejuni* enteritis.[75]

The observation that *C. jejuni* enteritis can be associated with episodes of loose or watery diarrhea has suggested that some isolates may produce an enterotoxin similar to either cholera toxin (CT) or *E. coli* heat-labile enterotoxin (LT). Several laboratories have reported the detection of a *C. jejuni* enterotoxin that is similar in effects and antigenic characteristics to CT and LT.[67–73] A 65- to 70-kDa *C. jejuni* protein has been purified that promotes cAMP accumulation in HeLa cells and/or fluid accumulation in rabbit ileal loops, and reacts with anti-CT antibodies.[68,72] However, other laboratories have failed to find any evidence for the presence of a CT-like enterotoxin using biologic, biochemical, and immunologic assays,[75,76] or by using LT- and CT-specific gene probes.[77,78] Patients with *C. jejuni* enteritis fail to produce antibodies that react with CT or LT.[75] It is, however, possible that the difficulties encountered in detecting enterotoxin activities in *C. jejuni* cultures are related to failure to establish optimal *in vitro* growth conditions for toxin production or to both temporal and strain-specific variability in the production of enterotoxins.

6. CONCLUDING REMARKS

In developed nations, *C. jejuni* enteritis is often a self-limiting disease in older children and adults. The disease may be treated effectively with antibiotics, which lessens the need for a vaccine. The hyperendemic nature of *Campylobacter* infection in underdeveloped countries suggests that this pathogen makes an important contribution to infant morbidity and mortality related to diarrheal disease. In these countries, an inexpensive and easily administered vaccine that is safely tolerated in a pediatric setting would be desirable. Recent data indicate that effective protection against *Campylobacter* disease might be achieved by oral immunization with killed bacteria supplemented with oral adjuvants. This approach will require extensive knowledge of the ability of various serologic variants or groups to elicit protective responses to heterologous serotypes. The safety and efficacy of small doses of oral adjuvants, such as active cholera toxin, remain to be investigated. Exposure to even small amounts of such bioactive toxins may be undesirable, especially in pediatric populations. It is important to determine the precise mechanisms involved in enteritis caused by *C. jejuni*. Identification of invasion-related determinants and the significance of specific toxins in disease may permit the construction of attenuated vaccine strains or the design of subunit-based preparations for oral immunization.

REFERENCES

1. Skirrow, M. B., and Blaser, M. J., 1992, Clinical and epidemiologic considerations, in: *Campylobacter jejuni: Current Status and Future Trends* (I. Nachamkin, M. J. Blaser, and L. S. Tompkins, eds.), American Society for Microbiology, Washington, D.C., pp. 3–8.
2. Kuroki, S., Haruta, T., Yoshioka, M., Kobayashi, Y., Nukina, M., and Nakanishi, H., 1991, Guillain–Barré syndrome associated with *Campylobacter* infection, *Pediatr. Infect. Dis. J.* **10:** 149–151.

3. Yuki, N., Taki, T., Inagaki, F., Kasama, T., Takahashi, M., Saito, K., Handa, S., and Miyatake, T., 1993, A bacterium lipopolysaccharide that elicits Guillain–Barré syndrome has a GM1 ganglioside-like structure, *J. Exp. Med.* **178:**1771–1775.
4. Mekalanos, J. J., 1992, Environmental signals controlling expression of virulence determinants in bacteria, *J. Bacteriol.* **174:**1–7.
5. Nachamkin, I., 1993, *Campylobacter* infections, *Curr. Opin. Infect. Dis.:* **6:**72–76.
6. Goossens, H., and Butzler, J.-P., 1992, Isolation and identification of *Campylobacter* spp., in: *Campylobacter jejuni: Current Status and Future Trends* (I. Nachamkin, M. J. Blaser, and L. S. Tompkins, eds.), American Society for Microbiology, Washington, D. C., pp. 93–109.
7. Guerrant, R. L., Lahita, R. G., Winn, W. C., Jr., and Roberts, R. B., 1978, Campylobacteriosis in man: Pathogenic mechanisms and review of 91 bloodstream infections, *Am. J. Med.* **65:** 584–592.
8. Cover, T. L., and Blaser, M. J., 1989, The pathobiology of *Campylobacter* infections in humans, *Annu. Rev. Med.* **40:**269–285.
9. Caldwell, M. B., Guerry, P., Lee, E. C., Burans, J. P., and Walker, R. I., 1985, Reversible expression of flagella in *Campylobacter jejuni, Infect. Immun.* **50:**941–943.
10. Harris, L. A., Logan, S. M., Guerry, P., and Trust, T. J., 1987, Antigenic variation of *Campylobacter* flagella, *J. Bacteriol.* **169:**5066–5071.
11. Skirrow, M. B., 1977, *Campylobacter* enteritis: A "new" disease, *Br. Med. J.* **2:**9–11.
12. Goossens, H., De Boeck, M., Coignau, H., Vlaes, L., Van den Borre, C., and Butzler, J. P., 1986, Modified selective medium for isolation of *Campylobacter* spp. from feces: Comparison with Preston medium, a blood-free medium, and a filtration system, *J. Clin. Microbiol.* **24:** 840–843.
13. Endtz, H. P., Ruijs, G. J., Zwinderman, A. H., Van der Reijden, T., Biever, M., and Moulton, R. P., 1991, Comparison of six media, including a semisolid agar, for the isolation of various *Campylobacter* species from stool specimens, *J. Clin. Microbiol.* **29:**1007–1110.
14. Walker, R. I., Caldwell, M. B., Lee, E. C., Guerry, P., Trust, T. J., and Ruiz-Palacios, G. M., 1986, Pathophysiology of *Campylobacter* enteritis, *Microbiol. Rev.* **50:**81–94.
15. Goossens, H., Henocque, G., Kremp, L., Rocque, J., Boury, R., Alanio, G., Vlaes, L., Hemelhof, W., Van den Borre, C., Macart, M., and Butzler, J. P., 1986, Nosocomial outbreak of *Campylobacter jejuni* meningitis in newborn infants, *Lancet* **1:**146–149.
16. Haq, J. A., Rahman, K. M., and Akbar, M. S., 1985, Haemolytic-uraemic syndrome and *Campylobacter,* Med. J. Aust. **142:**662–663.
17. Ebright, J. R., and Ryan, L. M., 1984, Acute erosive reactive arthritis associated with *Campylobacter jejuni*-induced colitis, *Am. J. Med.* **76:**321–323.
18. Doyle, M. P., and Jones, D. M., 1992, Food-borne transmission and antibiotic resistance of *Campylobacter jejuni,* in: *Campylobacter jejuni: Current Status and Future Trends* (I. Nachamkin, M. J. Blaser, and L. S. Tompkins, eds.), American Society for Microbiology, Washington, D.C., pp. 45–48.
19. Stern, N. J., 1992, Reservoirs for *Campylobacter jejuni* and approaches for intervention in poultry, in: *Campylobacter jejuni: Current Status and Future Trends* (I. Nachamkin, M. J. Blaser, and L. S. Tompkins, eds.), American Society for Microbiology, Washington, D.C., pp. 49–60.
20. Blaser, M. J., Cravens, J., Powers, B. W., Laforce, F. M., and Wang, W.-L. L., 1979, *Campylobacter* enteritis associated with unpasteurized milk, *Am. J. Med.* **67:**715–718.
21. Taylor, D. N., Porter, B. W., Williams, C. A., Miller, H. G., Bopp, C. A., and Blake, P. A., 1982, *Campylobacter* enteritis: A large outbreak traced to commercial raw milk, *West. J. Med.* **137:** 365–369.

21a. Tauxe, R. V., Vogt, R. L., Sours, H. E., Barrett, T., Feldman, R. A., Dickinson, R. J., and Witherell, L., 1982, *Campylobacter* enteritis associated with contaminated water, *Ann. Intern. Med.* **96:**292–296.

22. Blaser, M. J., Taylor, D. N., and Echeverria, P., 1986, Immune response to *Campylobacter jejuni* in a rural community in Thailand, *J. Infect. Dis.* **153**:249–254.

22a. Patton, C. M., and Wachsmuth, I. K., 1992, Typing schemes: Are current methods useful? in: *Campylobacter jejuni: Current Status and Future Trends* (I. Nachamkin, M. J. Blaser, and L. S. Tompkins, eds.), American Society for Microbiology, Washington, D.C., pp. 110–128.

23. Patton, C. M., Barrett, T. J., and Morris, G. K., 1985, Comparison of the Penner and Lior methods for serotyping *Campylobacter* spp., *J. Clin. Microbiol.* **22**:558–565.

24. Dunn, B. E., Blaser, M. J., and Snyder, E. L., 1987, Two-dimensional gel electrophoresis and immunoblotting of *Campylobacter* outer membrane proteins, *Infect. Immun.* **55**:1564–1572.

25. Nachamkin, I., and Hart, A. M., 1985, Western blot analysis of human antibody response to *Campylobacter jejuni* cellular antigens during gastrointestinal infection, *J. Clin. Microbiol.* **21**: 33–38.

26. Wenman, W. M., Chai, J., Louie, T. J., Goudreau, C., Lior, H., Newell, D. G., Pearson, A. D., and Taylor, D. E., 1985, Antigenic analysis of *Campylobacter* flagellar protein and other proteins, *J. Clin. Microbiol.* **21**:108–112.

27. Russel, R. G., Blaser, M. J., Sarmiento, J. I., and Fox, J., 1989, Experimental *Campylobacter jejuni* infection in *Macaca nemestrina*, *Infect. Immun.* **57**:1438–1444.

28. Black, R. E., Levine, M. M., Clements, M. L., Hughes, T. P., and Blaser, M. J., 1988, Experimental *Campylobacter* infection in humans, *J. Infect. Dis.* **157**:472–479.

29. Mills, S. D., and Bradbury, W. C., 1984, Human antibody response to outer membrane proteins of *Campylobacter jejuni* during infection, *Infect. Immun.* **43**:739–743.

30. Black, R. E., Perlman, D., Clements, M. L., Levine, M. M., and Blaser, M. J., 1992, Human volunteer studies with *Campylobacter* jejuni, in: *Campylobacter jejuni: Current Status and Future Trends* (I. Nachamkin, M. J. Blaser, and L. S. Tompkins, eds.), American Society for Microbiology, Washington, D.C., pp. 207–215.

31. Pennie, R. A., Pearson, R. D., Barrett, L. J., Lior, H., and Guerrant, R. L., 1986, Susceptibility of *Campylobacter jejuni* to strain-specific bactericidal activity in sera of infected patients, *Infect. Immun.* **52**:702–706.

32. Lane, E. M., Batchelor, R. A., Bourgeois, A. L., Burr, D. H., and Olson, J. G., 1987, Urine and faecal IgA response during naturally acquired infection with *Campylobacter jejuni*, *Lancet* **1**: 1141.

33. Burr, D. H., Caldwell, M. B., Bourgeois, A. L., Morgan, H. R., Wistar, R., Jr., and Walker, R. I., 1988, Mucosal and systemic immunity to *Campylobacter jejuni* in rabbits after gastric inoculation, *Infect. Immun.* **56**:99–105.

34. McSweegan, E., Burr, D. H., and Walker, R. I., 1987, Intestinal mucus gel and secretory antibody are barriers to *Campylobacter jejuni* adherence to INT 407 cells, *Infect. Immun.* **55**: 1431–1435.

35. Rollwagen, F. M., Pacheco, N. D., Clements, J. D., Pavlovskis, O., Rollins, D. M., and Walker, R. I., 1993, Killed *Campylobacter* elicits immune response and protection when administered with an oral adjuvant, *Vaccine* **11**:1316–1320.

36. Aspinall, G. O., Fujimoto, S., McDonald, A. G., Pang, H., Kurjanczyk, L. A., and Penner, J. L., 1994, Lipopolysaccharides from *Campylobacter jejuni* associated with Guillain–Barré syndrome patients mimic human gangliosides in structure, *Infect. Immun.* **62**:2122–2125.

37. Fox, J. G., 1992, In vivo models of enteric campylobacteriosis: Natural and experimental infections, in: *Campylobacter jejuni: Current Status and Future Trends* (I. Nachamkin, M. J. Blaser, and L. S. Tompkins, eds.), American Society for Microbiology, Washington, D.C., pp. 131–138.

38. Blaser, M. J., Duncan, D. J., Warren, G. H., and Wang, W.-L. L., 1983, Experimental *Campylobacter jejuni* infection of adult mice, *Infect. Immun.* **39**:908–916.

39. Caldwell, M. B., Walker, R. I., Stewart, S. D., and Rogers, J. E., 1983, Simple adult rabbit model for *Campylobacter jejuni* enteritis, *Infect. Immun.* **42**:1176–1182.

40. Babakhani, F. K., Bradley, G. A., and Joens, L. A., 1993, Newborn piglet model for campylobacteriosis, *Infect. Immun.* **61:**3466–3475.
41. Russell, R. G., 1992, *Campylobacter jejuni* colitis and immunity in primates: Epidemiology of natural infection, in: *Campylobacter jejuni: Current Status and Future Trends* (I. Nachamkin, M. J. Blaser, and L. S. Tompkins, eds.), American Society for Microbiology, Washington, D.C., pp. 148–157.
42. Russell, R. G., O'Donnoghue, M., Blake, D. C., Jr., Zulty, J., and DeTolla, L. J., 1993, Early colonic damage and invasion of *Campylobacter jejuni* in experimentally challenged infant *Macaca mulatta, J. Infect. Dis.* **168:**210–215.
43. Finlay, B. B., and Falkow, S., 1989, Common themes in microbial pathogenicity, *Microbiol. Rev.* **53:**210–230.
44. Fauchere, J.-L., Rosenau, A., Veron, M., Moyen, E. N., Richard, S., and Pfister, A., 1986, Association with HeLa cells of *Campylobacter jejuni* and *Campylobacter coli* isolated from human feces, *Infect. Immun.* **54:**283–287.
45. De Melo, M. A., and Pechere, J.-C., 1990, Identification of *Campylobacter jejuni* surface proteins that bind to eukaryotic cells in vitro, *Infect. Immun.* **58:**1749–1756.
46. Fauchere, J.-L., Kervella, M., Rosenau, A., Mohanna, K., and Veron, M., 1989, Adhesion to HeLa cells of *Campylobacter jejuni* and *C. coli* outer membrane components, *Res. Microbiol.* **140:**379–392.
47. Pei, Z., and Blaser, M. J., 1993, PEB1, the major cell-binding factor of *Campylobacter jejuni*, is a homolog of the binding component in gram-negative nutrient transport systems, *J. Biol. Chem.* **268:**18717–18725.
48. Newell, D. G., McBride, H., and Dolby, J. M., 1985, Investigations on the role of flagella in the colonization of infant mice with *Campylobacter jejuni* and the attachment of *Campylobacter jejuni* to human epithelial cells, *J. Hyg.* **95:**215–227.
49. McSweegan, E., and Walker, R. I., 1986, Identification and characterization of two *Campylobacter jejuni* adhesins for cellular and mucous substrates, *Infect. Immun.* **53:**141–148.
50. Grant, C. C. R., Konkel, M. E., Cieplak, W., Jr., and Tompkins, L. S., 1993, Role of flagella in adherence, internalization, and translocation of *Campylobacter jejuni* in nonpolarized and polarized epithelial cell cultures, *Infect. Immun.* **61:**1764–1771.
51. Newell, D. G., McBride, H., Saunders, F., Dehele, Y., and Pearson, A. D., 1985, The virulence of clinical and environmental isolates of *Campylobacter jejuni, J. Hyg.* **95:**45–54.
52. Konkel, M. E., and Joens, J. A., 1989, Adhesion to and invasion of HEp-2 cells by *Campylobacter* spp., *Infect. Immun.* **57:**2984–2990.
53. Konkel, M. E., Hayes, S. F., Joens, L. A., and Cieplak, W., Jr., 1992, Characteristics of the internalization and intracellular survival of *Campylobacter jejuni* in human epithelial cell cultures, *Microb. Pathog.* **13:**357–370.
54. De Melo, M. A., Gabbiani, G., and Pechere, J.-C., 1989, Cellular events and intracellular survival of *Campylobacter jejuni* during infection of HEp-2 cells, *Infect. Immun.* **57:**2214–2222.
55. Oelschlaeger, T. A., Guerry, P., and Kopecko, D. J., 1993, Unusual microtubule-dependent endocytosis mechanisms triggered by *Campylobacter jejuni* and *Citrobacter freundii, Proc. Natl. Acad. Sci. USA* **90:**6884–6888.
56. Konkel, M. E., and Cieplak, W., Jr., 1992, Altered synthetic response of *Campylobacter jejuni* to cocultivation with human epithelial cells is associated with enhanced internalization, *Infect. Immun.* **60:**4945–4949.
57. Konkel, M. E., Mead, D. J., and Cieplak, W., Jr., 1993, Kinetic and antigenic characterization of altered protein synthesis by *Campylobacter jejuni* during cultivation with human epithelial cells, *J. Infect. Dis.* **168:**948–954.
58. Panigrahi, P., Losonsky, G., DeTolla, L. J., and Morris, J. G., Jr., 1992, Human immune response to *Campylobacter jejuni* proteins expressed in vitro, *Infect. Immun.* **60:**4938–4944.
59. Wassenaar, T. M., Bleumink-Pluym, N. M. C., and Van der Zeijst, B. A. M., 1991, Inactivation

of *Campylobacter jejuni* flagellin genes by homologous recombination demonstrates that flaA but not flaB is required for invasion, *EMBO J.* **10:**2055–2061.

60. Konkel, M. E., Mead, D. J., Hayes, S. F., and Cieplak, W., Jr., 1992, Translocation of *Campylobacter jejuni* across human polarized epithelial cell monolayer cultures, *J. Infect. Dis.* **166:**308–315.
61. Everest, P. H., Goossens, H., Butzler, J.-P., Lloyd, D., Knutton, S., Ketley, J. M., and Williams, P. H., 1992, Differentiated Caco-2 cells as a model for enteric invasion by *Campylobacter jejuni* and *C. coli*, *J. Med. Microbiol.* **37:**319–325.
62. Kiehlbauch, J. A., Albach, R. A., Baum, L. L., and Chang, K. P., 1985, Phagocytosis of *Campylobacter jejuni* and its intracellular survival in mononuclear phagocytes, *Infect. Immun.* **48:**446–451.
63. Purdy, D., and Park, S. F., 1994, Cloning, nucleotide sequence and characterization of a gene encoding superoxide dismutase from *Campylobacter jejuni* and *Campylobacter coli*, *Microbiology* **140:**1203–1208.
64. Pesci, E. C., Cottle, D. L., and Pickett, C. L., 1994, Genetic, enzymatic, and pathogenic studies of the iron superoxide dismutase of *Campylobacter jejuni*, *Infect. Immun.* **62:**2687–2694.
65. Guerrant, R. L., Wanke, C. A., Pennie, R. A., Barrett, L. J., Lima, A. A. M., and O'Brien, A. D., 1987, Production of a unique cytotoxin by *Campylobacter jejuni*, *Infect. Immun.* **55:**2526–2530.
66. Mahajan, S., and Rodgers, F. G., 1990, Isolation, characterization, and host-cell-binding properties of a cytotoxin from *Campylobacter jejuni*, *J. Clin. Microbiol.* **28:**1314–1320.
67. Belbouri, A., and Megraud, F., 1988, Enterotoxin-like activity produced by *Campylobacter jejuni* and *Campylobacter coli* isolated from patients and healthy controls in Algeria, *FEMS Microbiol. Lett.* **51:**25–28.
68. Daikoku, T., Kawaguchi, M., Takama, K., and Suzuki, S., 1990, Partial purification and characterization of the enterotoxin produced by *Campylobacter jejuni*, *Infect. Immun.* **58:** 2414–2419.
69. Goossens, H., Butzler, J.-P., and Takeda, Y., 1985, Demonstration of cholera-like enterotoxin production by *Campylobacter jejuni*, *FEMS Microbiol. Lett.* **29:**73–76.
70. Johnson, W. M., and Lior, H., 1986, Cytotoxic and cytotonic factors produced by *Campylobacter jejuni*, *Campylobacter coli*, and *Campylobacter laridis*, *J. Clin. Microbiol.* **24:**275–281.
71. Klipstein, F. A., and Engert, R. F., 1985, Purification of *Campylobacter jejuni* enterotoxin, *Lancet,* **1:**1123–1124.
72. McCardell, B. A., Madden, J. M., and Lee, E. C., 1986, *Campylobacter jejuni* and *Campylobacter coli* production of a cytotonic toxin immunologically similar to cholera toxin, *J. Food Prot.* **47:** 943–949.
73. Ruiz-Palacios, G. M., Torres, J., Torres, N. I., Escamilla, E., Ruiz-Palacios, B. R., and Tamayo, J., 1983, Cholera-like enterotoxin produced by *Campylobacter jejuni*, characterisation and clinical significance, *Lancet* **2:**250–253.
74. Cover, T. L., Perez-Perez, G. I., and Blaser, M. J., 1990, Evaluation of cytotoxin activity in fecal filtrates from patients with *Campylobacter jejuni* or *Campylobacter coli* enteritis, *FEMS Microbiol. Lett.* **58:**301–304.
75. Perez-Perez, G. I., Cohn, D. L., Guerrant, R. L., Patton, C. M., Reller, L. B., and Blaser, M. J., 1989, Clinical and immunologic significance of cholera-like toxin and cytotoxin production by *Campylobacter* species in patients with acute inflammatory diarrhea in the USA, *J. Infect. Dis.* **160:**460–468.
76. Wadstrom, T., Baloda, S. B., Krovacek, K., Faris, A., Bengtson, S., and Walder, M., 1983, Swedish isolates of *Campylobacter jejuni/coli* do not produce cytotonic or cytotoxic enterotoxins, *Lancet* **2:**911.

77. Konkel, M. E., Lobet, Y., and Cieplak, W., Jr., 1992, Examination of multiple isolates of *Campylobacter jejuni* for evidence of cholera toxin-like activity, in: *Campylobacter jejuni: Current Status and Future Trends* (I. Nachamkin, M. J. Blaser, and L. S. Tompkins, eds.), American Society for Microbiology, Washington, D.C., pp. 193–198.
78. Baig, B. H., Wachsmuth, I. K., Morris, G. K., and Hill, W. E., 1986, Probing *Campylobacter jejuni* DNA coding for *Escherichia coli* heat-labile enterotoxin, *J. Infect. Dis.* **154:**542.

8

Clostridia and Bacteroides in Enteric Infections

DAVID M. LYERLY, ROGER L. VAN TASSELL, and TRACY D. WILKINS

1. INTRODUCTION

The ability of anaerobic bacteria to cause disease in humans and animals has been recognized for many years. The clostridia, for example, are well known for their ability to cause disease by producing a variety of toxins ranging from the potent neurotoxins of *Clostridium tetani* and *Clostridium botulinum* to the tissue-damaging toxins of the gas gangrene clostridia. *Bacteroides fragilis*, on the other hand, represents the major anaerobic pathogen in clinical specimens from patients with abscesses and soft-tissue infections. More recently, the importance of members of these genera in gastrointestinal infections has become recognized. *Clostridium difficile* has emerged as the major cause of nosocomial diarrhea because of its ability to produce diarrhea and pseudomembranous colitis in patients treated with antibiotics. *Clostridium perfringens*, which is the major cause of gas gangrene, produces a diverse group of toxins that are involved in enteric disease. In industrialized countries, outbreaks of *C. perfringens* food poisoning occur because of the production of a spore-associated enterotoxin. In certain regions in the Pacific rim, especially in Papua New Guinea, a disease known as pigbel, which is prevalent in persons who eat improperly cooked pork, results from infection with *C. perfringens* type C strains that produce β-toxin. In addition to these

DAVID M. LYERLY • TechLab, Inc., Corporate Research Center, Blacksburg, Virginia 24060. ROGER L. VAN TASSELL • Virginia Polytechnic Institute and State University, Blacksburg, Virginia 24061. TRACY D. WILKINS • Fralin Biotechnology Center, Virginia Polytechnic Institute and State University, Blacksburg, Virginia 24061.

Enteric Infections and Immunity, edited by Lois J. Paradise *et al.* Plenum Press, New York, 1996.

maladies, type E strains of *C. perfringens* and certain strains of *Clostridium spiroforme* produce a toxin (iota) that results in another distinctive type of enterotoxemia in animals.

We tend to think of the normal commensal organisms in the colon as nonpathogenic bacteria that do not pose a threat unless the integrity of the intestine has been altered. *B. fragilis,* for example, represents a normal inhabitant of the colonic microflora not typically thought of as an enteric pathogen, although its role in soft-tissue infections and septicemias has been well established. The recent discovery of the enterotoxigenic nature of certain strains of *B. fragilis,* however, shows that this commensal also can be an intestinal pathogen. In addition, the production of a toxin may help to explain why this organism is so efficient at causing extraintestinal infections once it disseminates from the colon.

Our knowledge on the role of anaerobes as enteric pathogens is still at an early stage but progress is being made in areas such as molecular mechanisms and improved diagnostic methods. Much work remains, however, in areas such as how our immune systems respond to these pathogens and how we can develop suitable vaccines. It is likely that there are other intestinal illnesses caused by other anaerobes that have not yet been recognized; therefore, it is important that efforts continue to be directed toward the identification of "new" anaerobic pathogens. In this chapter, we discuss the pathogenic properties of the anaerobic bacteria known to cause enteric disease. Although food-borne and infant botulism related to *C. botulinum* initially develop in a manner analogous to other types of enteric diseases, the disease results from systemic intoxication and not from the direct action of toxin within the intestine. Therefore, botulism is not covered herein, and the reader may refer to the suggested reading list at the end of the chapter for discussion of this topic.

2. PSEUDOMEMBRANOUS COLITIS AND ANTIBIOTIC-ASSOCIATED DIARRHEA

2.1. Description

C. difficile causes gastrointestinal disease ranging from mild diarrhea to life-threatening pseudomembranous colitis (PMC). The disease, which almost always is nosocomial in origin, occurs primarily in elderly patients following the disruption of the protective microflora by antibiotics and antineoplastic drugs. The clinical criteria typically include (1) diarrhea in which no other cause has been established, (2) loose or unformed stools, (3) three or more bowel movements per day, (4) duration of at least 2 days, and (5) prior history of antibiotic or antineoplastic agents in the preceding 4 to 6 weeks.

One of the first reports of PMC occurred before the turn of the century and the toxigenic nature of *C. difficile* in animals was described in reports from the mid-1930s. The role of this organism in PMC did not develop, however, until the onset of the antibiotic era when powerful antibiotics capable of drastically changing the colonic microflora became available and widely used. The role of *C. difficile* in the disease first became established in the late 1970s following reports of PMC and death in patients receiving clindamycin. We now know that the disease may occur following the use of any antibiotic or drug that alters the microflora of the gut.

The ability of this organism to form spores and to be spread by a highly efficient mode of transmission, diarrhea, makes *C. difficile* a formidable opportunistic pathogen. In a hospital setting involving debilitated elderly patients who are compromised, the organism may become a major health-care problem. These circumstances often lead to outbreaks of the disease, particularly in settings such as Veterans Administration medical centers and hospitals that have an unusually high percentage of elderly persons undergoing antibiotic therapy.

Hamsters are the primary animal model used to study *C. difficile* disease.(1) Treatment with antibiotics leaves the animal highly susceptible, and under these conditions, there is a rapid onset of the disease. In hamsters, the disease causes enterocolitis with extensive involvement of the cecum. Unless treated, the animals develop severe diarrhea and lethargy, and die from the infection. Mice and rats treated with antibiotics also develop *C. difficile* disease, but in these animals, the disease does not develop consistently and is less severe.

2.2. Epidemiology

C. difficile disease occurs most often in elderly persons undergoing antibiotic therapy in nosocomial settings. Typing methods such as serotyping, restriction endonuclease mapping, bacteriophage and plasmid typing, and protein analysis have been developed and used for epidemiological purposes. All of these methods have been used to demonstrate the prevalence of *C. difficile* in the hospital and transmission of the organism from patient to patient by the healthcare worker. Healthcare workers usually are not at risk although precautions should be taken if the worker is on antibiotic therapy.

The organism is not considered a member of the normal flora in adults. Only a low percentage (3% or less) of healthy adults have detectable levels of *C. difficile* in their stool specimens. On admission to the hospital, however, the number of asymptomatic adult carriers may increase significantly.[2] AIDS patients tend to be susceptible to *C. difficile* infections, an observation that is not surprising based on the compromised status and drug therapy that many of these individuals are receiving. Infants typically are resistant to the disease even though up to 50% of them may be positive for *C. difficile* and carry high levels of the toxins in their stools. Occasionally, however, infants who are severely debilitated (e.g., premature birth, sepsis) and on antibiotic therapy may develop the disease.

2.3. Pathogenesis

C. difficile produces two large single polypeptide toxins, A and B. The toxins can be produced in high quantities using a dialysis flask system previously developed for the large-scale production of botulinum toxin.[3] The system works well for other toxins of the clostridia, including iota toxin described below. Toxin A (M_r 308,000) is the largest bacterial toxin known. Toxin B (M_r 269,000) is slightly smaller. A description of the two toxins is provided in Fig. 1 and Table I. Both are cytotoxic and cause a similar type of rounding of eukaryotic cells, suggesting an effect on the cytoskeletal system. Toxin B is one of the most potent bacterial cytotoxins known, and may cause cell rounding by altering the action of factor *rho*, which is believed to be involved in the regulation of the cytoskeleton.[4] However, toxin B has little effect on the intestinal mucosa. Toxin A, on the other hand, is a potent tissue-damaging enterotoxin that is weakly cytotoxic. The action of toxin A on the gut mucosa leads many to believe that this toxin causes most of the clinical symptoms. There is evidence suggesting some synergistic action between the toxins, indicating that toxin B may act in conjunction with toxin A.

Toxigenic strains produce both toxins. Nontoxigenic strains, on the other hand, are negative for both toxins and do not carry the toxin genes. Only one strain has been described that produces one toxin without the other. In this instance, toxin B is produced. Interestingly, however, this strain carries the 5′-portion of the toxin A gene, but it is not known if this codes for the production of a toxin A fragment. Among the toxigenic strains, there is a wide variation in the level of toxin produced. Highly toxigenic strains produce high levels of both A and B whereas weakly toxigenic strains produce low levels of both toxins. There also are toxigenic strains that produce intermediate levels of toxin. The highly toxigenic strains may produce 100,000-fold more toxin than the weakly toxigenic strains. Curiously, however, no obvious correlation exists between the level of toxin production and severity of the disease.

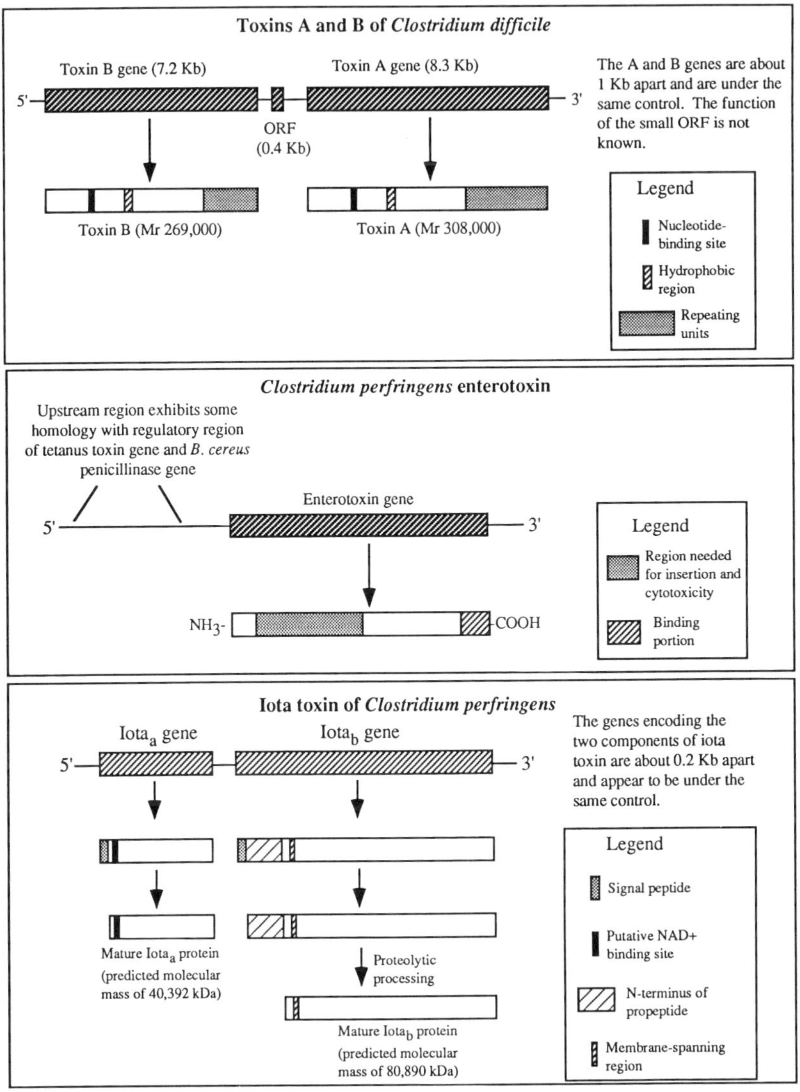

FIGURE 1. Genetics and molecular properties of the toxins of *Clostridium* species involved in enteric disease.

TABLE I
Properties of Toxins from Anaerobic Bacteria That Cause Enteric Disease

Toxin	Mechanism	Biological activity		Cross-reactivity
		Toxin A	*Toxin B*	
Toxins A and B of *C. difficile* (A, M_r 308,000; B, M_r 269,000)	Affects cytoskeletal system, resulting in cell rounding and cell death; toxin B alters factor rho, which is a GTP-binding protein	Enterotoxic Cytotoxic Lethal Hemagglutinin	Cytotoxic Lethal	Toxin A cross-reacts with *C. sordellii* toxin HT; toxin B cross-reacts with *C. sordellii* toxin LT
C. perfringens enterotoxin (M_r 35,000)	Toxin or toxic fragment inserts into membrane, affecting cell permeability and resulting in cell death	Cytotoxic Enterotoxic		No known cross-reactivity with other toxins or proteins; possesses superantigen properties similar to those of *S. aureus* heat-stable enterotoxins
C. perfringens beta toxin (M_r 35,000)	Unknown	Dermonecrotic Lethal Hemolytic		Some homology with alpha-toxin, gamma-toxin, and leukocidin of *S. aureus*
Iota toxin of *C. perfringens* (M_r iota$_a$, 40,000; M_r iota$_b$, 81,000) and *C. spiroforme*	A binary toxin that catalyzes the transfer of the ADP-ribose moiety of NAD onto the Arg residue of monomeric actin, thereby inhibiting actin polymerization	Cytotoxic Lethal Dermonecrotic Enterotoxic		The active and binding components of iota toxin from *C. perfringens* cross-react with the components of *C. spiroforme* iota toxin and are interchangeable; functionally, iota toxin and toxin C2 from *C. botulinum* are similar but they do not cross-react
B. fragilis toxin (M_r 20,000)	Biological activity may result from protease activity	Enterotoxic Cytotoxic		Cross-reacts with membrane-associated components, possibly inactive toxin

The A and B genes appear to be the result of gene duplication. The toxins exhibit an overall homology of 45% at the amino acid level, and they share several structural features, including a putative nucleotide-binding site, hydrophobic region, four cysteine residues, and a series of contiguous repeating units.[5,6] The repeating units, which probably function as the binding portion, exhibit homology with several glucan-binding proteins from *Streptococcus* species. Toxins A and B exhibit extensive homology with the large cytotoxins of *Clostridium sordellii*.[7] Toxin A is highly related to toxin HT (hemorrhagic toxin) of *C. sordellii*, which also is a tissue-damaging toxin. Toxin B, on the other hand, is highly related to toxin LT (lethal toxin) of *C. sordellii*. Interestingly, toxin LT is about 1000-fold less cytotoxic than toxin B but at least 10-fold more lethal. Antitoxin against the *C. sordellii* toxins neutralizes the *C. difficile* toxins and still is used in some medical institutions for diagnostic purposes. *C. sordellii*, however, is *not* associated with gastrointestinal disease and is more typically implicated in wound infections in animals.

2.4. Immunity

In the experimental hamster model, vaccination against both toxins is sufficient to completely protect the animal against the onset of *C. difficile* disease. In germfree mice, which develop a more chronic form of the disease, passive immunization with a neutralizing monoclonal antibody against toxin A is completely protective against the onset of the disease.[8] Based on these results, research on active and passive immunization against *C. difficile* disease is under way and may lead to the development of vaccines that are warranted in high-risk situations.

Little is known about the immune response to the toxins during the disease. Some preliminary evidence indicates that patients with *C. difficile* disease develop serum antibodies against both toxins and low levels of neutralizing secretory antibodies against toxin A. A potential problem in the detection of anti-A and anti-B antibodies, however, is the fact that both toxins bind certain subclasses of antibodies in a nonimmune manner. This nonimmune binding may result in false-positive readings and incorrect interpretation of the immune status of the individual. The nature of the nonimmune binding of immunoglobulin to the toxins has not been determined.

2.5. Treatment and Prevention

In some instances, *C. difficile* disease is self-limiting and the episodes of diarrhea may stop simply by discontinuing the inciting agent. If the condition of the patient improves following this simple step, additional therapy other than supportive measures to correct for fluid loss may not be warranted. More commonly, the method of treating the disease is oral administration of vancomycin or metronidazole. Because more extensive clinical evaluations have been performed with vancomycin, it tends to be used more often than metronidazole. Vancomycin is very effective against the organism because it is poorly adsorbed in the intestine and high concentrations typically are achieved. A typical dose of vancomycin ranges from 125 to 500 mg delivered at 6-hr intervals. The higher doses tend to be used in patients who are critically ill.[1]

A major problem is the increasing incidence of relapse and the development of vancomycin-resistant enterococci. In some locations, relapse rates of 20 to 25% are being reported. As a result, research into other types of treatments is being pursued and includes: (1) treatment with anion-exchange resins such as cholestyramine and colestipol, which bind the toxins without exposing the patient to additional antibiotic therapy, (2) treatment with probiotics such as *Saccharomyces boulardii*, a nonpathogenic yeast used to treat gastrointestinal illness in Europe, and (3) passive immunoprophylaxis with gamma globulin in persons at high risk.

The spread of spores by the healthcare worker is the primary mode of transmission. The use of preventive measures such as vinyl disposable gloves, laboratory coats, and proper washing procedures is indicated, and all items (thermometers, commodes, endoscopes, etc.) that come into contact with the patient should be properly cleaned and disinfected. The isolation of patients, if possible, also is recommended in institutions that have endemic levels of the disease. Institutions that report high incidences of the disease should consider the use of sporicidal agents and disinfectants for cleaning the rooms of patients. It is important that healthcare workers be educated concerning the manner in which the disease is spread and precautions that should be used to minimize transmission.

3. *CLOSTRIDIUM PERFRINGENS* FOOD POISONING

3.1. Description

C. perfringens is a common inhabitant of the soil and can be isolated from a variety of sources, including uncooked meat, water sources, and the intestinal tract of humans and animals. *C. perfringens* food poisoning results from the ingestion of contaminated foods and the subsequent production of a spore-associated enterotoxin. The disease was first reported before the turn of the century and has become increasingly recognized since the 1940s. The organism grows rapidly over temperatures ranging from room temperature to 50°C, and under optimal conditions, the generation time may be only 10 min. Therefore, there is the potential to produce large amounts of the enterotoxin in infected persons. Fortunately, in most instances, the disease is mild and hospitalization seldom is required. There are several isolated reports of persons who apparently died of the disease, but those persons were elderly debilitated individuals. Clinically, there is an incubation period of 8 to 12 hr after contaminated food is ingested before the onset of diarrhea, abdominal cramping, and occasionally nausea. Clues to the diagnosis, which usually become apparent after the episode, include a short incubation time period, a suspected meat source, diarrhea and intestinal cramping, and short duration of the illness. Because of the short duration, the diagnosis usually is made after the patient recovers and consists of detecting $>10^5$ *C. perfringens* per gram in the suspected food source and $>10^6$ of the same type of *C. perfringens* per gram of feces. The enterotoxin can be detected by several different types of assays, including tissue culture assay, passive reversed hemagglutination, and enzyme immunoassay. The tissue culture assay is used primarily for research purposes whereas the enzyme immunoassay is being evaluated for clinical use.

3.2. Epidemiology

From 1973 through 1987, more than 12,000 cases of *C. perfringens* food poisoning were reported to the Centers for Disease Control, accounting for 11.2% of the total number of cases of food poisoning. From 1989 through 1991, 2889 cases were reported, and in 1991, food poisoning related to *C. perfringens* represented 23.1% of the total number of cases of food poisoning that were reported. Outbreaks typically have been associated with commercial food services for institutions and restaurants, and usually have been traced to meat or meat products, including beef and poultry. *C. perfringens* represents the third leading bacterial cause of food poisoning in the United States behind *Salmonella* and *Staphylococcus aureus*. Some clinicians suspect that the incidence of *C. perfringens* food poisoning is higher because in many instances, the etiologic agent responsible for the outbreak often is not identified. However, not all strains isolated during food poisoning outbreaks contain the enterotoxin gene, and the isolation of *C. perfringens* from persons suspected of having *C. perfringens* food poisoning does not definitively show that the isolate produces enterotoxin. For example, in one study involving 245 strains isolated from food and feces from 186 confirmed outbreaks of *C. perfringens* food poisoning in the United Kingdom, only 59% were positive when analyzed using gene probes specific for the enterotoxin gene.[(9)]

In addition to causing food poisoning, in some instances enterotoxigenic strains have been associated with antibiotic-associated diarrhea in which the diarrhea lasts much longer and fecal blood and mucus are present. The incidence of antibiotic-associated diarrheas resulting from enterotoxigenic *C. perfringens* has not been established. However, some scientists have suggested that this organism may be responsible for the majority of the cases in some outbreaks. The role of enterotoxigenic *C. perfringens* in antibiotic-associated diarrheas is an important consideration because *C. difficile*, which causes essentially all cases of pseudomembranous colitis, is found in only about 25% of the reported cases of antibiotic-associated diarrheas.

3.3. Pathogenesis

C. perfringens food poisoning results from the production of a spore-associated enterotoxin (Fig. 1, Table I). At one time, it was suggested that the enterotoxin actually represents a structural component of the spore, but this possibility now is being questioned. Studies on the physicochemical properties of the toxin have progressed more rapidly since the development by Duncan and Strong[10] of a medium, which contains yeast extract (0.4%), proteose peptone (1.5%), soluble starch (0.4%), sodium thioglycolate (0.1%), and $Na_2HPO_4 \cdot 7H_2O$ (1.0%), that promotes the onset of sporulation and results in high yields of toxin. The enterotoxin is a single polypeptide with a molecular weight of 35,000. The toxin is hydrophobic and forms aggregates. Limited proteolysis results in the removal of the N-terminal 25 amino acids and a mature 32,500-kDa toxin that is two to three times more cytotoxic than uncleaved toxin. The C-terminus is the binding portion that recognizes the receptor.[11] The size of the toxin and its superantigen properties have led to speculation that it may be related to the enterotoxins of *S. aureus* that cause similar clinical symptoms. However, the *C. perfringens* enterotoxin is inactivated by heating at 60°C for 5 min, distinguishing it from the heat-stable enterotoxins of *S. aureus*. In addition, there is no significant homology between the *C. perfringens* enterotoxin and the enterotoxins of *S. aureus* at the DNA level.

The toxin is produced once the organism begins to sporulate in the intestinal tract. After being released during cell lysis, the toxin appears to be proteolytically processed to a more active form. After binding to a proteinaceous receptor, the toxin (or a toxic fragment) inserts into the membrane and affects the permeability of cells, resulting in the leakage of intracellular components and, ultimately, cell death.[11] This mechanism likely is responsible for the damage to the gut mucosa, stimulating the loss of fluid and electrolytes from the intestinal tract.

The enterotoxin gene, which is located on the chromosome, has been cloned and the entire sequence determined.[12] There are indications that some minor heterogeneity exists at the DNA level among different enterotoxigenic strains. These minor variations in gene sequence, however, do not appear to result in any significant differences at the protein level although additional studies are needed to confirm these preliminary observations. A region upstream of the gene has significant homology with the -35 and -10 promoter sequences for both the *C. tetani* tetanus toxin gene and the *Bacillus cereus* penicillinase gene, but there is no homology in the structural gene itself. There is some evidence that the transcriptional factors that regulate the genes involved in sporulation also increase the level of enterotoxin. Thus, there is a direct correlation in the level of sporulation and production of enterotoxin.

3.4. Immunity

Monoclonal antibodies against the enterotoxin have been developed and used to map epitopes on the molecule.[11] One of the antibodies that binds to the C-terminus blocks the binding of the toxin to its receptor. This blocking effect completely inhibits the action of the enterotoxin, thus neutralizing its cytotoxic activity. Little is known about the immune response to the enterotoxin during the infection.

3.5. Treatment and Prevention

Because of the mild and self-limiting nature of *C. perfringens* food poisoning, the typical therapy includes rehydration by the intake of fluids. Antibiotic therapy usually is not recommended. It is important, however, that the patient seek medical attention if complications develop. The disease is not spread by patient-to-patient contact and isolation precautions are not necessary. However, any suspected cases should be reported to the local public health authorities so that any contaminated foods can be identified and improper food preparation and sanitation methods can be corrected. The disease results primarily from the contamination of meats. All meats should be properly cooked to destroy the spores and foods should be refrigerated to minimize bacterial growth. Meats that are rewarmed should be heated to 75°C internally to destroy the bacteria.

In instances in which antibiotic-associated diarrheas are traced to *C. perfringens*, preventive measures such those described for *C. difficile* may be warranted. Additional studies are needed, however, to document the role of this organism in antibiotic-associated intestinal illnesses.

4. ENTERITIS NECROTICANS (PIGBEL)

4.1. Description

Enteritis necroticans, more commonly known as pigbel, is caused by type C strains of *C. perfringens* that produce α- and β-toxin, both of which are lethal and necrotizing. The β-toxin appears to be the more important of the two toxins in pigbel, although α-toxin (phospholipase C) is probably the major virulence factor in gas gangrene infections. Pigbel occurs following the consumption of improperly cooked pork. In areas such as the Highlands of New Guinea, hogs or large slabs of pork are cooked between layers of bread, fruit, and leaves at low temperatures (80°C) that do not kill the organism. When the meat is eaten, the organism grows in the intestinal tract, releasing the α- and β-toxins.

The disease develops in several forms and is similar to veterinary enterotoxemias caused by *C. perfringens* and to Darmbrand, the epidemic enteritis necroticans that occurred in regions of Europe following World War II. The acute stage of pigbel, which may be fulminating and result in the death of the patient within 24 hr, usually begins within several days after a high-protein meal that typically includes pork. Clinically, there may be mild diarrhea or constipation, and nausea. Children may complain of being hungry, yet develop severe intestinal pain when they eat. The milder form of the disease is difficult to distinguish from gastroenteritis, and the patient usually recovers without complications. Chronic pigbel, which develops in some patients with past histories of either the acute or milder form of the disease, results from intestinal obstruction.

The diagnosis of the disease is based on the patient history and the clinical features. Isolating the organism may be difficult because type C strains are overshadowed by the abundance of type A strains that are indistinguishable morphologically and culturally. Patients with pigbel often develop antibodies against the organism and toxins, but the detection of these antibodies by serological testing typically is not performed for diagnostic purposes.

4.2. Epidemiology

In the Highlands of Papua New Guinea where pigbel has been studied in detail, the disease is the most common cause of death in children older than 1 year of age. The annual incidence has been estimated at 33.6 per 10,000 in children under the age of 15 and at 48.3 per 10,000 in the adult population. The organism is present in human and pig feces and in soil samples from the villages that report a high incidence. Therefore, in addition to eating improperly cooked meat, the large slabs of pork are contaminated even further by handling the meat in dirty environments. Part of the national pastime in the Highlands region includes pig feasting activities. When these activities occur, the incidence increases and the epidemiology can be studied by following the consumption of pork. In addition to New Guinea, the disease occurs in other regions in southeast Asia and parts of China and Africa.

4.3. Pathogenesis

Once the contaminated meats are ingested, the organism grows in the intestinal tract and releases α- and β-toxins which can be absorbed systemically. The genes for these toxins have been identified and the β-toxin gene has been found only in type B and C strains. The β-toxin exhibits homology with alpha-toxin, gamma-toxin, and leukocidin of *S. aureus*, and exists as monomeric and multimeric forms[13] (Table I). The α- and β-toxins are both lethal and necrotizing, and are rapidly destroyed by endogenous proteases such as trypsin. Thus, the progression of the disease appears to be affected by the level of endogenous protease activity. It has been suggested that the conditions needed for the disease to develop include the presence of type C strains that produce β-toxin, consumption of high-protein nutrients for rapid growth of the organism, low levels of endogenous trypsin, and the presence of trypsin inhibitors in the diet such as those found in sweet potatoes or those from parasites.

4.4. Immunity

Vaccination against β-toxoid results in the development of protective antibodies. In one study, vaccination with β-toxoid decreased the incidence of the disease from 33 cases to 4 cases per 10,000, and the results from other vaccination studies have been similar. It is recommended that children in high-risk regions be vaccinated with β-toxoid at 4, 6, and 12 months and that boosters be administered to increase the level of protective antibody.

4.5. Treatment and Management

Once a diagnosis has been made in patients with the acute form of the disease, the decision of when to operate and start intravenous feeding must be made quickly. If surgery is delayed, the disease can progress rapidly and become fatal. This consideration must be balanced, however, with the possibility that in some instances, children recover with proper management and without surgery. The guidelines for the treatment of pigbel have been established and are described in detail elsewhere.[14]

5. IOTA TOXIN-MEDIATED ENTEROTOXEMIA

5.1. Description

In addition to its role in food poisoning and pigbel, *C. perfringens* has been implicated in enterotoxemia in animals. In this instance, the disease results from infection with type C strains that produce iota toxin. *C. spiroforme*, which has a characteristic shape resembling a helical coil, produces a cross-reactive iota toxin responsible for a similar disease in rabbits. The disease may be induced by antibiotic treatment and can become a major concern in farm animals and in animal care facilities.

5.2. Pathogenesis

The disease results from the production of iota toxin, which is a binary toxin comprised of two nonlinked polypeptides. One of the polypeptides ($iota_a$) represents the active enzyme. The other polypeptide ($iota_b$) is believed to be the binding component (Fig. 1, Table I). Individually, neither polypeptide is toxic. When mixed together, however, they become dermonecrotic and are lethal when injected into experimental animals.[15–17] The production of both components of iota toxin is enhanced by the addition of divalent cations to the growth medium.[18] The *C. perfringens* iota toxin has been studied in more detail than the *C. spiroforme* iota toxin, and more is known about its properties and gene sequence. However, it has been established that the $iota_a$ component of both toxins is an enzyme that transfers the ADP-ribosyl moiety of NAD onto monomeric actin, thus inhibiting the polymerization of actin from its globular form to the filamentous form. In this regard, iota toxin is similar to another binary toxin, C2 produced by some strains of *C. botulinum* type C, although an association between C2 and enteric disease has not been as well defined. An unusual strain of *C. difficile* has been identified that produces an ADP-ribosylating enzyme which cross-reacts with $iota_a$. However, the strain does not produce an $iota_b$ component and the $iota_a$ component, by itself, is not toxic.[19]

The iota toxins of *C. perfringens* and *C. spiroforme* are very similar in size and they exhibit extensive cross-reactivity and cross-neutralization. The components can be interchanged and continue to be active. Similarly, the ADP-ribosylating enzyme of *C. difficile* is toxic when mixed with the binding component of either iota toxin. The C2 binary toxin of *C. botulinum*, however, does not cross-react and is not neutralized by antiserum against iota toxin. In addition, its active and binding components cannot be used interchangeably with those of iota toxin.

The $iota_a$ and $iota_b$ genes of *C. perfringens* have been sequenced and both components appear to be regulated by the region upstream of the $iota_a$ gene.[20] Both components have a signal sequence that is cleaved during the secretory process. The $iota_b$ component may require additional processing to become fully functional. There are several additional cleavage sites that are candidates for further processing of the $iota_b$ propeptide to a mature fully active peptide. After binding, $iota_b$ enhances the uptake of $iota_a$, resulting in the full expression of the toxic activity. The ADP-ribosylation of monomeric actin by $iota_a$ is believed to be the mechanism by which the toxin causes cell rounding.

5.3. Immunity

Animals that are vaccinated against crude or highly purified preparations of iota toxin develop antibodies against both components. The antibodies are protective against iota toxin-mediated disease. No information is available on the immune response to iota toxin in animals with *C. perfringens* or *C. spiroforme* enterotoxemia. It has been established, however, that vaccination against iota toxin protects animals against the disease, indicating that iota toxin functions as an important virulence factor. Therapeutic measures against iota toxin-mediated disease have not been established other than treatment for the symptoms.

6. *BACTEROIDES FRAGILIS* ENTEROTOXIN

6.1. Description

B. fragilis, which is the anaerobic species most commonly isolated from human clinical specimens, causes abscesses, soft-tissue infections, and bacteremias. This organism also is a normal inhabitant of the human colonic flora comprising about 1 to 2% of the flora, and this is the source from which it infects tissues. The most studied virulence factor of *B. fragilis* is a thin capsular polysaccharide layer that appears to protect the organism from phagocytosis,[9] enhancing the ability of the organism to cause soft-tissue infections. Recent evidence demonstrating the production of an enterotoxin raises new clues as to the virulence mechanisms of this organism and its possible role in diarrheal illness. Thus, enterotoxigenic *B. fragilis* may soon represent a "new" addition to the growing list of enteric pathogens.

The organism has been associated with diarrheal disease in farm animals, including lambs, calves, piglets, and foals, and evidence is accumulating that shows a correlation between the presence of *B. fragilis* in stool specimens and outbreaks of diarrhea in humans.[21] It is estimated from studies on hundreds of isolates from clinical material, intestinal contents, and raw sewage that about 10% of all *B. fragilis* strains are enterotoxigenic.

Several animal models such as the injection of organisms into ligated ceca and oral challenge have been used to study *B. fragilis*-associated intestinal disease in rabbits and lambs. These animal models, however, are expensive, complicated to perform, and tedious. More recently, it has been shown that *B. fragilis* enterotoxin is cytotoxic for certain colon carcinoma cell lines, particularly HT29/C1 cells, which develop morphological changes that include loss of cell-to-cell attachment, rounding, swelling, and pyknosis.[22] The tissue culture assay makes it much easier to detect and quantitate the toxin.

6.2. Epidemiology

Little is known about the epidemiology of *B. fragilis* diarrheal disease. However, there have been numerous studies on *B. fragilis* as part of the normal microflora of the colon and some inferences may be made from these studies. In adults, the growth of enterotoxigenic *B. fragilis* and production of high levels of toxin in the colon likely is affected by competition with other major colonic species, including the more numerous *B. vulgatus*, *B. uniformis*, and *B. stercoris*. In neonates and infants, the story may be quite different. Initial colonization of the large bowel typically involves facultative organisms followed within days by high levels of *Bifidobacterium* species. In some infants, however, high levels of *Bacteroides* species become established. Therefore, it is possible that colonization by enterotoxigenic strains of *B. fragilis* prior to the establishment of a balanced protective flora may lead to the onset of intestinal disease.

6.3. Pathogenesis

The *B. fragilis* enterotoxin has been purified to homogeneity from culture filtrates by sequential ammonium sulfate precipitation, anion-exchange chromatography, and hydrophobic chromatography, and has been partially characterized[23] (Table I). The secreted toxin, which is a single polypeptide with a molecular weight of about 20,000, induces the same type of fluid accumulation in ligated lamb ileal loops and morphological changes in colon carcinoma cell lines as crude culture filtrates from enterotoxigenic strains. In culture filtrate, the enterotoxin is highly susceptible to inactivation by proteases and is protected by inhibitors of serine proteases. In its pure form, the enterotoxin undergoes autolysis but is completely protected by the incorporation of chelating agents or dithiothreitol. Recently, neutralizing antisera to crude and purified toxin were prepared against the enterotoxin and have been used to develop a cytotoxicity neutralization assay as a confirmatory test. In addition, the antisera have been used to develop an indirect enzyme immunoassay specific for the toxin. Together, these tests have been used to show that enterotoxigenic strains produce a high amount of membrane-associated components that are immunologically related to the enterotoxin. Studies are under way to determine if these cross-reactive components are inactive enterotoxin. The enzyme immunoassay offers some advantages over the tissue culture assay because it is capable of detecting inactive or membrane-associated toxin. Thus, although additional studies are needed, the immunoassay may be more broadly applicable for the clinical detection of the enterotoxin and enterotoxigenic *B. fragilis*, and for epidemiological studies.

The gene encoding the enterotoxin has been sequenced in part and sequence analysis has revealed that the enterotoxin exhibits high levels of homology with zinc metalloproteases. These observations have now been extended to show that the highly purified enterotoxin cleaves certain types of proteins, including gelatin and actin. These results indicate that the enterotoxin is a metalloprotease and suggest a possible mechanism of action.[24]

7. CONCLUSIONS

The ability of diverse anaerobic pathogens such as *C. difficile*, *C. perfringens*, *C. spiroforme*, and *B. fragilis* to cause enteric disease results from the ability of the organisms to grow in the complex ecosystem of the intestine and produce toxins capable of attacking and damaging the gut mucosa. The toxins produced by these bacteria represent a diverse group of biologically active proteins that vary widely in their physicochemical and functional properties. We now are beginning to understand the mechanisms of some of these toxins at the molecular level. Some are ADP-ribosylating enzymes, some affect eukaryotic factors that are involved in the regulation of the cytoskeletal system, and with others, the mechanism is unknown. Very little is known about the immune response to the toxins. Continued research into the mechanism of action of these toxins and how our bodies respond immunologically during these diseases may lead to better vaccines and treatments, and to a better understanding of how these unusual enteric pathogens cause disease.

The following references are recommended for additional information on the role of clostridia and their toxins in intestinal disease:

Borriello, S. P., 1985, *Clostridia in Gastrointestinal Disease*, CRC Press, Boca Raton, FL.

Finegold, S. M., and George, W. L., 1989, *Anaerobic Infections in Humans*, Academic Press, San Diego.

Hatheway, C. L., 1990, The toxigenic clostridia, *Clin. Microbiol. Rev.* **3:**66–98.

McClane, B. A., Hann, P. C., and Wnek, A. P., 1988, *Clostridium perfringens* enterotoxin, *Microb. Pathog.* **4:**317–323.

Rolfe, R. D., and Finegold, S. M., 1988, *Clostridium difficile: Its Role in Intestinal Disease*, Academic Press, San Diego.

Rood, J. I., and Cole, S. T., 1991, Molecular genetics and pathogenesis of *Clostridium perfringens*, *Microbiol. Rev.* **55:**621–648.

Sebald, M., 1992, *Genetics and Molecular Biology of Anaerobic Bacteria*, Springer-Verlag, Berlin.

REFERENCES

1. Rolfe, R. D., and Finegold, S. M., 1988, *Clostridium difficile: Its Role in Intestinal Disease*, Academic Press, San Diego.
2. McFarland, L. V., Mulligan, M. E., Kwok, R. Y. Y., and Stamm, W. E., 1991, Nosocomial acquisition of *Clostridium difficile* infection, *N. Engl. J. Med.* **320:**204–210.
3. Lyerly, D. M., and Wilkins, T. D., 1991, Toxins of anaerobes, in: *Anaerobic Microbiology, a Practical Approach* (P. N. Levett, ed.), IRL Press, New York, pp. 163–181.
4. Just, I., Fritz, G., Aktories, K., Giry, M., Popoff, M., Boquet, P., Hegenbarth, S., and von Eichel-Streiber, C., 1994, *Clostridium difficile* toxin B acts on the GTP-binding protein *rho*, *J. Biol. Chem.* **269:**10706–10712.
5. Dove, C. H., Wang, S.-Z., Price, S. B., Phelps, C. J., Lyerly, D. M., Wilkins, T. D., and Johnson, J. L., 1990, Molecular characterization of the *Clostridium difficile* toxin A gene, *Infect. Immun.* **58:**480–488.
6. Barroso, L. A., Wang, S.-Z., Phelps, C. J., Johnson, J. L., and Wilkins, T. D., 1990, Nucleotide sequence of *Clostridium difficile* toxin B gene, *Nucl. Acids Res.* **18:**4004.
7. Martinez, R. D., and Wilkins, T. D., 1992, Comparison of *Clostridium sordellii* toxins HT and LT with toxins A and B of *Clostridium difficile*, *J. Med. Microbiol.* **36:**30–36.
8. Lyerly, D. M., Phelps, C. J., Toth, J., and Wilkins, T. D., 1986, Characterization of toxins A and B of *Clostridium difficile* with monoclonal antibodies, *Infect. Immun.* **54:**70–76.
9. Van Damme-Jongsten, M., Rodhouse, J., Gilbert, R. J., and Notermans, S., 1990, Synthetic DNA probes for detection of enterotoxigenic *Clostridium perfringens* strains isolated from outbreaks of food poisoning, *J. Clin. Microbiol.* **28:**131–133.
10. Duncan, C. L., and Strong, D. H., 1968, Improved medium for sporulation of *Clostridium perfringens*, *Appl. Microbiol.* **16:**82–89.
11. Hanna, P. C., Wieckowski, E. U., Mietzner, T. A., and McClane, B. A., 1992, Mapping of functional regions of *Clostridium perfringens* type A enterotoxin, *Infect. Immun.* **60:**2110–2114.

12. Granum, P. E., and Stewart, G. S. A. B., 1992, Molecular biology of *Clostridium perfringens* enterotoxin, in: *Genetics and Molecular Biology of Anaerobic Bacteria* (M. Sebald, ed.) Springer-Verlag, Berlin, pp. 235–247.
13. Sakurai, J., and Fujii, U., 1987, Purification and characterization of *Clostridium perfringens* beta toxin, *Toxicon* **25:**1301–1310.
14. Finegold, S. M., and George, W. L., 1989, *Anaerobic Infections in Humans,* Academic Press, San Diego.
15. Stiles, B. G., and Wilkins, T. D., 1986, *Clostridium perfringens* iota toxin: Synergism between two proteins, *Toxicon* **24:**767–773.
16. Stiles, B. G., and Wilkins, T. D., 1986, Purification and characterization of *Clostridium perfringens* iota toxin: Dependence on two nonlinked proteins for biological activity, *Infect. Immun.* **54:**683–688.
17. Popoff, M. R., and Bouquet, P., 1988, *Clostridium spiroforme* toxin is a binary toxin which ADP-ribosylates cellular actin, *Biochem. Biophys. Res. Commun.* **152:**1361–1368.
18. Carman, R. J., Van Tassell, R. L., and Wilkins, T. D., 1987, Production of iota toxin by *Clostridium spiroforme*: A requirement for divalent cations, *Vet. Microbiol.* **15:**115–120.
19. Popoff, M. R., Rubin, E. J., Gill, D. M., and Boquet, P., 1988, Actin-specific ADP-ribosyltransferase produced by a *Clostridium difficile* strain, *Infect. Immun.* **56:**2299–2306.
20. Perelle, S., Gibert, M., Boquet, P., and Popoff, M. R., 1993, Characterization of *Clostridium perfringens* iota-toxin genes and expression in Escherichia coli, *Infect. Immun.* **61:**5147–5156.
21. Sack, R. B., Albert, M. J., Alam, K., Neogi, P. K. B., and Akbar, M. S., 1994, Isolation of enterotoxigenic *Bacteroides fragilis* from Bangladeshi children with diarrhea: A controlled study, *J. Clin. Microbiol.* **32:**960–963.
22. Weikel, C. S., Grieco, F. D., Reuben, J., Myers, L. L., and Sack, R. B., 1992, Human colonic epithelial cells, HT29/C1, treated with crude *Bacteroides fragilis* enterotoxin dramatically alter their morphology, *Infect. Immun.* **60:**321–327.
23. Van Tassell, R. L., Lyerly, D. M., and Wilkins, T. D., 1992, Purification and characterization of an enterotoxin from *Bacteroides fragilis, Infect. Immun.* **60:**1343–1350.
24. Moncrief, J. S., and Van Tassell, R. L., personal communication.

9

The Staphylococcal Toxins in Human Disease

MERLIN S. BERGDOLL

1. INTRODUCTION

The staphylococci are involved in many human illnesses and infections, including staphylococcal food poisoning, enterocolitis, pneumonia, septicemia, osteomyelitis, diarrhea, empyema, scalded skin syndrome (SSS), toxic shock syndrome (TSS), boils, infected wounds, and infected incisions. However, most of these illnesses have not been associated with any particular product of the staphylococci, although the staphylococci produce many biologically active products: the enterotoxins (SE), exfoliatin or epidermolytic toxin (ET), toxic shock syndrome toxin (TSST), hemolysins, proteases, coagulases, nucleases, lipase, phosphatase, and others. Only food poisoning[1] and pseudomembranous enterocolitis (SE),[2] SSS (ET),[3] and TSS (TSST)[4,5] have been associated with particular staphylococcal toxins.

2. THE STAPHYLOCOCCAL ENTEROTOXINS (SE)

Although the SEs are potent toxins, their involvement in staphylococcal diseases has been inadequately investigated. This may be because their reactions in food poisoning are relatively mild and of short duration, a few hours to 1 day in most cases. Food poisoning differs from other staphylococcal diseases because it results from ingestion of the SE with food instead of resulting from staphylococcal infections.

MERLIN S. BERGDOLL • Food Research Institute, University of Wisconsin, Madison, Wisconsin 53706.

Enteric Infections and Immunity, edited by Lois J. Paradise *et al.* Plenum Press, New York, 1996.

Table I
Symptoms of Staphylococcal Food Poisoning in 122 Cases[a]

Symptom	Cases[b]	No reaction	Mild reaction	Severe reaction
Vomiting	122	15	12	95
Pains in abdomen	122	6	40	76
Diarrhea	103	13	75	15
Headache	101	29	59	13
Muscular cramping	113	41	58	14

[a]Includes 94 students, 8 lunchroom supervisors who took some cream puffs home, and 20 from cream puffs sold at three cafes and from the bakery truck.
[b]Cases from which this information was available.

2.1. Staphylococcal Food Poisoning

2.1.1. Food Poisoning Symptoms

The most observable symptoms are vomiting and/or diarrhea, with most victims experiencing both symptoms (Table I).[(6)] The more severe cases may result in prostration and hospitalization. Deaths do occur occasionally in children or in older individuals. Normally, recovery occurs within a few hours with no sequelae. Because of the relative mildness of this disease, the SEs generally have been considered relatively impotent. However, the amount of SE required to cause illness in sensitive individuals indicates that these toxins are very potent. Chocolate milk that was the cause of several hundred school children becoming ill contained 100 to 200 ng/half-pint, the amount of milk that the majority of the children drank (Table II).[(7)] Although it is expected that a smaller amount of SE would be required to

TABLE II
Enterotoxin A Analysis in 2% Chocolate Milk

	Enterotoxin A	
Sample	ng/ml	ng/$\frac{1}{2}$ pint
1	0.40	94
2	0.73	172
3	0.48	113
4	0.63	149
5	0.78	184
6	0.65	153
Av.	0.61	144

cause illness in children than in adults, some of the victims were 19 years of age. Those who drank more than one half-pint of the milk were more seriously ill than those who drank only one half-pint.

Staphylococcal food poisoning is a worldwide disease, but it is not possible to determine how frequently it occurs because in many countries the illnesses are not reported and records are not kept. For example, only in recent years has attention been given to this disease in Brazil. Although there was interest in the staphylococci and their production of SEs, there was little interest in the investigation of food poisoning outbreaks. The investigations of these outbreaks were undertaken with the assistance of M. S. Bergdoll.[8] The major foods involved in Brazil were cream-filled cake and a type of white cheese made on farms and by small producers. Any food that provides a good medium for the growth of staphylococci may be involved in this type of food poisoning. Refrigeration of vulnerable foods is the best preventive measure.

2.1.2. *Intravenous Potency of the SE*

The emetic symptoms in monkeys require about 1 to 2 μg/kg when given intragastrically to rhesus monkeys, whereas only 20 ng/kg is required to produce an emetic reaction when administered intravenously.[1] However, much larger intravenous doses in monkeys produce a multitude of symptoms (Table III). The symptoms are similar to those observed in TSS.[9]

2.1.3. *SE in TSS*

Although most TSS cases are caused by another toxin, TSST, there is evidence that the SEs, particularly enterotoxin B (SEB), can cause TSS, primarily in nonmenstrually related cases (Table IV).[10] All types of staphylococcal infections can result in TSS. Although known toxins are not responsible for these infections, if the staphylococci associated with the infection produce a toxin such as SE, TSS can result. Pneumonia from staphylococci producing SEB resulted in TSS in a number of patients

TABLE III
Activity of Enterotoxin B Intravenously Given to Monkeys

Fever, Vomiting, Diarrhea, Hypotension, Hyperfibrinogenemia, Pulmonary edema, Increase in heart rate, Thrombocytopenia, Initial leukopenia followed by a neutrophilic leukocytosis, Decrease in urine output, Pooling of blood in vascular beds, Elevation of blood urea nitrogen, Gradual decrease in serum proteins, Increase in serum creatinine, Increase in glutamic oxaloacetic transaminase, Evidence of endothelial cell degeneration, Shock and death

TABLE IV
Production of Toxins by *S. aureus* in TSS Patients

	Menstural TSS (300)		Nonmenstrual TSS (80)	
Toxin	No.	%	No.	%
TSST (total)	280	93.3	68	85.0
TSST	115	38.3	22	27.5
TSST, SEA	142	47.3	33	41.3
TSST, SEC	12	4.0	13	16.3
TSST, SEA, SEC	8	2.7	0	0.0
TSST, other	3	1.0	0	0.0
SEB	10	3.3	9	11.3
SEC	1	0.3	1	1.3
Negative	9	3.0	2	2.5

originally ill with influenza.[11] Apparently this was not a new experience because in the late 1950s a number of deaths occurred among elderly people with staphylococcal pneumonia following influenza. Many of the staphylococcal strains isolated from these patients produced SEB, which could have been the cause of death. At that time TSS was an unknown disease. Enterotoxin C (SEC) also has been involved in respiratory disease. The staphylococci from six of eight patients produced SEC and TSST, and all six died; the other two patients did not die even though the staphylococci from one produced SEA and TSST and the other produced TSST.

2.1.4. *SE in Pseudomembranous Enterocolitis*

Surgalla and Dack[2] in the early 1950s did associate SE production by staphylococcal strains isolated from the stools of surgical patients ill with pseudomembranous enterocolitis. The majority of these strains (51 of 58) produced identified SEs with 80.9% of the enterotoxigenic ones producing SEA and SEB; a few produced only SEA. SEB was produced in relatively large amounts by many of the strains and was the major cause of the many deaths that resulted from this illness. The symptoms were typical of those of severe staphylococcal food poisoning: vomiting, diarrhea, collapse, and death.

This illness is not common, requiring a special set of circumstances for it to occur. Although a few similar illnesses have subsequently been identified, the ones mentioned above occurred as a result of the advent of antibiotic use. These were administered to patients undergoing surgery to prevent infection; however, it was not recognized at the time that staphylococci could become resistant to antibiotics. The antibiotics eliminated the natural flora of the intestinal tract without destroying the antibiotic-resistant staphylococci. Under normal conditions, staphylococci will not grow in the intestinal tract because they cannot compete with the natural flora. Once the latter was destroyed, the staphylococci (isolated in pure culture in many cases) grew uninhibited and produced large amounts of SE. That SE was the causative agent was demonstrated in dogs and chinchillas. Antibiotic-treated animals were administered both enterotoxigenic and nonenterotoxigenic staphylococci, but only those that were given the enterotoxigenic staphylococci developed enteritis.(12) Also, enteritis could be produced by administering pure SEB intragastrically, the severity depending on the amount of SEB administered.(13)

Although enterocolitis does occur during staphylococcal food poisoning, it is less severe because the amount of toxin is small and occurs in one dose rather than continuously. It is rare that the conditions exist for the staphylococci to grow in the intestinal tract; hence, cases similar to the earlier ones are seldom seen. There have been cases of food poisoning where a secondary reaction occurred some hours later. This may have been related to the staphylococci from the food poisoning growing in the intestinal tract and producing SE. How this could have happened is not known. However, if antibiotics had been administered in treating the food poisoning, this could be the cause. One fatality did occur when a young man experiencing food poisoning was treated with antibiotics.

2.1.5. *SE in Staphylococcal Diarrhea*

Involvement of the SEs in other staphylococcal diseases is difficult to verify because in most cases the symptoms of the infections vary and cannot be correlated with SE production by the staphylococci responsible for the infections. However, many cases of diarrhea in hospitals, particularly among babies, are the result of staphylococcal infections. Little information is available about SE production of the staphylococci involved, primarily because in the United States very few strains have been examined for SE production. Investigators in other countries have examined staphylococci isolated from stools from diarrheal patients for SE production. Zak *et al.*(14) examined 365 *S. aureus* strains collected mostly from diarrheal patients in the late 1960s in Poland. Over 75% of these strains produced SE with the majority (75.7%) producing SEB. In a similar study in Hungary in the 1970s, Osvath-Marton *et al.*(15) reported that 72% of the staphylococcal strains isolated from diarrheal patients (mostly children) were enterotoxigenic, with the majority producing SEA. Although the diarrhea may be related to more than one factor, SE is likely one of the factors because diarrhea is a common symptom in food poisoning.

2.1.6. *SE in Staphylococcal Sepsis*

Staphylococcal sepsis is normally a serious disease with death resulting in many cases. In this instance any products produced by the staphylococci do not need to diffuse into the circulation as they are already there. The disease probably is caused by a combination of factors rather than any particular one. However, if the staphylococci produce either an SE or TSST, the effects of the illness may be more severe than otherwise. Studies of staphylococci isolated from sepsis patients in Madison (Wisconsin) and Germany show that approximately 38% of the strains isolated were enterotoxigenic. In addition, two of the strains in the Madison study and four in the German study produced TSST. There is no information available to relate SE production to any of the symptoms observed in the sepsis patients, but if SE is produced, it is expected to contribute to the symptoms. In one hospital, two sepsis patients, from whom staphylococcal strains that produced SEA and TSST were isolated, died. Because of the early death, it was not possible to record the symptoms that might have indicated the patient was suffering from TSS.

2.1.7. *SE in Hospital Staphylococcal Infections*

It is not unusual for patients entering hospitals to develop staphylococcal infections because these organisms are common inhabitants of hospitals. Examination of staphylococci isolated from outbreaks of infections in three different hospitals showed 27 of 28 of these cultures to produce SEA, the SE produced most widely by human strains of staphylococci. It is likely that a common strain of staphylococci was responsible for each of the outbreaks.

3. STAPHYLOCOCCAL TOXIC SHOCK SYNDROME TOXIN (TSST)

Another toxin known to be involved in a particular staphylococcal disease is TSST, the toxin responsible for the majority of cases of TSS.[4,5] Many TSS cases are associated with the use of tampons during menstruation, although TSS can result from all types of staphylococcal infections.[16] Even though TSST is not responsible for the infections, if it is produced by the staphylococci involved, TSS may result.

It was not until 1980 that TSST was isolated and identified as a new toxin,[4,5] despite having been produced by the staphylococci decades earlier. The disease came into prominence when it was identified as the cause of illness in young women using tampons during menstruation.[17] Although the identification of TSS as a new disease was a result of staphylococcal illnesses among young people with a common set of symptoms, it was not related to any particular type of staphylococcal infection.[18] Undoubtedly, this disease had been around for a long time, but was not recognized as a specific disease. Today, essentially half of all TSS cases are the result of nonmenstrual staphylococcal infections.

3.1. Symptoms of TSS

3.1.1. Major Criteria for TSS

One of the major early symptoms of TSS is fever with a temperature of at least 38.9°C, which is one of the four major criteria of the strict definition of TSS.[9] The other criteria are: (1) diffuse macular erythroderma; (2) desquamation 1–2 weeks after onset of illness, primarily of the palms, soles, fingers, and toes; and (3) hypotension: systolic blood pressure $\leqslant$ 90 mm Hg for adults and < 5th percentile for those under 16 years of age; syncope or orthostatic dizziness. In addition to these four criteria, involvement of three or more of the following organ symptoms is required: (1) gastrointestinal: vomiting or diarrhea at onset of illness; (2) muscular: severe myalgia or creatinine phosphokinase level greater than twice the upper limit of normal; (3) mucous membrane: vaginal, oropharyngeal, or conjunctival hyperemia; (4) renal: BUN or serum creatinine greater than twice the upper limit of normal, or equal to or greater than 5 white blood cells per high-power field in the absence of a urinary tract infection; (5) hepatic: total bilirubin, SGOT, or SPGT greater than twice the upper limit of normal; (6) hematologic: platelets equal to or greater than 100,000/mm^3; and (7) central nervous system: disorientation or alterations in consciousness without focal neurologic signs when fever and hypotension are absent.

3.1.2. Early Symptoms of TSS

In addition to the high temperature, early symptoms also include chills, headache, nausea, possibly vomiting, sore and tender mouth and throat, generalized myalgias, muscle tenderness and weakness, abdominal pain, malaise, and profuse diarrhea.[19] A decrease in vasomotor tone as measured by a low systemic vascular resistance appears early and results in the pooling of blood in the peripheral vasculature with vascular congestion in organs and tissues, probable relaxation of the microcirculation, and poor venous return.

3.1.3. Later Symptoms of TSS

Over the course of the next 24 to 48 hr, edema of the face, hands, and feet, neuralgic pain in the joints, particularly in the wrists, knees, ankles, fingers, and toes, erythroderma, conjunctival injection, cough, dizziness, syncope, incontinence, and oliguria may develop. In the case of severe disease, diffuse encephalopathy, with confusion, agitation, disorientation, irritability and combativeness, lethargy, hallucinations, and possibly unresponsiveness may be observed. Because TSS is a toxin-mediated disease, the sequence of appearance and the progression of symptoms are predictable.

All of the oral mucous membranes including the tongue will be intensely erythematous and about 25% of patients will have a strawberry tongue. In menstruating patients, the cervicovaginal mucous membranes will be intensely red, with a purulent menstrual discharge in half of the patients. The decrease in circulating blood volume, interstitial edema compressing small vessels, and organ congestion result in ischemia and multisystem organ involvement, with ultimate failure. It is not clear whether this multisystem organ involvement is related exclusively to ischemia or to a direct toxin or mediator-induced damage as well.

3.1.4. *Appearance of Symptoms of TSS*

One remarkable aspect of TSS is the suddenness with which the symptoms appear and progress in previously healthy individuals. In menstrual cases, the symptoms do not occur before the second or third day after the beginning of menstruation, primarily because it takes that amount of time for sufficient growth of staphylococci to occur in the vagina to produce TSST. However, regarding surgical procedures in which the incision becomes contaminated with toxin-producing staphylococci, the onset of the symptoms occurs as early as 8 hr. This is because the organisms grow much more rapidly in the incision than in the vagina.

3.1.5. *Postoperative TSS*

Apparently, contamination of incisions with staphylococci during or following the operation is not unusual as cases of TSS have been reported from all types of operations.[20] It is difficult to determine the source of the staphylococci because adequate culturing of hospital personnel is seldom done. In some cases, particularly when more than one patient was operated on by the same surgeon, it has been demonstrated that the surgeon was colonized with the same staphylococcal strain isolated from the incisions.[21]

In postoperative cases of TSS, the wound invariably appears normal and uninflamed despite the presence of toxin-producing bacteria. Traditional findings of swelling, erythema, tenderness, and pus are rarely present. In nonmenstrual, and nonpostoperative patients, a rigorous search should be made for the source of the infection with attendant toxin production. The importance of identifying and draining the focus of infection is obvious, because even small undrained residual foci of infection can perpetuate the course.

3.1.6. Emesis and Diarrhea in TSS

The fact that the early symptoms may include vomiting and/or diarrhea is of interest because TSST does not produce an emetic reaction in monkeys as do the SEs, even when given intravenously. One explanation for the emetic reaction is that many strains produce both an enterotoxin and TSST and it is the enterotoxin that is producing the emetic reaction. This also may be true of the diarrheal reaction, although the injection of purified TSST into baboons resulted in diarrhea in all five animals, whereas only two of the animals vomited.(22)

3.1.7. Similarity of Symptoms of TSS to Other Diseases

The initial symptoms of fever, vomiting, and diarrhea are similar to those of influenza and sometimes the illness is mistakenly diagnosed as influenza. Also, the gastrointestinal symptoms are similar to those of staphylococcal food poisoning; however, the fever and severe myalgias are not associated with food poisoning. Although diarrhea is characteristic of food poisoning, that accompanying TSS is a profuse secretory diarrhea and of several days' duration.

4. EXFOLIATIN OR EPIDERMOLYTIC TOXIN (ET)

Another toxin produced by the staphylococci related to a particular disease is ET, which is the cause of SSS,(3) primarily a disease of very young children. Although the disease is caused by ET, the staphylococci (primarily belonging to phage group II) also produce an unidentified SE. Attempts to purify this toxin were unsuccessful; it appeared to be more labile than the identified SEs. Whether it has a role in SSS is not known. The laboratory workers producing the ET experienced desquamation of the palms of the hands, a similar experience of those purifying the SE. The effects of this disease are primarily external.

Staphylococcal Scalded Skin Syndrome (SSS)

The term SSS is applied to a combination of staphylococcal skin infections such as Ritter's disease or toxic epidermal necrolysis, bullous impetigo, and a scarlatiniform rash.(3) Winternitz(23) was the first to show that Ritter's disease was a staphylococcal disease, with numerous investigators providing additional evidence that SSS is a staphylococcal disease.

SSS Symptoms

The symptoms observed vary with the different forms of the disease. Ritter's disease is characterized by erythema that begins around the mouth and spreads over the entire body within 2 days,[3] which includes exudative lesions with yellow crusty deposits. The disease may begin with a purulent conjunctivitis or upper respiratory infection. After gentle stroking, the skin becomes irreversibly wrinkled and can be displaced with slight pressure (positive Nikolsky sign). Large flaccid bullae filled with clear fluid appear, with the result that large areas of the epidermis peel revealing moist red glistening corium beneath. The exposed areas dry and develop into large seborrhea-like flakes, which continues for several days with rapid replacement of epithelium. Recovery is complete within 7 to 10 days.

Staphylococcal bullous impetigo differs from Ritter's disease in that it is characterized by flaccid localized bullous lesions on normal-appearing skin. A positive Nikolsky sign is absent. Lesions rupture spontaneously revealing a moist red surface. Exfoliation is limited to the margin of the original bulla.

Staphylococcal scarlatiniform rash produces a generalized erythema similar to Ritter's disease, but with no bullae or exfoliation. After 1 to 2 days, cracks appear in the skin creases that become covered with thick flakes of dried skin that nearly cover the entire epidermis. Desquamation of these flakes reveals healed skin beneath.

5. UNKNOWN STAPHYLOCOCCAL TOXINS

The fact that the staphylococci are involved in many diseases and infections indicates that unidentified, biologically active compounds may be produced by these organisms. For example, the identification of TSST as late as 1981, even though it was being produced many years before, indicates that other toxins may be produced by the staphylococci. Only when they can be associated with some specific biological activity will they be isolated and identified.

6. STAPHYLOCOCCAL TOXINS AS SUPERANTIGENS

It is well known that the SEs are among the most mitogenic of known compounds.[24] Other products of the staphylococci are also mitogenic, with

TSST being similar to the SEs in its lymphocyte mitogenicity.[(25)] This activity was observed when the toxins stimulated rabbit splenocytes to proliferate, as measured by incorporation of [^{3}H]thymidine into DNA. Lymphocytes isolated from both murine and rabbit spleens and human peripheral and cord blood proliferated nonspecifically *in vitro*. Proliferation in the mouse and human systems depended on the presence of macrophages in the lymphocyte populations. Proliferation was achieved with 0.1 ng TSST for murine and rabbit splenocytes and 0.01 ng TSST for human mononuclear cells (Table V). The reactivity resided solely in the T-cell populations.

The fact that the SEs and TSST reacted with many more T cells than did conventional mitogens, such as the plant lectins, phytohemagglutinin (PHA) and concanavalin (Con A), resulted in their designation as superantigens.[(26)] Further study revealed that unlike other antigens, the attachment of SEs and TSST to T cells does not require processing of the protein molecules for them to be presented to the T cells by major histocompatibility complex (MHC) class II molecules.[(27)] Neither do they attach to the T-cell receptor (TCR) in the same manner as conventional antigens, which attach in the groove between the α and β chains of the TCR. Instead, the superantigens attach to variable (V) portions of the β chain (V_β) at the side of the receptor, with each superantigen attaching to specific V_β elements of the TCR. Conventional antigen attachment to T cells is highly specific and very limited, whereas the superantigens attach to essentially all T cells that presented the V_β elements specific for the superantigen in question. Further, the superantigens stimulate both $CD4^+$ and $CD8^+$ T cells, whereas conventional antigens stimulate only $CD4^+$ T cells.

The superantigens, particularly the SEs and TSST, have been the subjects of intensive research in recent years, primarily to obtain a better

TABLE V
Proliferation of Lymphocytes by TSST in Mononuclear Cells from Human PBMs, Murine and Rabbit Splenocytes[25]

	[^{3}H]thymidine (cpm)		
TSST (ng)	Human PBMs	Murine splenocytes	Rabbit splenocytes
10.0	77,708 ± 4,783	367,717 ± 17,050	256,302 ± 16,325
1.0	91,274 ± 3,252	225,257 ± 20,465	252,836 ± 14,774
0.1	58,933 ± 5,119	43,536 ± 6,182	171,493 ± 41,003
.01	17,702 ± 813	3,749 ± 684	9,059 ± 1,143
0.001	3,442 ± 297	4,265 ± 621	10,212 ± 1,761
0.0	1,003 ± 232	2,386 ± 329	6,389 ± 449

understanding of the immune system. Review of this research is beyond the scope of this presentation, except where it possibly involves understanding the role of the staphylococcal toxins in disease. Some research has been done in this area, but at the present time the function of the SEs and TSST in their respective diseases is inadequately understood.

6.1. Superantigen Effect on Antibody Production

In contrast to the finding that tetanus toxoid, mumps, and herpes zoster activity was limited to $CD4^+$ cells,[28] TSST also stimulated $CD8^+$ cells; approximately 25% of all blasts were $CD8^+$.[29] The binding to $CD8^+$ cells is of particular importance because the functional activation of these cells inhibits immunoglobulin production. Apparently, the $CD8^+$ cells are responsible for TSST-induced suppression of the response of immunoglobulin-secreting cells to pokeweed mitogen.[30] This may be one explanation why a high percentage of TSS patients do not develop antibodies to TSST during convalescence.[16] Also, SEB can induce multiple T-suppressor cell populations that inhibit both primary and secondary plaque-forming cell responses.[31]

6.2. Cytokine Production by Superantigens

Most of the early research on the production of cytokines by the staphylococcal toxins was done with TSST; the work reported here will be that done with this toxin, although the results can apply to the SE as well. In this case the disease used as the example is TSS because it has been extensively studied. This could also apply to any disease resulting from the SEs contaminating the blood; however, it is questionable whether it would apply to staphylococcal food poisoning, although it has been speculated that it might.[27]

6.2.1. Cytokine Production by TSST

The cause of TSS has not been completely elucidated, primarily because no animal model has been found that elicits all of the signs and symptoms observed in TSS. Although TSST appears to be a very potent toxin, it is doubtful that it is directly responsible for the signs and symptoms observed in TSS. It is more likely that the cytokines that it stimulates the production and/or release of in animals and humans—such as interleukin 1 (IL-1),[32,33] IL-2,[34,35] tumor necrosis factor (TNF),[36,37] interferon γ (INF-γ),[36,37] and others—produce the signs and symptoms observed in TSS.

6.2.2. *Activity of IL-1*

The evidence for involvement of IL-1 in TSS appears to be adequate as IL-1 given to animals produces a number of signs and symptoms similar to those observed in TSS.[38] Activities of IL-1 include the modulation of lymphokine production and lymphocytic reactivity caused by IL-2 release, neutrophil mobilization from the bone marrow, hypoferremia, muscle proteolysis, and acute-phase protein synthesis.[38] These activities of IL-1 may explain much of the pathogenesis of TSS. The action of IL-1 on various cells may be mediated by an increase in arachidonic acid, followed by increases in prostaglandins and leukotrienes, which are highly active substances that may contribute to other aspects of the syndrome, such as diarrhea and hypotension.

Culture filtrate fluids from TSS strains produced biphasic fevers in rabbits similar to those induced by other exogenous pyrogens.[39] The major fever occurring several hours later was prolonged, with IL-1 being detectable at its peak. An earlier febrile peak (1½ to 2 hr after injection) may be related to the direct effect of TSST on the thermoregulatory center. Plasma taken from rabbits at this point produced no febrile response in recipient rabbits, indicating the toxin had been cleared from the circulation. Other exogenous pyrogens are rapidly cleared after intravenous injection and can be found in the reticuloendothelial system, the major source of IL-1. It would appear from the IL-1 production by staphylococci isolated from TSS patients that it may be an essential factor in the development of TSS.

The amount of IL-1 generated by staphylococcal culture filtrate fluids is directly related to the amount of TSST produced.[40] If this is also valid for the SEs, one might expect that SEB-producing strains would be more frequently involved in TSS than are strains producing the other SEs, because SEB is produced in much larger amounts than SEA or SED. SEC also is produced in relatively large amounts, but is seldom involved in TSS because of the high percentage of individuals who have protective antibody titers to this toxin.[10]

6.2.3. *Activity of IL-2*

Although IL-2's main activity appears to be that of a growth factor for activated T cells, it may cause T cells to secrete a variety of other lymphokines. One such substance is IFN-γ, which has a marked effect on a variety of immune functions including natural killer (NK) cell activity, generation of cytolytic T cells, macrophage activation, and modulation of the expression of histocompatibility antigens. IL-2 is capable of inducing T cells to secrete a B-cell growth factor which apparently can contribute to both cell cycle transitions and lymphokine secretion. Thus, IL-2 appears to be important in the modulation of immune responses. The fact that TSST can stimulate the production of large amounts of IL-2 in human peripheral blood lymphocytes may be important in the induction of harmful responses. Because many activities associated with IL-2 *in vitro* are similar to those observed in TSS, IL-2 may be an important contributor to the signs and symptoms observed in TSS.

6.2.4. *Activity of TNF*

The role that TNF may play in TSS, if any, is not understood. One major effect of TNF is as a pyrogen; an intravenous injection of 1 μg/kg rTNF-α into rabbits produced monophasic fever with peak elevation after 45–55 min.[41] The only other well-defined substance of leukocytic origin to produce a rapid increase in core temperature is IL-1. Because TNF is produced by stimulating leukocytes, it must be considered an endogenous pyrogen. Other roles that TNF may play in TSS are unknown, but one possibility is the inhibition of migratory functions of PMN.[42] The failure of staphylococci to elicit a purulent response from the host in surgical wound infection resulting in TSS may be related to this effect. Many of these infections were not detected until after the patient developed TSS. Incubation of supernatant fluids from mononuclear leukocytes with TSST resulted in potent inhibition of both PMN random and chemotactic migratory functions. Neutralization of the activity with anti-TNF antiserum identified the factor as TNF.

TNF and IL-1 (2 μg/kg TNF and 2–5 μg/kg IL-1) can induce a sharp drop in mean arterial pressure (MAP) in the rabbit within 10–15 min.[43] Associated with this decrease in the systemic vascular resistance (SVR) and central venous pressure (CVP) is an increase in cardiac output and heart rate. Intravenous injection of 2 μg/kg TSST into rabbits induced significant decreases in MAP and SVR and an increase in CVP. A dose of 5 μg/kg resulted in a shocklike state followed by death. High levels of TNF were detectable in the serum. Significant levels of TNF could be demonstrated in both rabbit and human mononuclear cells after treatment with as little as 0.1 ng/ml TSST. Combined administration of TNF and IL-1 (1 μg/kg each bolus and 5 ng/kg per min infusion for 2 hr) into the rabbit resulted in a shocklike state with a more profound and sustained fall in MAP than when either was given singly.

7. EPILOGUE

The staphylococci are very common microorganisms, and can be found almost everywhere. The genus is composed of over 30 species, most of which are coagulase-negative and less likely to be involved in disease. Although the major species, *S. aureus*, is the most virulent of the species, it is a common inhabitant of the human body without causing any problems. Only when it finds a suitable environment, such as a cut, wound, or incision, in which to grow and produce its toxic proteins, does it cause problems. Its adaptability to antibiotics has made it one of the most difficult organisms to control, particularly in a hospital environment. It likely is the most studied organism and will continue to be so because of its involvement in many diseases and infections.

REFERENCES

1. Bergdoll, M. S., 1989, *Staphylococcus aureus*, in: *Foodborne Bacterial Pathogens* (M. P. Doyle, ed.). Dekker, New York, pp. 463–523.
2. Surgalla, M. J., and Dack, G. M., 1955, Enterotoxin produced by micrococci from cases of enteritis after antibiotic therapy, *J. Am. Med. Assoc.* **158**:649–650.
3. Rogolsky, M., 1979, Nonenteric toxins of *Staphylococcus aureus*, *Microbiol. Rev.* **43**:320–360.
4. Bergdoll, M. S., Crass, B. A., Reiser, R. F., Robbins, R. N., and Davis, J. P., 1981, A new staphylococcal enterotoxin, enterotoxin F, associated with toxic-shock syndrome *Staphylococcus aureus* isolates, *Lancet* **1**:1017–1021.
5. Schlievert, P. M., Shands, K. N., Dan, B. B., Schmid, G. P., and Niahirnura, R. D., 1981, Identification and characterization of an exotoxin from *Staphylococcus aureus* associated with toxic shock syndrome, *J. Infect. Dis.* **143**:509–516.
6. Dennison, G. A., 1936, Epidemiology and symptomatology of staphylococcus food poisoning, *Am. J. Public Health* **26**:1168–1175.
7. Evenson, M. L., Hinds, M. W., Bernstein, R. S., and Bergdoll, M. S., 1987, Estimation of human dose of staphylococcal enterotoxin A from a large outbreak of food poisoning involving chocolate milk, *Int. J. Food Microbiol.* **7**:311–316.
8. do Carmo, L. S., and Bergdoll, M. S., 1990, Staphylococcal food poisoning in Belo Horizonte (Brazil), *Rev. Microbiol. Sao Paulo* **21**:320–323.
9. Centers for Disease Control, 1980, Follow-up on toxic-shock syndrome, *Morbid. Mortal. Weekly Rep.* **29**:441–445.
10. Crass, B. A., and Bergdoll, M. S., 1986, Involvement of staphylococcal enterotoxins in nonmenstrual toxic shock syndrome, *J. Clin. Microbiol.* **23**:1138–1139.
11. Schlievert, P. M., 1986, Staphylococcal enterotoxin B and toxic-shock syndrome toxin-l are significantly associated with nonmenstrual TSS, *Lancet* **1**:1149–1150.
12. Prohaska, J. V., Jacobson, M. J., Drake, C. T., and Tan, T.-L., 1959, Staphylococcal enterotoxin enteritis, *Surg. Gynecol. Obstet.* **108**:73–77.
13. Prohaska, J. V., 1963, Role of staphylococcal enterotoxin in the induction of experimental ileitis, *Ann. Surg.* **158**:492–497.
14. Zak, C., Jeljaszewicz, J., and Stochmal, I., 1973, Serological types of enterotoxins produced by strains of *Staphylococcus aureus* isolated from feces, in: *Staphylococci and Staphylococcal Infections* (J. Jeljaszewicz and W. Hryniewicz, eds.), Polish Medical Publishers, Warsaw pp. 526–528.
15. Osvath-Marton, A., Ban, E., and Molnar, L., 1976, Occurrence of enterotoxin-producing *Staphylococcus aureus* strains in enterocolitis in children, in: *Staphylococci and Staphylococcus Diseases* (J. Jeljaszewicz, ed.), Gustav Fischer Verlag, Stuttgart, pp. 589–592.

16. Crass, B. A., and Bergdoll, M. S., 1986, Toxin involvement in toxic shock syndrome, *J. Infect. Dis.* **153**:918–926.
17. Shands, K. N., Schmidt, G. P., Dan, B. B., Blum, D., Guidotti, R. J., Hargrett, N. T., Anderson, R. L., Hill, D. L., Broome, C. V., Band, J. D., and Fraser, D. W., 1980, Toxic-shock syndrome in menstruating women: Association with tampon use and *Staphylococcus aureus* and clinical features in 52 cases, *N. Engl. J. Med.* **303**:1436–1442.
18. Todd, J., Fishaut, M., Kapral, F., and Welch, T., 1978, Toxic-shock syndrome associated with phage-group I staphylococci, *Lancet* **2**:1116–1118.
19. Chesney, P. J., 1991, Clinical spectrum and therapy, in: *Toxic Shock Syndrome* (M. S. Bergdoll and P. J. Chesney, eds.), CRC Press, Boca Raton, FL, pp. 33–49.
20. Bergdoll, M. S., 1990, Toxic shock syndrome from surgical infections, in: *Pathogenesis of Wound and Biomaterial-Associated Infections* (T. Wadstrom, I. Eliasson, I. Holder, and A. Ljungh, eds.), Springer-Verlag, Berlin, pp. 121–128.
21. Kreiswirth, B. N., Kravitz, G. R., and Novick, R. P., 1986, Nosocomial transmission of toxic shock syndrome, *Ann. Intern. Med.* **105**:704–707.
22. Quimby, F., and Nguyen, H. T., 1985, Animal studies of toxic shock syndrome, *Crit. Rev. Microbiol.* **12**(1):1–44.
23. Winternitz, R., 1898, Ein beitrag zur kenntnis der dermatitis exfoliative neonatorum (Ritter), *Arch. Dermatol. Syphilol.* **44**:397–416.
24. Langford, M. P., Stanton, G. J., and Johnson H. M., 1978, Biological effects of staphylococcal enterotoxin A on human peripheral lymphocytes, *Infect. Immun.* **22**:62–68.
25. Poindexter, N. J., and Schlievert, P. M., 1985, Toxic-shock-syndrome toxin 1-induced proliferation of lymphocytes: Comparison of the mitogenic response of human, murine, and rabbit lymphocytes, *J. Infect. Dis.* **151**:65–72.
26. White, J., Herman, A., Pullen, A. M., Kubo, R., Kappler, J. W., and Marrack, P., 1989, The V_β-specific superantigen staphylococcal enterotoxin B: Stimulation of mature T cells and clonal deletion in neonatal mice, *Cell* **56**:27–35.
27. Marrack, P., and Kappler, J., 1990, The staphylococcal enterotoxins and their relatives, *Science* **248**:705–711.
28. Reinherz, E. L., Morimoto, C., Penta, A. C., and Schlossman, S. F., 1981, Subpopulations of the T4+ inducer T cell subset in man: Evidence for an amplifier population preferentially expressing Ia antigen upon activation, *J. Immunol.* **126**:67–70.
29. Calvano, S. E., Quimby, F. W., Antonacci, A. C., Reiser, R. F., Bergdoll, M. S., and Dineen, P., 1984, Analysis of the mitogenic effects of toxic shock toxin on human peripheral blood mononuclear cells *in vitro*, *Clin. Immunol. Immunopathol.* **33**:99–110.
30. Poindexter, N. J., and Schlievert, P. M., 1986, Suppression of immunoglobulin-secreting cells from human peripheral blood by toxic shock syndrome toxin-1, *J. Infect. Dis.* **153**:772–779.
31. Taub, D. D., Lin, Y.-S., and Rogers, T. J., 1990, Characterization and genetic restriction of suppressor effector cells induced by staphylococcal enterotoxin B, *J. Immunol.* **144**:456–462.
32. Parsonnet, J., Hickman, R. K., Eardley, D. D., and Pier, G. B., 1985, Induction of human interleukin-1 by toxic shock syndrome toxin-1, *J. Infect. Dis.* **151**:514–522.
33. Ikejima, T., Dinarello, C. A., Gill, D. M., and Wolff, S. M., 1984, Induction of human interleukin-1 by a product of *Staphylococcus aureus* associated with toxic shock syndrome, *J. Clin. Invest.* **73**:1312–1320.
34. Micusan, V. V., Mercier, G., Bhatti, A. R., Reiser, R. F., Bergdoll, M. S., and Oth, D., 1986, Production of human and murine interleukin-2 by toxic shock syndrome toxin-1, *Immunology* **58**:203–208.
35. Uchiyama, T., Kamagata, Y., Wakai, M., Yoshioka, M., Fujikawa, H., and Igarashi, H., 1986, Study of the biological activities of toxic shock syndrome toxin-1. I. Proliferative response and interleukin 2 production by T cells stimulated with the toxin, *Microbiol. Immunol.* **30**:469–483.

36. Jupin, C., Anderson, S., Damais, C., Alouf, J. E., and Parant, M., 1988, Toxic shock syndrome toxin 1 as an inducer of human necrosis factors and interferon, *J. Exp. Med.* **167**:752–761.
37. Anderson, S., Jupin, C., de Azavedo, J., Damais, C., Parant, M., and Alouf, J. E., 1989, TSST-1 induces production of tumor necrosis factor by human monocytes and of interferon-gamma, interleukin 2, and lymphotoxin by lymphocytes (Abstract), *Rev. Infect. Dis.* **11**(Suppl. 1):S331–S332.
38. Dinarello, C. A., 1984, Interleukin-1, *Rev. Infect. Dis.* **6**:51–95.
39. Ikejima, T., Dinarello, C. A., Gill, D. M., and Wolff, S. M., 1984, Induction of human interleukin-1 by a product of *Staphylococcus aureus* associated with toxic shock syndrome, *J. Clin. Invest.* **73**:1312–1320.
40. Parsonnet, J., Gillis, Z. A., and Pier, G. B., 1986, Induction of interleukin-1 by strains of *Staphylococcus aureus* from patients with nonmenstrual toxic shock syndrome, *J. Infect. Dis.* **154**:55–63.
41. Dinarello, C. A., Cannon, J. G., Wolff, S. M., Bernheim, H. A., Beutler, B., Cerami, A., Figari, I. S., Palladino, M. A., Jr., and O'Connor, J. V., 1986, Tumor necrosis factor (cachectin) is an endogenous pyrogen and induces production of interleukin 1, *J. Exp. Med.* **163**:1433–1450.
42. Fast, D. J., Schlievert, P. M., and Nelson, R. D., 1988, Nonpurulent response to toxic shock syndrome toxin 1-producing *Staphylococcus aureus*, *J. Immunol.* **140**:949–953.
43. Ikejima, T., Okusawa, S., Van Der Meer, J. W. M., and Dinarello, C. A., 1988, Induction by toxic-shock-syndrome toxin-1 of a circulating tumor necrosis factor-like substance in rabbits and of immunoreactive tumor necrosis factor and interleukin-1 from human mononuclear cells, *J. Infect. Dis.* **158**:1017–1025.

10

Immunological Aspects of Viral Gastroenteritis

SHUNZO CHIBA

1. VIRAL GASTROENTERITIS AND ITS ETIOLOGICAL AGENTS

During the two decades since the discovery of Norwalk virus in 1972[1] and rotavirus in 1973,[2] the viral etiology of acute nonbacterial gastroenteritis has been established and a considerable number of viruses are now recognized pathogens, responsible for medically important diarrheal illness.

Major categories of human gastroenteritis viruses include rotavirus, enteric adenovirus, Norwalk virus and its related agents, calicivirus and astrovirus[3] (Table I). Among these viruses, rotavirus is the most important cause of severe diarrhea in infants and young children, whereas the Norwalk group of viruses are common causes of epidemic gastroenteritis associated with contaminated water and foods in older children and adults. After the successful gene cloning of Norwalk virus,[4] it was classified as a member of the Caliciviridae family. Classification of Norwalk virus-like small round structured viruses (SRSVs) is still under investigation. Enteric adenovirus is the second most common pathogen after rotavirus, causing diarrhea in infants and young children. Calicivirus and astrovirus are less important pathogens that occur in sporadic gastroenteritis, but they occasionally cause institutional or food-borne outbreaks of acute gastroenteritis.

Human rotavirus was first discovered by electron microscopy in 1973 and rotaviruses are now known to exist in a wide variety of animals as well. Group A rotaviruses of serotypes 1–4 commonly prevail in humans everywhere. In view of the high morbidity and mortality of rotavirus gastroenter-

SHUNZO CHIBA • Department of Pediatrics, Sapporo Medical University, School of Medicine, Sapporo, Hokkaido 060, Japan.

Enteric Infections and Immunity, edited by Lois J. Paradise *et al.* Plenum Press, New York, 1996.

TABLE I
Human Gastroenteritis Viruses

Virus	Diameter of virion	Nucleic acid	*In vitro* replication	Serotypes
Rotavirus				
Group A	70–75 nm	Double-stranded segmented RNA	Yes	2 subgroups; 7 serotypes (1–4, 8, 9, 12)
Group B	70–75 nm	Double-stranded segmented RNA	No	Unknown
Group C	70–75 nm	Double-stranded segmented RNA	Yes	Unknown
Adenovirus	70–80 nm	Double-stranded DNA	Yes	2 serotypes (40, 41)
Norwalk virus	27–32 nm	Single-stranded RNA, positive sense	No	Unknown
Calicivirus	27–39 nm	Presumably single-stranded RNA	No	Unknown
Astrovirus	27–32 nm	Single-stranded RNA	Yes	5 serotypes (1–5)
SRSVs	26–40 nm	Presumably single-stranded RNA	No	Unknown

itis, development of rotavirus vaccines is listed as the highest priority by WHO and field trials of some promising vaccine candidates are now in progress.

However, mechanisms of immune resistance against viral gastroenteritis including rotavirus disease are still controversial and not fully understood.

In this chapter, immunological aspects of viral gastroenteritis mainly resulting from rotavirus infection will be discussed on the basis of the major findings so far reported.

2. IMMUNITY AGAINST ROTAVIRUS GASTROENTERITIS

2.1. Pathogenesis of Rotavirus Gastroenteritis

Rotavirus gastroenteritis is characterized by an abrupt onset of watery diarrhea that is not characteristically associated with blood or mucus in the stool. Vomiting is often present before or after onset of diarrhea. Rotavirus is excreted in very high concentrations early in the course of the illness, with as many as 10^{11} particles per gram of feces.

Study[(5,6)] of biopsied specimens of the jejunal mucosa of infants and young children hospitalized with rotavirus gastroenteritis has identified the following lesions: shortening and blunting of the villi, mononuclear cell infiltration in the lamina propria, distended cisternae of the endoplasmic reticulum and in lysosomes of columnar epithelial cells. Virus particles were also observed in goblet cells and phagocytes in the lamina propria. The destroyed infected enterocytes were replaced by immature cells, resulting in impaired absorption.

Information relating to the pathogenesis of rotavirus infection has also been obtained from experimentally infected animals that developed diarrheal illness following administration of human rotavirus. Functional alterations in small intestinal villous epithelial cells were demonstrated in piglets(7) that developed diarrhea following administration of human rotavirus. Glucose-coupled sodium transport was impaired, sucrase activity was decreased, and thymidine kinase activity was increased. However, adenylate cyclase and cyclic AMP were not stimulated.

Thus, diarrhea in rotavirus gastroenteritis is thought to result from decreased epithelial surface as well as from impaired absorption secondary to the altered ion transport and depressed disaccharidases in replaced immature cells.

2.2. Antigenic Characterization of Rotavirus

As members of the Reoviridae family, rotaviruses possess a segmented, double-stranded RNA genome and are nonenveloped icosahedrons with diameters of about 70 nm. The virion has an inner core and both inner and outer capsids.(8)

There are several levels of classification for rotaviruses based on antigenic determinants of the capsid proteins. The first division of rotaviruses is into groups A–F and is based on inner capsid protein VP6 differences. Members of group A are the ones most frequently associated with gastroenteritis in children throughout the world.(9) Group A rotaviruses are divided into two subgroups, I and II, according to nonneutralizing epitopes of VP6. Within subgroup II (which contains most of the human group A strains), the third level of classification, the serotype, is designated according to epitopes on proteins VP4 and VP7 which are in the outer capsid. The serotype epitopes are detectable in neutralization assays.(10–12) Immunogenically, VP7 is more important than VP4.(13) Thus far, 14 serotypes of viruses from various animals and humans have been recognized for VP7. The seven isolated from humans are serotypes 1, 2, 3, 4, 8, 9, and 12(14): 1, 3, 4, and 9 are in subgroup II and 2, 8, and 12 in subgroup I. Classification of serotypes according to VP4 is less well advanced.(15) Serotypes 1–4 are found throughout the world and account for the vast majority (about 95%) of human infections.(16) According to epidemiologic studies on serotype distribution in various regions of the world, several strains of rotavirus belonging to different serotypes cocirculate each year. Every 1 or 2 years, the predominating serotypes will be replaced by different ones, which then cocirculate. This alternating process, which repeats itself, has been found for each area studied. These studies have demonstrated variation of serotype frequency both between and among geographic regions as well as among countries.[9]

2.3. Systemic Antibody Response and Its Role in Protective Immunity

As already described, considerable progress has been made in identifying and characterizing the viral proteins. However, the factors that determine susceptibility to rotavirus infection and illness in humans are still unclear. The mechanisms responsible for immunity to human rotavirus infections and illness are poorly understood.

The viral proteins important in protection against rotavirus infection are VP7 and VP4. This has been demonstrated by using specific viral reassortants in genetic studies and by using specific monoclonal antibodies directed at epitopes of either of these two proteins. Antigenic determinants of other viral proteins were not able to neutralize virions or to induce protection in susceptible hosts.[17–19] During rotavirus infection in humans, IgM, IgG, and IgA antibody responses to both viral proteins, VP4 and VP7, have been demonstrated in serum and intestinal secretions.[20,21] One of the characteristics of response to human rotavirus infections is that antibodies produced after an episode of gastroenteritis are not directed solely to the major protective antigens of the viral protein. Heterotypic antibodies, specific to other VP7 serotype antigens, also are synthesized. This seems to be the case even following a primary infection.[22–24] Rotavirus infections in animal species tend to be predominantly homotypic, however.[14] The neutralization test detects the total reaction against neutralization epitopes on both VP4 and VP7. Recently, the relative contribution of VP4 and VP7 antibodies to the immune response has been examined by the use of reassortant rotaviruses containing either one genome segment for these two neutralization proteins or epitope-blocking assays with monoclonal antibodies directed at an epitope of VP4 or VP7.[21,25–29] It has been suggested that the VP7 response predominates after parenteral vaccination and VP4 response predominates after oral infection.[30,31] However, the opposite response with a dominant VP7 response after oral infection has also been reported.[27]

Inconsistency of the relative strength of immunogenicity of VP4 and VP7 may not be explained by the route of infection. The immunogenicity of both of these proteins may also be modified by the molecular properties of the virus strain or by properties of the infected host. A couple of recent reports support the latter possibility. In general, antibody responses to VP7 epitopes of the infecting serotype of virus were observed at a high frequency in children naturally infected or vaccinated with rotavirus. However, antibody responses to VP4 and heterotypic VP7 were observed only when the individuals possessed antibodies to any serotype of rotavirus in their acute-phase or prevaccination sera.[25,26] Further studies are necessary to determine whether such a pattern of antibody response is common and whether these heterotypic antibodies play a role in heterotypic protection.

Newborns asymptomatically infected with rotavirus appear to have less severe disease after a reexposure to rotavirus later in life than children with no early exposure.[32] The role of preexisting homotypic and heterotypic serum and secretory antibodies in immune resistance is still unclear. Because of the multiplicity of rotavirus serotypes, knowing the potentials of homotypic and/or heterotypic antibodies for providing protection against infection would facilitate development of effective vaccine(s).

There is a controversy regarding the existence of systemic immunity and its serotype specificity. Both secretory and humoral antibodies are formed in rotavirus infections. The relative importance of each of these in specific, protective immunity is not known. Intestinal secretory antibodies appear to be more relevant than serum antibodies with respect to protection against disease in experimental animals, but whether significant protection against human disease is exerted by mucosal immunity is not clear.[24,33–40] The concentration of serum antibodies specific for serotypes 1, 2, or 6 in groups of adult human volunteers correlated with protection against development of illness following experimental infection with human serotype 1 strain rotavirus.[35] Examination of the preinfection sera, using monoclonal antibodies in competitive epitope-blocking immunoassays, indicated that a specific site in antigen "A" on VP7 might participate as part of a major protective antigen.[35] Volunteer serum antibodies specific for this site at significant levels corresponded to decreased virus shedding and/or illness after experimental infection.

On the basis of clinical and serovirological findings obtained during outbreaks of gastroenteritis caused by serotype 3 rotavirus in a welfare nursery, we found that a critical level of serotype-specific neutralizing antibody (≥128) in the serum was associated with homotypic resistance to illness (Table II).[24] Protection against diarrhea was serotype specific, however, since correlation between heterotypic antibody levels and resistance to illness was lacking. More recent findings in the same nursery were obtained during an outbreak of gastroenteritis caused by serotype 1 rotavirus.[33] Eight infants had preexisting, neutralizing, anti-serotype 1 antibodies with titers ≥ 64. Two of these infants showed mild clinical disease; the remainder showed none. Of 41 infants with titers ≤ 32, 29 (71%) developed gastroenteritis. These results were less clear-cut than in the previous, serotype 3, study. Nevertheless, enhancement of resistance to clinical disease by significant levels of preexisting, homotypic antibodies was observed again.

These findings indicated that either serum neutralizing antibodies

TABLE II
Relation between Preexisting Titers of Neutralizing Antibody to Homotypic Rotavirus and Infection or Clinical Illness during Two Outbreaks of Gastroenteritis Caused by Serotype 3 Rotavirus[a]

Antibody titer	No. of infants tested	No. of infants infected	No. of infants with clinical illness
<1:16	7	7	6
1:16	13	13	12
1:32	6	6	6
1:64	12	9	8
1:128	6	5	1
1:256	8	2	0
≥1:512	9	2	0

[a]Reprinted with permission from Ref. 24.

were protective against rotavirus infection and illness or the titers of these antibodies correlated with other factors involved in protection. Serum antibodies may enter the lumen of the intestinal tract by transudation or may be just a reflection of intestinal immunity.(41) It is of interest that circulating rotavirus antibodies can appear in the gastrointestinal tract of neonatal calves if the level of circulating antibodies is sufficiently high and that these mucosal antibodies can provide protection against experimentally induced infection and diarrhea.(42,43)

As already described, various studies on natural and experimental rotavirus infection in human subjects have revealed correlation between the level of homotypic serum antibodies and protection against rotavirus infection and illness. However, in one adult volunteer study, the level of homotypic serum antibodies did not correlate with protection.(44,45) On the other hand, several studies in experimental animals have found poor correlation between protection against disease and serum neutralizing antibody titers. Calves infected with avirulent strains of rotavirus were protected from challenge with an apparently unrelated (by neutralization assay) virulent strain.(46) In another experimental study, gnotobiotic calves vaccinated with B223 bovine strain demonstrated the presence of serotype-specific resistance to challenge but also cross-protection against other bovine serotype strains such as B641, but the homo- or heterologous reverse cross-reactivity was not present after B641 vaccination with respect to levels of neutralizing antibody developed.(47) Knowing the antibody titer in serum or intestinal fluids did not enable prediction of level of resistance to challenge administration of virus. Other studies in neonatal mice provided evidence that active protection following oral immunization with live rotavirus is not dependent on serum and intestinal neutralizing antibodies.[48]

A case–control study on rotavirus diarrhea in children carried out in Bangladesh indicated that the correlation of protection with serum neutralizing antibody titers was not serotype-specific.[48] Furthermore, in vaccine trials with children, protection has been reported between RIT 4237 or WC3 vaccine derived from bovine rotaviruses and natural infections with human rotaviruses, although these vaccine viruses do not belong to any of the four main serotype groups of human rotavirus.[49,50] However, other reports have shown a lack of protection against rotavirus infection in RIT-vaccinated children[51] and immunity to repeat infections appears to be serotype dependent.

Thus, the basis of immunity to rotavirus, particularly the role of neutralizing antibody in serotype-specific protection, needs to be clarified. Possible explanations for the differences in the results include the criteria accepted for evidence of protection in the various studies and different background of vaccines with regard to the history of rotavirus infections. One of the difficulties in evaluation of rotavirus vaccine efficacy is that unknown quantity, lack of knowledge as to previous rotavirus infection. That is, vaccination may not be the primary exposure of the person to rotaviral protective antigens.[47] Then, of course, anamnestic antibody responses to vaccination in some individuals would confound analysis of relationships of postvaccinal serum or secretory antibody levels to protection against challenge infection.[52] Intestinal secretory antibodies often were homotypic only, even when serum antibodies were heterotypic with respect to infecting strains. Failure of parallel responses in circulating and local antibody specificities may result from several causes, including (1) technical (neutralization test sensitivity may differ), (2) loss of immunoglobulins because of intestinal enzyme action, and (3) variation in the abilities of macrophages to present or lymphocytes to recognize different immunogens in spleen versus mucosa.[52]

As for persistence of immunity, we found that even homotypic protection was of short duration, lasting for half a year but not beyond a year.[24] This lack of persistence may explain recurrent attacks of gastroenteritis caused by rotavirus of one serotype. In a study with adult volunteers, specific strain protection lasted for 9 to 12 months, or possibly longer in some persons, after oral infection with live virus. Serum IgG antibody levels correlated with protection against oral challenge.[39] A prospective study of natural rotavirus infection in a cohort of infants suggested that protection against reinfection with homotypic virus persisted over 2 years.[36]

The reasons for the discrepancy in the duration of immunity between studies are not clear but may have resulted from the level of reexposure to homotypic or heterotypic virus in different epidemiological settings.

We extended our studies to examine the possibility of heterotypic protection induced by natural rotavirus infection in the previously described welfare nursery.(24) Table III presents our findings on heterotypic antibody responses among 44 infants infected with serotype 3 rotavirus. More than half of the infants developed specific protective antibody titers (⩾128). In addition, neutralization tests revealed heterotypic responses to types 1, 2, and 4 in 77, 23, and 62% of the patients, respectively. We examined the neutralizing antibody response to rotavirus serotypes 1–4 in 28 infants inoculated with rhesus rotavirus vaccine (RRV-1 strain) which is antigenically similar to the serotype 3 human rotavirus.(33) The results indicated a heterotypic antibody response whose sequence was similar to that seen after natural infection with serotype 3 rotavirus. These observations suggested that immunity to heterotypic virus might also be induced by a rotavirus vaccine. As already mentioned, rotavirus has two neutralization antigens, VP4 and VP7, in the outer layer of the particle. In our recent study, examination of the serotype-specific epitope on VP7 and cross-reactive epitopes on VP4 after vaccination and after natural infection revealed that the incidence of a VP7 homotypic response was higher than that of the heterotypic response. For heterotypic responses to VP7 to occur, it appeared that patients or vaccinees had to have preexisting antibodies. The serotype specificity of these antibodies was irrelevant. This situation was the same for cross-reacting antibodies to VP4. Therefore, repeated infections or booster administration of vaccine might be necessary to induce antibodies to cross-reactive epitopes on VP4. We had a chance to assess the efficacy of the vaccine against the disease caused by a heterotypic virus.(53) During the follow-up period after booster vaccination with RRV-1 (serotype 3), an outbreak of gastroenteritis caused by serotype 1 rotavirus occurred in the group of infants studied.

TABLE III
Concomitant Increase in Titers of Neutralizing Antibody to Rotaviruses of Serotypes 1, 2, and 4 among Infants Infected with Serotype 3[a]

Serotype (strain) specificity of antibody	No. of serotype 3-infected infants tested for antibody	No. (%) of infants with antibody response
1 (KU)	44	34 (77)
2 (S2)	44	10 (23)
4 (HOCHI)	42	26 (62)

[a]Reprinted with permission from Ref. 33.

Although almost all infants were infected and shed rotavirus in their stools, clinical disease developed in only 20% of infants receiving the booster dose, whereas such disease was apparent in 50 and 94% of infants who received one dose of vaccine and placebo, respectively.[53] The findings were correlated with the level of preoutbreak neutralizing antibody to homotypic (serotype 1) virus and may suggest the induction of heterotypic protection by booster inoculation with rhesus rotavirus vaccine. It was also reported that inoculation of WC 3 bovine rotavirus vaccine (serotype 6) could boost neutralizing antibody titers to human serotypes 1–4 in subjects primed by a previous natural infection.[27]

2.4. Mucosal Immunity

In contrast to results supporting the importance of the systemic antibody responses in natural rotavirus infection in humans, experimental studies in animals suggested that local intestinal antibody prevents disease following oral inoculation with rotaviruses but that systemic antibody is not protective.[54,55] It was observed that newborn calves frequently developed rotavirus diarrhea despite a moderately high level of circulating rotavirus antibodies derived from colostrum.[56] This observation was confirmed experimentally when five calves were fed colostrum containing rotavirus antibodies within 24 hr of birth and then challenged 7 days later with calf rotavirus. These calves developed a diarrheal illness despite the presence of a high level of serum rotavirus antibodies at the time of challenge. In contrast, two calves that were fed colostrum containing rotavirus antibodies 4 hr before challenge and 4 and 24 hr after challenge with calf rotavirus failed to develop diarrhea within the normal incubation period.

In another experiment,[55] a group of 1-day-old gnotobiotic lambs was fed colostrum that contained rotavirus antibodies and two other groups were fed either (1) the same preparation of colostrum on days 1 through 4 or (2) serum containing rotavirus antibodies on days 2 through 4. Following challenge with lamb rotavirus at 2 days of age, the first group of lambs developed diarrhea despite the presence of serum antibodies to rotavirus acquired from the colostrum, whereas the other two groups of lambs did not.

Using a suckling mouse model for rotavirus infection, a team of investigators hyperimmunized dams by oral administration of a SA-11 simian virus strain.[57] Milk fractions containing IgG or IgA specific antibodies were protective when orally administered to suckling mice challenged with SA-11; IgA had greater protective capacity than IgG. In contrast, the IgG fraction of milk from dams parenterally vaccinated showed strongest activity, although mice suckled on these dams were also protected from challenge. When newborn mice of immunized dams were foster-nursed by specific-antibody-negative dams, transplacentally acquired rotavirus-specific antibodies, even in high titers, were unable to protect the neonates against oral challenge with SA-11.

These observations suggest that antibody in the lumen of the small intestine is the major determinant of resistance to rotavirus illness. In addition, passively acquired gastrointestinal rotavirus antibodies, but not circulating antibodies, have been shown to protect newborn mice and calves against experimental rotavirus disease. From the findings of other investigators,[58] such protection against North Carolina porcine rotavirus was induced in agammaglobulinemic piglets by orally administered rotavirus IgG derived from colostrum of cows immunized with heterologous and heterotypic rotavirus SA-11. This observation indicated the possible induction of the heterotypic immunity in the gut.

Although there are many reports concerning the development of local antirotavirus antibodies,[44,59–62] little information exists as to the duration of this antibody or its role in protection against rotavirus infection and illness in humans. Only a few studies provided evidence for the protective effect of locally produced antibodies in rotavirus infection in humans. One report correlated preexisting jejunal antibodies with protection of adults challenged with a human rotavirus.[63] Obviously, these experiments cannot be performed in children for ethical reasons, so the role of jejunal antibody levels cannot be confirmed in the equivalent human situation.

More recently, another study suggested the effective role of coproantibody in clinical protection of children against reinfection with rotavirus.[64] For 14–31 months after occurrence of severe rotaviral illness, 35 children were tested for specific coproantibody levels. These titers were compared with the incidence of naturally occurring reinfection. Ninety-four percent of the children became infected again; 38% retained high coproantibody levels ("plateaus"). In this latter group, reinfection was more common. In addition, the ratio of symptomatic children to the total number infected was lower in "plateau" children than in the others. Multiple exposure to rotavirus evidently stimulated production of persistently high levels of specific copro-IgA and this correlated with protection against clinical disease.

Except for experimental infection in adult volunteers, where intestinal fluid can be obtained for antibody determination, coproantibody titer is usually measured to detect locally produced antibody. Copro-IgA conversions are shown to be sensitive markers of primary and recurrent rotavirus infection. However, measurement of this antibody caused problems related to high background levels and possible degradation.[65] There are several reports indicating that serum and salivary IgA antibody to rotavirus is a good indicator of local antibody responses.[66–69]

2.5. Cellular Immunity

The lack of convincing evidence of serotype-specific systemic and local humoral immunity has led investigators to consider the potential role of cellular immunity in rotavirus infection. Results of several studies indicate that rotavirus-specific cytotoxic T lymphocytes may be important for protection against the disease.

Cytotoxic immunity was shown to be induced after rotavirus infection in mice and immune cytotoxic T cells could prevent or clear rotavirus infection in animal models.[70,71] Protection against rotavirus disease after adoptive transfer of virus-specific cytotoxic T lymphocytes was demonstrated to occur in the absence of neutralizing antibodies in the sera of suckling mice.

Mice with severe combined immunodeficiency could recover from chronic rotavirus infection after adoptive transfer of immune $CD8^+$ T lymphocytes.[72] In a more recent study in gnotobiotic calves, depletion of $CD8^+$ or $CD4^+$ T lymphocytes by administration of monoclonal antibodies showed that $CD8^+$ cells are involved in restriction of rotavirus replication and $CD4^+$ cells are required for generation of mucosal and humoral specific antibodies.[73] Thus, experimental animal studies suggest that immune responses mediated by cytotoxic T cells play a direct role in protection and recovery from rotavirus infection. However, the role of cellular immunity in humans has not been adequately studied. Information is limited to the facts that rotavirus-specific lymphoproliferative activity can be detected only in peripheral blood mononuclear cells from subjects with previous exposure to rotavirus and that the number of positive responders increases gradually with age from 0% in newborns to 80–90% in older children and adults.[74–77]

2.6. Maternal–Infantile Immunity

The roles of transplacentally transmitted maternal antibody and the role of breast-feeding in preventing rotaviral disease have been described.

Children under the age of 3 months are quite resistant to rotaviral disease; this may be related to levels of maternal antibody.

By immunizing mouse dams either orally or parenterally with SA-11 simian rotavirus, passively acquired gastrointestinal but not circulating antibodies were found to protect suckling mice against diarrhea induced by challenge with SA-11 virus. In addition, passively acquired protection in suckling mice against rotavirus challenge was dependent on both serotype and titer of antibody.[57] More recent studies used colostrum-deprived cynomolgus monkeys, which were offspring of dams naturally exposed to rotaviruses, as demonstrated by serum antibody titers of 2000 to >64,000 in ELISA.[78] Newborn and infant monkeys were challenged orally with simian virus SA-11. Lack of protection indicated that circulating antibodies of maternal origin do not enhance resistance to rotaviral disease. Such observations may support the following two possibilities: the disease in the human newborn is mild because of (1) immaturity of the enterocytes or (2) passively acquired secretory IgA antibodies obtained from breast milk, rather than because of maternal antibodies acquired transplacentally.

Presence of passively acquired gastrointestinal immunity through breast-feeding has been supported by experimental animal studies, the interfering effect of breast-feeding on take rates of rotavirus vaccines in infants, and the protective effects of infant formulas containing anti-rotavirus milk antibodies from hyperimmunized cows.

Cordle *et al.*(79) reported a quantitative evaluation of the ability of bovine milk antibodies to prevent gnotobiotic piglets from diarrhea caused by human rotavirus. Results of passive protection studies for human rotavirus serotype 1 are shown in Fig. 1. The dose–response relationship between the amount of neutralizing IgG antibody fed and protection of piglets from diarrhea was clear. Inhibition of virus shed in feces also corresponded to IgG administered.(79,80)

Bovine colostrum containing neutralizing antibodies against four different serotypes of human rotavirus produced no clinical benefit when used to treat children with rotavirus gastroenteritis, but it was effective in preventing diarrhea caused by infection with human rotavirus.(81–83)

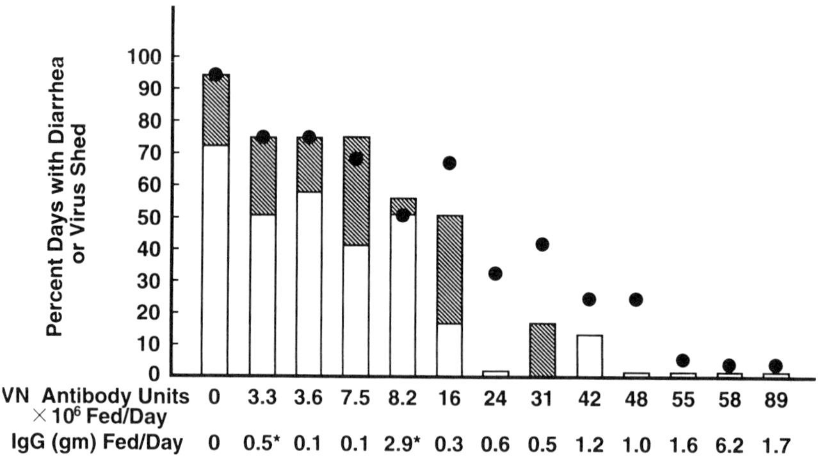

FIGURE 1. Antibody dose–diarrhea/virus shedding response for HRV serotype 1(Wa)-challenged piglets. Clear bars, mild diarrhea; hatched bars, profuse diarrhea; ●, virus shedding; *, nonimmune. Reprinted with permission from Ref. 79.

In contrast to the above-mentioned benefit of breast-feeding or an infant formula containing rotavirus antibodies, several studies indicated a significant adverse effect of breast-feeding with respect to rotavirus vaccine take. In a meta-analysis of three studies in which rhesus rotavirus vaccine was administered as a single dose, it was shown that 48% of breast-fed infants seroconverted versus 70% of bottle-fed infants ($p < 0.005$).[84] In another meta-analysis of 16 groups of infants who received one of three different vaccines, the authors concluded that breast-feeding inhibited the immunogenicity of the vaccines to a small but statistically significant extent.[85]

3. IMMUNITY AGAINST GASTROENTERITIS CAUSED BY NORWALK VIRUS AND HUMAN CALICIVIRUS

To date, there has been only our report on immunity against gastroenteritis caused by the morphologically distinctive human calicivirus.[86] We prospectively followed up infants in a welfare nursery who were exposed to an outbreak of calicivirus gastroenteritis. Forty-one paired pre- and post-outbreak sera from patients and from healthy contacts involved in the outbreak were tested for anticalicivirus antibodies by a radioimmunoassay blocking test. The 23 infants who had no preexisting antibody all became infected as revealed by a sharp rise in antibody titer and 18 (78%) of them developed gastroenteritis. Fifteen of eighteen infants with preexisting antibody also became infected, but only 3 of the 18 became ill. The data support the development of humoral immunity against calicivirus gastroenteritis.

Although Norwalk virus is currently classified as a member of the family Caliciviridae, immunity against this virus is somewhat mysterious. According to results from a series of volunteer studies conducted in the United States, challenge of volunteers with this virus produced disease in some persons and not in others.[87] Repeat challenge with a homologous virus within several months of the original infection did not produce clinical illness.[88] Immunity to challenge with a heterologous virus, i.e., Norwalk virus followed by Hawaii agent, did not develop.[88] Subsequent challenge with a homologous virus 2 to 4 years later produced disease in the same volunteers who had symptomatic infection with the first contact with the virus.[87] Those individuals who were asymptomatic and did not have infection with the first challenge, did not have infection with subsequent challenge (Fig. 2).[87] Neither serum nor local intestinal antibody correlated with resistance to Norwalk virus challenge. On the contrary, high levels of serum or local jejunal antibody to the Norwalk virus correlated with an increased risk of developing illness, whereas individuals with low levels of circulating antibody appeared less likely to develop illness.[87] The explanation for such paradox-

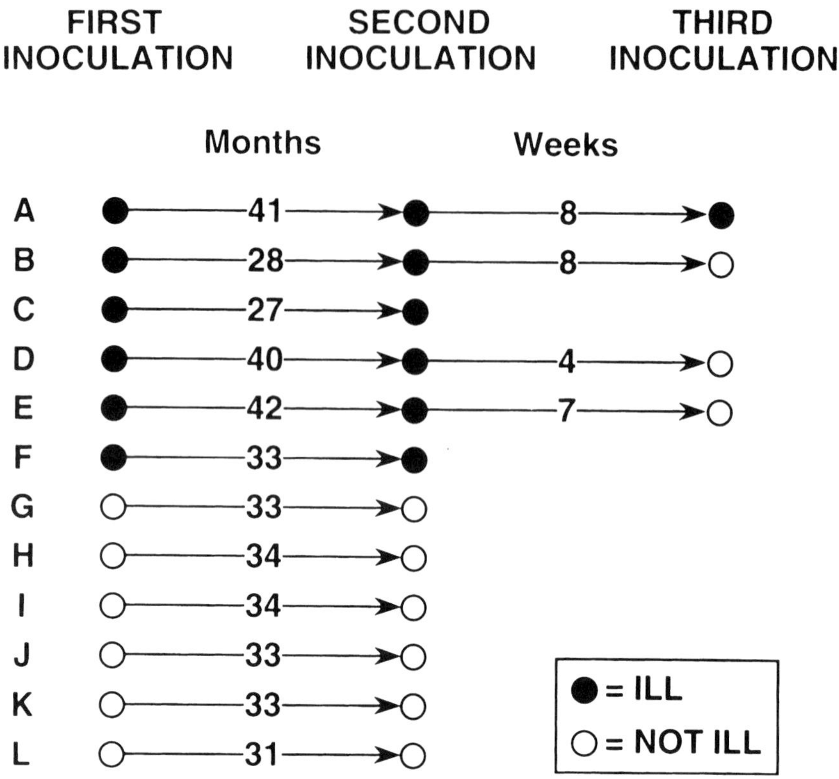

FIGURE 2. Sequence of Norwalk-agent inoculation studies in 12 volunteers (shown individually as A–L). The numbers indicate the months or weeks between inoculations. Solid circles represent volunteers who experienced clinical illness, and open circles those in whom clinical illness failed to develop. Reprinted with permission from Ref. 87.

ical findings is unclear, but it is possible that some individuals are resistant to these viruses because of genetically determined factors.[89] Since antibody was measured by radioimmunoassay in the previous volunteer studies, determination of neutralizing antibody in the same serum and jejunal fluid may provide more valuable information regarding the immunity to Norwalk virus gastroenteritis in the future.

4. CONCLUSION

Despite controversies about mechanisms of immunity against viral gastroenteritis, current strategy of developing effective rotavirus vaccines is based on the serotype-specific nature of immune protection against the disease. Thus, combined polyvalent vaccines or genetic reassortants having epitopes of four major human serotypes are being evaluated by field trials.

REFERENCES

1. Kapikian, A. Z., Wyatt, R. G., Dolin, R., Thornhill, T. S., Kalica, A. R., Chanock, R. M., 1972, Visualization by immune electron microscopy of a 27-nm particle associated with acute infectious nonbacterial gastroenteritis, *J. Virol.* **10:**1075–1081.
2. Bishop, R. F., Davidson, G. P., Holmes, I. H., and Ruck, B. J., 1973, Virus particles in epithelial cells of duodenal mucosa from children with acute non-bacterial gastroenteritis, *Lancet* **2:**1281–1283.
3. Blacklow, N. R., and Greenberg, H. B., 1991, Viral gastroenteritis, *N. Engl. J. Med.* **325:** 252–264.
4. Jiang, X., Graham, D. Y., Wang, K., and Estes, M. K., 1990, Norwalk virus gene cloning and characterization, *Science* **250:**1580–1583.
5. Holmes, I. H., Ruck, B. J., Bishop, R. F., and Davidson, G. R., 1975, Infantile enteritis viruses: Morphogenesis and morphology, *J. Virol.* **16:**937–943.
6. Suzuki, H., and Konno, T., 1975, Reovirus-like particles in jejunal mucosa of a Japanese infant with acute infectious nonbacterial gastroenteritis, *Tohoku J. Exp. Med.* **115:**199–211.
7. Davidson, G. P., Gall, D. B., Petric, M., Butler, D. G., and Hamilton, J. R., 1977, Human rotavirus enteritis induced in conventional piglets. Intestinal structure and transport, *J. Clin. Invest.* **60:**1402–1409.
8. Estes, M. K., and Cohen, J., 1989, Rotavirus gene structure and function, *Microbiol. Rev.* **53:** 410–449.
9. Pickering, L. K., and O'Ryan, M., 1991, Serotypes of rotavirus that infect infants symptomatically and asymptomatically, in: *Immunology of Milk and the Neonate* (J. Mestecky et al., eds.), Plenum Press, New York, pp. 241–247.
10. Prasad, B. V. V., Burns, J. W., Marrietta, E., Estes, M. K., and Chiu, W., 1990, Localization of VP4 neutralization sites in rotavirus by three-dimensional cryo-electron microscopy, *Nature* **343:**476–479.
11. Taniguchi, K., Morita, Y., Urasawa, T., and Urasawa, S., 1987, Cross-reactive neutralization epitopes on VP3 of human rotavirus, analysis with monoclonal antibodies and antigenic variants, *J. Virol.* **61:**1726–1730.
12. Hoshino, Y., Sereno, M. M., Midthun, K., Flores, J., Kapikian, A. Z., and Chanock, R. M., 1985, Independent segregation of two antigenic specificities (VP3 and VP7) involved in neutralization of rotavirus infectivity, *Proc. Natl. Acad. Sci. USA* **82:**8701–8704.
13. Dyall-Smith, M. L., Lazdins, I., Tregear, G. W., and Holmes, I. H., 1986, Location of the major antigenic sites involved in rotavirus serotype-specific neutralization, *Proc. Natl. Acad. Sci. USA* **83:**3465–3468.
14. Beards, G. M., King, J. A., Mazhar, S., Landon, J., and Desselberger, U., 1993, Homotypic and heterotypic immune responses to group A rotaviruses in parenterally immunized sheep, *Vaccine* **11:**262–266.
15. Gorziglia, M., Larralde, G., Kapikian, A. Z., and Chanock, R. M., 1990, Antigenic relationships among human rotaviruses as determined by outer capsid protein VP4, *Proc. Natl. Acad. Sci. USA* **87:**7155–7159.
16. Beards, G. M., Desselberger, U., and Flewett, T. H., 1989, Temporal and geographical distribution of human rotavirus serotypes 1983–1989, *J. Clin. Microbiol.* **27:**2827–2833.

17. Matsui, S. M., Offit, P. A., and Vo, P. T., 1989, Passive protection against rotavirus-induced diarrhea by monoclonal antibodies to the heterotypic neutralization domain of VP7 and VP8 fragment of VP4, *J. Clin. Microbiol.* **27**:780–782.
18. Hoshino, Y., Saif, L. J., Sereno, M. M., Chanock, R. M., and Kapikian, A. Z., 1988, Infection immunity of piglets to either VP3 or VP7 outer capsid protein confers resistance to challenge with a virulent rotavirus bearing the corresponding antigen, *J. Virol.* **62**:744–748.
19. Offit, P. A., Clark, H. F., Blavat, G., and Greenberg, H. B., 1986, Reassortant rotaviruses containing structural proteins VP3 and VP7 from different parents induce antibodies protective against each parenteral serotype, *J. Virol.* **60**:491–496.
20. Coulson, B. S., Grimwood, K., Masendycz, P. J., Jennifer, S. L., Norman, M., Ruth, F. B., and Graeme, L. B., 1990, Comparison of rotavirus immunoglobulin A coproconversion with other indices of rotavirus infection in a longitudinal study in childhood, *J. Clin. Microbiol.* **28**:1367–1374.
21. Richardson, S. C., and Bishop, R. F., 1990, Homotypic serum antibody responses to rotavirus proteins following primary infection of young children with serotype 1 rotavirus, *J. Clin. Microbiol.* **28**:1891–1897.
22. Gerna, G., Sarasini, A., Torsellini, M., Torre, D., Parea, M., and Battaglia, M., 1990, Group- and type-specific serologic response in infants and children with primary rotavirus infections and gastroenteritis caused by a strain of known serotype, *J. Infect. Dis.* **161**:1105–1111.
23. Brüssow, H., Werchaw, H., Lerner, L., Mietens, C., Liedtke, W., Sidoti, J., and Sotek, J., 1988, Seroconversion patterns to four human rotavirus serotypes in hospitalized infants with acute rotavirus gastroenteritis, *J. Infect. Dis.* **158**:588–595.
24. Chiba, S., Yokoyama, T., Nakata, S., Morita, Y., Urasawa, T., Taniguchi, K., Urasawa, S., and Nakao, T., 1986, Protective effect of naturally acquired homotypic and heterotypic rotavirus antibodies, *Lancet* **2**:417–421.
25. Matson, D. O., O'Ryan, M. L., Pickering, L. K., Chiba, S., Nakata, S., Raj, P., and Estes, M. K., 1992, Characterization of serum antibody responses to natural rotavirus infections in children by VP7-specific epitope blocking assays, *J. Clin. Microbiol.* **30**:1056–1061.
26. Taniguchi, K., Urasawa, T., Kobayashi, N., Ahmed, M. U., Adachi, N., Chiba, S., and Urawasa, S., 1991, Antibody response to serotype-specific and cross-reactive neutralization epitopes on VP4 and VP7 after rotavirus infection or vaccination, *J. Clin. Microbiol.* **29**:483–487.
27. Ward, R. L., Knowlton, D. R., Greenberg, H. B., Schiff, G. M., and Bernstein, D. I., 1990, Serum-neutralizing antibody to VP4 and VP7 proteins in infants following vaccination with WC3 bovine rotavirus, *J. Virol.* **64**:2687–2691.
28. Brüssow, H., Offit, P. A., Gerna, G., Bruttin, A., and Sidoti, J., 1990, Polypeptide specificity of antiviral serum antibodies in children naturally infected with human rotavirus, *J. Virol.* **64**: 4130–4136.
29. Shaw, R. D., Fong, K. J., Losonsky, G. A., Levine, M. M., Maldonado, Y., Yolken, R., Flores, J., Kapikian, A. Z., Vo, P. T., and Greenberg, H. B., 1987, Epitope-specific immune responses to rotavirus vaccination, *Gastroenterology* **93**:941–950.
30. Flores, J., Prez-Schael, I., Blanco, M., Vilar, M., Garcia, D., Perez, M., Daoud, N., Midthun, K., and Kapikian, A. Z., 1989, Reactions to and antigenicity of two human–rhesus rotavirus reassortant vaccine candidates of serotypes 1 and 2 in Venezuelan infants, *J. Clin. Microbiol.* **27**:512–518.
31. Svensson, L., Sheshberadaran, H., Vene, S., Norrby, E., Gradien, M., and Wadell, G., 1987, Serum antibody responses to individual viral polypeptides in human rotavirus infections, *J. Gen. Virol.* **68**:643–651.
32. Bishop, R. F., Barnes, G. L., Cipriani, E., and Lund, J. S., 1983, Clinical immunity after neonatal rotavirus infection, *N. Engl. J. Med.* **309**:72–76.

33. Chiba, S., Nakata, S., Ukae, S., and Adachi, N., 1993, Virological and serological aspects of immune resistance to rotavirus gastroenteritis, *Clin. Infect. Dis.* **16:**S117–S121.
34. Clemens, J. D., Ward, R. L., Rao, M. R., Sack, D. A., Knowlton, D. R., van Loon, F. P. L., Huda, S., McNeal, M., Ahmed, F., and Schiff, G. M., 1992, Seroepidemiologic evaluation of antibodies to rotavirus as correlates of the risk of clinically significant rotavirus diarrhea in rural Bangladesh, J. Infect. Dis. **165:**161–165.
35. Green, K. Y., and Kapikian, A. Z., 1992, Identification of VP7 epitopes associated with protection against human rotavirus illness or shedding in volunteers, *J. Virol.* **66:**548–553.
36. Bernstein, D. I., Sander, D. S., Smith, V. E., Schiff, G. M., and Ward, R. L., 1991, Protection from rotavirus reinfection: 2-year prospective study, *J. Infect. Dis.* **164:**277–283.
37. Green, K. Y., Taniguchi, K., Mackow, E. R., and Kapikian, A. Z., 1990, Homotypic and heterotypic epitope-specific antibody responses in adult and infant rotavirus vaccinees: Implication for vaccine development, *J. Infect. Dis.* **161:**667–679.
38. Bernstein, D. I., Smith, V. E., Sander, D. S., Pax, K. A., Schiff, G. M., and Ward, R. L., 1990, Evaluation of WC3 rotavirus vaccine and correlates of protection in healthy infants, *J. Infect. Dis.* **162:**1055–1062.
39. Ward, R. L., Bernstein, D. I., Shukla, R., McNeal, M. M., Sherwood, J. R., Young, E. C., and Schiff, G. M., 1990, Protection of adults rechallenged with a human rotavirus, *J. Infect. Dis.* **161:**440–445.
40. Kapikian, A. Z., Wyatt, R. G., Levine, M. M., Yolken, R. H., VanKirk, D. H., Dolin, R., Greenberg, H. B., and Chanock, R. M., 1983, Oral administration of human rotavirus to volunteers: Induction of illness and correlates of resistance, *J. Infect. Dis.* **147:**95–106.
41. Edelman, R., 1987, Perspective on the development and deployment of rotavirus vaccines, *Pediatr. Infect. Dis. J.* **6:**704–710.
42. Besser, T. E., McGuire, T. C., Gay, C. C., and Pritchett, L. C., 1988, Transfer of functional immunoglobulin G (IgG) antibody into the gastrointestinal tract accounts for IgG clearance in calves, *J. Virol.* **62:**2234–2237.
43. Besser, T. E., Gay, C. C., McGuire, T. C., and Evermann, J. F., 1988, Passive immunity to bovine rotavirus infection associated with transfer of serum antibody into the intestinal lumen, *J. Virol.* **62:**2238–2242.
44. Bernstein, D. I., Ziegler, J. M., and Ward, R. L., 1986, Rotavirus fecal IgA response in adults challenged with human rotavirus, *J. Med. Virol.* **20:**297–304.
45. Ward, R. L., Bernstein, D. I., Young, E. C., Sherwood, J. R., Knowlton, D. R., and Schiff, G. M., 1986, Human rotavirus studies in volunteers: Determination of infectious dose and serological response to infection, *J. Infect. Dis.* **154:**871–880.
46. Bridger, J. C., and Oldham, G., 1987, Avirulent rotavirus infections protect calves from disease with and without inducing high levels of neutralizing antibody, *J. Gen. Virol.* **68:** 2311–2317.
47. Woode, G. N., Zheng, S., Rosen, B. I., Knight, N., Gourley, N. E. K., and Ramig, R. F., 1987, Protection between different serotypes of bovine rotavirus in gnotobiotic calves: Specificity of serum antibody and coproantibody responses, *J. Clin. Microbiol.* **25:**1052–1058.
48. Ward, R. L., McNeal, M. M., and Sheridan, J. F., 1992, Evidence that active protection following oral immunization of mice with live rotavirus is not dependent on neutralizing antibody, *Virology* **188:**57–66.
49. Vesikari, T., Isolauri, E., D'Hondt, E., Delem, A., Andre, F. E., and Zissis, G., 1984, Protection of infants against rotavirus diarrhea by RIT4237 attenuated bovine rotavirus strain vaccine, *Lancet* **1:**977–981.
50. Clark, H. F., Borian, F. E., Bell, L. M., Modesto, K., Gouvea, V., and Plotkin, S. A., 1988,

Protective effect of WC3 vaccine against rotavirus diarrhea in infants during a predominantly serotype 1 rotavirus season, *J. Infect. Dis.* **158:**570–587.
51. DeMol, P., Zissis, G., Butzler, J. P., Mutwewingabo, A., and Andre, F. E., 1986, Failure of live, attenuated oral rotavirus vaccine, *Lancet* **2:**108.
52. Conner, M. E., Gilger, M. A., Estes, M. K., and Graham, D. Y., 1991, Serologic and mucosal immune response to rotavirus infection in the rabbit model, *J. Virol.* **65:**2562–2571.
53. Ukae, S., Nakata, S., Adachi, N., Kogawa, K., and Chiba, S., 1994, Efficacy of rhesus rotavirus vaccine MMU-18006 against gastroenteritis due to serotype 1 rotavirus, *Vaccine* **12:**933–939.
54. Saif, L. J., Redman, D. R., Smith, K. L., and Theil, K. W., 1983, Passive immunity to bovine rotavirus in newborn calves fed colostrum supplements from immunized or nonimmunized cows, *Infect. Immun.* **41:**1118–1131.
55. Snodgrass, D. R., and Wells, P. W., 1976, Rotavirus infection in lambs: Studies on passive protection, *Arch. Virol.* **52:**201–205.
56. Woode, G. N., Jones, J., and Bridger, J., 1975, Levels of colostral antibodies against neonatal calf diarrhoea virus, *Vet. Res.* **97:**148–149.
57. Offit, P. A., and Clark, H. F., 1985, Protection against rotavirus-induced gastroenteritis in a murine model by passively acquired gastrointestinal but not circulating antibodies, *J. Virol.* **54:**58–64.
58. Leece, J. G., Leary, H. L., Clarke, D. A., and Batema, R. P., 1991, Protection of agammaglobulinemic piglets from porcine rotavirus infection by antibody against simian rotavirus SA-11, *J. Clin. Microbiol.* **29:**1382–1386.
59. Shinozaki, T., Araki, K., Ushijima, H., Kim, B., Tajima, T., and Fujii, R., 1986, Coproantibody response to rotavirus in an outbreak in a day-care nursery, *Eur. J. Pediatr.* **144:**515–516.
60. Hjelt, K., Grauballe, P. C., Schiotz, P. O., Andersen, L., and Krasilnikoff, P. A., 1985, Intestinal and serum immune response to a naturally acquired rotavirus gastroenteritis in children, *J. Pediatr. Gastroenterol. Nutr.* **4:**60–66.
61. Inouye, S., Matsuno, S., and Yamaguchi, Y., 1984, Efficient coating of the solid phase with rotavirus antigens for enzyme linked immunosorbent assay of immunoglobulin A antibody in feces, *J. Clin. Microbiol.* **19:**259–263.
62. Riepenhoff-Talty, M., Bogger-Goren, S., Li, P., Carmody, P. J., Barrett, H. J., and Ogra, P. L., 1981, Development of serum and intestinal antibody response to rotavirus after acquired rotavirus infection in man, *J. Med. Virol.* **8:**215–222.
63. Ward, R. L., Bernstein, D. I., Shukla, R., Young, E. C., Sherwood, J. R., McNeal, M. M., Walker, C., and Schiff, G. M., 1989, Effects of antibody to rotavirus on protection of adults challenged with a human rotavirus, *J. Infect. Dis.* **159:**79–88.
64. Coulson, B. S., Grimwood, K., Hudson, I. L., Barnes, G. L., and Bishop, R. F., 1992, Role of coproantibody in clinical protection of children during reinfection with rotavirus, *J. Clin. Microbiol.* **30:**1678–1684.
65. Hjelt, K., and Grauballe, P. C., 1990, Protective levels of intestinal rotavirus antibodies, *J. Infect. Dis.* **161:**352–353.
66. Ward, R. L., Pax, K. A., Sherwood, J. R., Young, E. C., Schiff, G. M., and Bernstein, D. I., 1992, Salivary antibody titers in adults challenged with a human rotavirus, *J. Med. Virol.* **36:** 222–225.
67. Losonsky, G. A., and Reymann, M., 1990, The immune response in primary asymptomatic and symptomatic rotavirus infection in newborn infants, *J. Infect. Dis.* **161:**330–332.
68. Aiyar, J., Bhan, M. K., Bhandari, N., Kumar, R., Raj, P., and Sazawal, S., 1990, Rotavirus-specific antibody response in saliva of infants with rotavirus diarrhea, *J. Infect. Dis.* **162:**1383–1384.

69. Bernstein, D. I., McNeal, M. M., Schiff, G. M., and Ward, R. L., 1989, Induction and persistence of local rotavirus antibodies in relation to serum antibodies, *J. Med. Virol.* **28:**90–95.
70. Offit, P. A., Boyle, D. B., Both, G. W., Hill, N. L., Svoboda, Y. M., Cunningham, S. L., Jenkins, R. J., and McCrae, M. A., 1991, Outer capsid glycoprotein VP7 is recognized by cross-reactive, rotavirus-specific, cytotoxic T lymphocytes, *Virology* **184:**563–568.
71. Offit, P. A., and Dudzik, K. I., 1990, Rotavirus-specific cytotoxic T lymphocytes passively protect against gastroenteritis in suckling mice, *J. Virol.* **64:**6325–6328.
72. Dharakul, T., Rott, L., and Greenberg, H. B., 1990, Recovery from chronic rotavirus infection in mice with severe combined immunodeficiency: Virus clearance mediated by adoptive transfer of immune CD 8+ T lymphocytes, *J. Virol.* **64:**4375–4382.
73. Oldham, G., Bridger, J. C., Howard, C. J., and Parsons, K. R., 1993, In vivo role of lymphocyte subpopulations in the control of virus excretion and mucosal antibody responses of cattle infected with rotavirus, *J. Virol.* **67:**5012–5019.
74. Offit, P. A., Hoffenberg, E. J., Pia, E. S., Panackal, P. A., and Hill, N. L., 1992, Rotavirus-specific helper T cell responses in newborns, infants, children, and adults, *J. Infect. Dis.* **165:**1107–1111.
75. Elaraby, I., El-Sharkawy, S., Abbassy, A., and Hussein, M., 1992, A study on delayed hypersensitivity to rotavirus in infancy and childhood, *Ann. Trop. Paediatr.* **12:**83–86.
76. Yasukawa, M., Nakagomi, O., and Kobayashi, Y., 1990, Rotavirus induces proliferative response and augments non-specific cytotoxic activity of lymphocytes in humans, *Clin. Exp. Immunol.* **80:**49–55.
77. Totterdell, B. M., Banatvala, J. E., Chrystie, I. L., Ball, G., and Cubitt, W. D., 1988, Systemic lymphoproliferative responses to rotavirus, *J. Med. Virol.* **25:**37–44.
78. Petschow, B. W., Litov, R. E., Young, L. J. T., and McGraw, T. R., 1992, Response of colostrum-deprived cynomolgus monkeys to intragastric challenge exposure with simian rotavirus strain SA11, *Am. J. Vet. Res.* **53:**674–678.
79. Cordle, C. T., Schaller, J. R., Winship, T. R., Candler, E. L., Hilty, M. D., Smith, K. L., Saif, L. J., Kohler, E. M., and Krakowa, S., 1991, Passive immune protection from diarrhea caused by rotavirus or E. coli: An animal model to demonstrate and quantitate efficacy, in: *Immunology of Milk and the Neonate* (J. Mestecky et al., eds.), Plenum Press, New York, pp. 317–321.
80. Schaller, J. P., Saif, L. J., Cordle, C. T., Candler, E., Winship, T. R., and Smith, K. L., 1992, Prevention of human rotavirus-induced diarrhea in gnotobiotic piglets using bovine antibody, *J. Infect. Dis.* **165:**623–630.
81. Turner, R. B., and Kelsey, D. K., 1993, Passive immunization for prevention of rotavirus illness in healthy infants, *Pediatr. Infect. Dis. J.* **12:**718–722.
82. Ebina, T., Ohta, M., Kanamaru, Y., Yamamoto-Osumi, Y., and Baba, K., 1992, Passive immunizations of suckling mice and infants with bovine colostrum containing antibodies to human rotavirus, *J. Med. Virol.* **38:**117–123.
83. Davidson, G. P., Whyte, P. B. D., Daniels, E., Franklin, K., Nunan, H., McCloud, P. I., Moore, A. G., and Moore, D. J., 1989, Passive immunization of children with bovine colostrum containing antibodies to human rotavirus, *Lancet* **2:**709–712.
84. Pichichero, M. E., 1990, Effect of breast-feeding on oral rhesus rotavirus vaccine seroconversion: A metaanalysis, *J. Infect. Dis.* **162:**753–755.
85. Glass, R. I., Ing, D. J., Stoll, B. J., and Ing, R. T., 1991, Immune response to rotavirus vaccines among breast-fed and nonbreast-fed children, in: *Immunology of Milk and the Neonate* (J. Mestecky et al., eds.), Plenum Press, New York, pp. 249–254.
86. Nakata, S., Chiba, S., Terashima, H., Yokoyama, T., and Nakao, T., 1985, Humoral immunity in infants with gastroenteritis caused by human calicivirus, *J. Infect. Dis.* **152:**274–279.

87. Parrino, T. A., Schreiber, D. S., Trier, J. S., Kapikian, A. Z., and Blacklow, N. R., 1977, Clinical immunity in acute gastroenteritis caused by Norwalk agent, *N. Engl. J. Med.* **297:**86–89.
88. Wyatt, R. G., Dolin, R., Blacklow, N. R., Dupont, H. L., Buscho, R. F., Thornhill, T. S., Kapikian, A. Z., and Chanock, R. M., 1974, Comparison of three agents of acute infectious nonbacterial gastroenteritis by cross-challenge in volunteers, *J. Infect. Dis.* **129:**709–714.
89. Johnson, P. C., Mathewson, J. J., Dupont, H. L., and Greenberg, H. B., 1990, Multiple-challenge study of host susceptibility to Norwalk gastroenteritis in US adults, *J. Infect. Dis.* **161:**18–21.

11

Immunopathogenesis of *Entamoeba histolytica*

SERGIO ARIAS-NEGRETE and KRIS CHADEE

Entamoeba histolytica is an enteric protozoan parasite that infects up to 10% of the world's population resulting in 100,000 deaths per year from amebic colitis and liver abscess.[1] The parasite's distribution is worldwide. Humans are infected by two morphologically identical species of amoebae: *Entamoeba histolytica* Schaudinn, 1903, which is pathogenic and is capable of causing invasive amebiasis (but remains as a harmless commensal in the majority of infections), and *Entamoeba dispar* Brumpt, 1925, which is not pathogenic. Biochemical, immunological, genetic, and clinical characteristics support the existence of the two different and well-defined species.[2] This important finding has clarified the long-standing issue of interconverting amoebae and has explained the low percentage of individuals with clinical symptoms.

1. THE LIFE CYCLE OF *E. HISTOLYTICA*

During the life cycle of *E. histolytica*, three stages are recognized: the trophozoite or invasive form, the cyst which is the infective and transmission form, and the metacyst. When the cyst reaches the small intestine, it excysts releasing a quadrinucleated amoeba which divides twice and gives rise to eight uninucleated daughter trophozoites. The motile trophozoites (10–40 μm) colonize the large intestine, where they multiply by binary fission and may cause lesions or form typical tetranucleated cysts (8–20 μm diameter) to complete the life cycle.[3] The trophozoite surface is highly pleomorphic

SERGIO ARIAS-NEGRETE and KRIS CHADEE • Institute of Parasitology, McGill University, Ste-Anne de Bellevue, Quebec H9X 3V9, Canada.

Enteric Infections and Immunity, edited by Lois J. Paradise *et al.* Plenum Press, New York, 1996.

exhibiting lobopodia, endocytotic stomata, filopodia, and posterior uroid. Trophozoites lack mitochondria, peroxisomes, and a typical Golgi apparatus and endoplasmic reticulum; they also lack a structured cytoskeleton and cytoplasmic microtubules (only observed during nuclear division). The cytoplasm contains numerous digestive vacuoles and a conspicuous nucleus.[3] *E. histolytica* is considered to be a very primitive eukaryote and its metabolic pathways deviate from those of complex eukaryotes.[4]

2. PATHOGENESIS OF *E. HISTOLYTICA*

Pathogenesis is defined as the origination and development of a disease. In *E. histolytica* infections, disease causation involves a series of distinct events: (1) colonization of the colon, (2) disruption or dissolution of nonspecific host defenses (e.g., mucus glycoproteins) by enzymes or toxins (secretagogues), (3) attachment of amoebae to colonic epithelial cells, and (4) lysis of enterocytes and inflammatory cells leading to amebic invasion resulting in colonic ulceration and/or dissemination of the parasites to soft organs causing amebic abscesses.[5]

Adherence Mechanisms

E. histolytica adherence in the gut is important not only for colonization but also for invasion of the large bowel. Colonic biopsies of patients with intestinal amebiasis have shown amebic trophozoites in the lumen and adherent to the mucus layer in areas devoid of ulceration. Areas of focal ulceration were characterized by amoebae in the lesion and mucus depletion.[6] Animal models of the disease have confirmed that the sequence of events involves (1) trophozoite colonization of the mucus layer (Fig. 1A), (2) depletion of goblet cell mucin, (3) amebic adherence to enterocytes (Fig. 1B) followed by cytolysis of the cells, and finally (4) trophozoite invasion into the lamina propria and crypts.[7,8] *In vitro* studies with trophozoites and target cells have elucidated the molecular basis of amebic adherence mechanisms. Even though a variety of surface proteins that mediate adherence to target cells (Table I) have been described in *E. histolytica*, the galactose lectin[9,10] is the most important and is central in pathogenesis. Evidence for this is suggested by the inhibition of adherence and cytolysis of target cells in the presence of galactose, *N*-acetyl-D-galactosamine (Gal/GalNAc), and purified colonic mucins.[11,18] Indeed, the Gal/GalNAc lectin is involved in amebic colonization of the colonic mucus blanket[19] and in adherence and cytolysis of a variety of target cells. The nonreduced lectin (260 kDa) is composed of two subunits of 170 and 35 Kda.[20] The 170-kDa

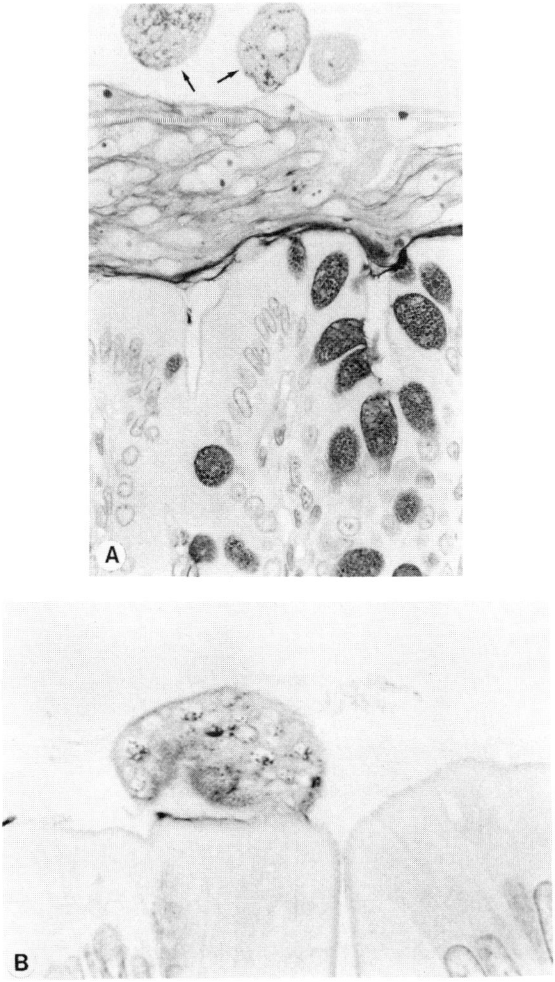

FIGURE 1. (A) Colonization of *Entamoeba histolytica* trophozoites (arrows) in and on the mucus layer of the colon 2 days following intracecal inoculation with amoebae. Periodic acid–Schiff reagent/Alcian blue, × 165. (B) Adherence of *E. histolytica* trophozoites to the interglandular epithelium following mucus depletion 4 days after infection. Periodic acid–Schiff reagent/Alcian blue, × 413.

TABLE I
Adherence Proteins of *Entamoeba histolytica*

Protein (molecular mass)	Receptor	Function(s)	Refs.
Gal/GalNAc lectin (260 kDa: 170 and 35 kDa)	Galactose, *N*-acetylgalactosamine	Adhesion to mucins and diverse cells	9–11
Lectin (220 kDa)	Chitotriose and hyaluronate	Adhesion to red blood cells and MDCK cells	12
Adhesin (112 kDa)	Unknown	Adhesion to erythrocytes and MDCK cells	13
Fibronectin receptor (37 kDa)	Fibronectin and laminin	Mediates adhesion to extracellular matrix	14
Serine-rich *E. histolytica* protein (47/52 kDa) (SREHP)	Unknown	Adhesion to CHO cells	15
LPG (lipophosphoglycoconjugate) (45 and 110 kDa; polydisperse)	Unknown	Adhesion to CHO cells	16, 17

heavy subunit contains the adherence domains but both subunits have to be present in their native form for functional lectin activity.[21]

Secretagogues from E. histolytica

The secretagogues from *E. histolytica* (Table II) constitute secretory/excretory products or are integral parasite components. They may play a role in the pathophysiology of amebic diarrhea or in pathogenesis. Recent studies[29] have shown that *E. histolytica* exhibit potent mucus secretagogue activity by enhancing the secretion of preformed and newly synthesized mucin as well as stimulating colonic glycoprotein biosynthesis. Hypersecretion of mucus contributes to the depletion and alteration of the protective mucus layer and facilitates amebic adherence to mucosal epithelial cells and, thus, amebic invasion. The mechanism whereby amoebae may induce and regulate mucin secretion is a complex phenomenon involving both parasite and host-induced modulators.[19] Intestinal secretions, other than mucins, have been demonstrated with amebic lysates for ileal and colonic electrolyte transport.[26] Intestinal amebic secretagogues may therefore play a role in modulating intestinal secretory mechanisms independent from parasite contact-dependent cell damage.

Contact-Dependent Cytolysis

While it is clear that adherence is important for amebic killing of target cells, it is not known how amoebae actually kill the cells following contact. *In vitro* studies have shown that amoebae can kill target cells immediately following contact or that contacted cells are destined to die at a later

TABLE II
Secretagogues from *Entamoeba histolytica*

Secretagogue	Target tissue/cell	Effect	Refs.
Proteinases (mostly neutral)	Connective tissue	Cytotoxic Enterotoxic?	22
Pore-forming protein	Lipid bilayers	Increased cell permeability to ions	23, 24
Serotonin	Gut	Electrolyte transport	25
Substance P	Gut	Electrolyte transport	26
Neurotensin (vasoactive peptide)	Gut, blood vessels	Increased vascular permeability, electrolyte transport	26
Prostaglandin, fatty acids	Gut, all cells	Mucus secretion, electrolyte transport, immunomodulation	26–28

time.[30] It is possible that following amebic contact via the Gal/GalNAc adherence lectin, diacylglycerol production directly activates protein kinase C (PKC) resulting in the phosphorylation of putative regulatory substances that are involved in the cytolytic process. PKC from *E. histolytica* has recently been described and activators of PKC increase amebic cytolytic activity.[31] The major parasite factors that are presumed to be involved in amebic killing include acid pH vesicles, pore-forming proteins, phospholipases, and secreted proteinases.

Acid pH Vesicles

E. histolytica does not contain cytoplasmic granules, but does possess numerous acidic (pH 5.1) cytoplasmic (endocytotic) vesicles. Increasing vesicle pH (> 5.7) with weak bases (NH_4Cl, primaquine, chloroquine) inhibited amebic killing. This inhibitory effect persisted for at least 2 hr in control medium following prolonged elevation of amebic vesicle pH with NH_4Cl, despite a rapid return of vesicle pH to normal values.[32] Amebic cytolytic activity was also reduced by inhibition of protein synthesis via cycloheximide treatment of trophozoites. An immediate increase in vesicle pH can reduce activity of acid optimal enzymes in vesicles (e.g., calcium-independent phospholipase A) or may be required for the solubility of the amebic pore-forming protein (see below). The effect of NH_4Cl and cycloheximide may be analogous in that a sustained increase in vesicle pH may cause mistargeting of enzymes (or proteins) normally routed to lysosomes which are important for amebic cytolysis. In fact, the increased vesicle exocytosis that occurs during amebic killing of target cells was inhibitable by calcium antagonists TMB-8 and bepridil and the microfilament inhibitor cytochalasin D.[33] However, attempts to measure [^{125}I]dextran (previously loaded into acid pH vesicles of amoebae) release onto dead target cells have been unsuccessful. Morphologic studies have demonstrated that amebic cytoplasmic vesicles are in close proximity to the site of interaction of the parasite and target cells; however, there is no evidence of membrane fusion or exocytosis at these sites. At present, indirect evidence remains for the participation of vesicle constituents in amebic cytolysis.

Pore-Forming Proteins

An amebic ion channel or pore-forming protein (amebapore) has been described.[23,24] *E. histolytica* secretes a high-speed sedimentable protein known as amebapore (AP; 13–15 kDa by SDS–PAGE analysis). AP spontaneously inserts into artificial lipid bilayers and cell membranes and causes a rapid collapse of membrane potential because of this channel-forming ability, which is moderately cation (Na or K) selective. AP was found associated with highly dense, small-diameter particles (M_r 28 kDa). Two forms of AP exist differing in their pI (6.8 for α-AP and 5.3 for β-AP),[34] and their incorporation into membranes is dependent on pH. The cDNA encoding the precursor of the pore-forming peptide has been cloned and the deduced amino acid sequence reveals a putative signal peptide and a mature peptide of 77 amino acids, including six cysteine residues. It is similar to melittin, the membranolytic peptide of bee venom.[35,36] At present, it is not known how AP contributes to cytolysis.

Phospholipases

E. histolytica contains two phospholipase A_2 enzymes: a calcium-independent enzyme maximally active at acid pH and a calcium-dependent plasma membrane enzyme with an optimal pH of 7.5.[37] Phospholipase A_2 hydrolyzes diacylphospholipids such as phosphatidylcholine, with the production of cytolytic lysodiacylphospholipids and free fatty acids which have detergentlike effects on cell membranes and could kill cells. The evidence suggesting that amebic phospholipase activity is important in cytolysis of target cells includes the inhibition of amebic killing by calcium chelators and phospholipase A_2 pharmacologic antagonists.[38]

Proteinases

E. histolytica proteinases have traditionally been implicated in pathogenesis even though direct evidence for this is scarce. In general, the proteinases secreted by amoebae consist of a major thiol neutral proteinase,[39] cathepsin B,[40] a collagenase,[41] and a cysteine proteinase.[42] Recent studies[43] have also demonstrated proteolytic activity on the surface of glutaraldehyde-fixed trophozoites that resides in a 70-kDa plasma membrane protein. The neutral proteinase causes cell rounding and detachment from substrates but is not cytotoxic.[39] The enzyme also degrades fibronectin, laminin, and type 1 collagen, as well as a model of connective tissue extracellular matrix, suggesting a role for this enzyme in interrupting cell-to-cell adhesion which may participate in intestinal invasion. Similar enzyme activities including cytotoxic/enterotoxic effects have been reported for other amebic proteinases[22] (Table II). There is also a good correlation between proteinase production and amebic virulent potential suggesting a role for amebic proteinases in pathogenesis even though direct evidence for this is lacking.[44]

It is clear that the pathogenesis of *E. histolytica* infection involves multifactorial mechanisms. While the molecular events of amebic adherence to its relevant host receptor, colonic mucin, are known to occur via the amebic Gal/GalNAc adherence lectin for colonization and initiation of invasion, the events following adherence to target cells are unclear. *E. histolytica* infections may or may not lead to disease; however, it is quite likely that local pathophysiologic events may occur as a direct result of the numerous secretagogues released by the parasites that may predispose the host to colonic invasion.

3. PATHOGENESIS OF INVASIVE AMEBIASIS

Following depletion or dissolution of the protective mucus barrier, *E. histolytica* trophozoites attach to and indiscriminately kill the mucosal epithelial cells.[7,8,19] Systemic infections occur when trophozoites gain entry into the blood vessels and lymphatics and are disseminated to soft organs (most notably the liver because of the hepatic portal vein drainage from the intestine) following colonic invasion. One of the hallmarks of trophozoites lodged in the hepatic sinusoids is a massive neutrophilic [polymorphonuclear leukocytes (PMNs)] response.[45,46] *E. histolytica* trophozoites contain a membrane-bound heat-stable PMN chemoattractant that may be responsible for the recruitment of these cells toward amoebae in tissues.[47] The PMNs are rapidly killed on direct contact with the parasite and the released PMN contents (e.g., lysosomal enzymes) cause lysis of adjacent hepatocytes that establishes the initial lesions of liver amebiasis. *In vitro* studies have shown that lysis of human PMNs by amoebae enhances destruction of Chang liver cells by nonoxidative product,[48] and in *in vivo* animal models of the disease, massive PMN infiltration and killing of the cells results in destruction of hepatocytes adjacent to and at a distance from the amoebae.[45,46] Thus, PMNs are of importance in the establishment of the initial hepatic lesions and in the acute stages of infection, whereas in colonic amebiasis, PMNs may serve to aggravate the pathologic response.

4. IMMUNE RESPONSE IN AMEBIASIS

The immune system is sensitized by *E. histolytica* antigens during invasion. Suppression of immune functions by steroid therapy, thymectomy, splenectomy, and with anti-immune cell antibodies results in exacerbation of amebiasis in animal models. Moreover, a low recurrence of reinfection by *E. histolytica* is indicative of the development of a protective immune response. However, primary invasive infection by amoebae is characterized by a transient state of immunosuppression affecting both antibody- and cell-mediated immunity.[49,50]

4.1. Humoral Immunity

Specific antiamebic antibodies are detected in 95% of patients with amebic liver abscess.[51] Invasive liver or intestinal amebiasis induces a long-lasting immune response which consists of specific IgG and IgM antiamebic antibodies.[52] Coproantibodies have been shown in 80% of cases of amebic dysentery patients to be of the IgA class. However, these antibodies only persist for a short time.[53,54] It is not known whether secretory IgA against amoebae might play a protective role against amebic colonization or invasion. Even in individuals with asymptomatic *E. histolytica* infections, there are serum antiamebic antibodies present. Sera from patients with invasive amebiasis recognize defined and well-studied amebic proteins such as the Gal lectin of *E. histolytica.*[55] The anti-Gal lectin antibodies have predictive values in the early stages of invasive amebiasis, and these have been utilized to monitor patients with amebic liver abscess.[56] Anti-Gal lectin antibodies may play a role in protective immunity as salivary secretory IgA anti-Gal lectin antibodies inhibited amebic adherence to epithelial cells *in vitro.*[57] Protective immunity in experimental hepatic amebiasis has been induced by using the native[58] or recombinant[59] 170-kDa Gal lectin or a serine-rich *E. histolytica* protein (SREHP).[60] Interestingly, passive immunization with anti-SREHP antibodies also prevents the development of amebic liver abscesses in SCID mice.[61] At present, there is no evidence to suggest that antiamebic antibody titer correlates with the clinical outcome of the disease or plays a role in protective immunity. *E. histolytica* trophozoites evade humoral response by capping and shedding antibodies.[62] In addition, they are resistant to complement-mediated lysis[63] even though they can activate both the classical and alternative pathways of complement.[64] The Gal lectin binds C8 and C9 and thus inhibits the assembly of the membrane attack complex.[65]

4.2. Cell-Mediated Immunity

In contrast to the humoral immune response, there is a wealth of information indicating that cell-mediated immunity plays a central role in the control of and resistance to reinfection with *E. histolytica.* While macrophage-mediated effector mechanisms are the most important, their functions are regulated by lymphokines produced by T lymphocytes or macrophages themselves. Other phagocytes may play a role in host defense against amebiasis in immune individuals as neutrophils activated with IFN-γ alone or in combination with TNF-α were shown to kill amoebae *in vitro*.[66]

4.2.1. *Role of T Cells*

Immunity mediated by T lymphocytes is important in amebiasis since T cells can perform cytotoxic activity against *E. histolytica* trophozoites or regulate cell immune functions by producing lymphokines. Human T cells ($CD8^+$) isolated from healthy donors and stimulated with PHA kill *E. histolytica* trophozoites.[67] Likewise, supernatants derived from T cells stimulated with amebic proteins or mitogens activate monocyte-derived macrophages for oxidative and nonoxidative amebicidal activity.[68] Recently, it was reported that peripheral blood mononuclear cells from patients cured of invasive amebiasis proliferate and produce IL-2 and IFN-γ and kill amoebae in response to the Gal lectin.[69] In the gerbil model for intestinal amebiasis, development of an effective cytotoxic T-cell response *in vitro* and *in vivo* against *E. histolytica* was noticed only after 20 and 30 days postinfection; at the onset of infection, no differences with the control group were found.[70]

During acute hepatic amebiasis, patients were shown to have a low $CD4^+/CD8^+$ lymphocyte ratio relative to sex- and age-matched healthy controls.[68] Similarly, mice hyperimmunized with amebic antigens show a reduction in Thy-1.2$^+$ and Lyt-1$^+$ cells; this T-cell imbalance could cause a delay in the development of cell-mediated immunity.[71] During amebic invasion, both mouse and human T cells proliferate in response to amebic protein stimulation but not to concanavalin A (Con A), phytohemagglutinin A (PHA), or lipopolysaccharide (LPS).[68,71] This transient immunosuppression could be advantageous for *E. histolytica* invasion.

Interestingly, human sera containing antiamebic antibodies suppress proliferation and IFN-γ production by peripheral lymphocytes from patients cured of amebic liver abscess. This effect does not depend on antiamebic antibody titers as preabsorption with viable trophozoites did not abolish the suppressive activity.[72] It is evident that during invasion *E. histolytica* releases a variety of products that can sensitize different T-cell subpopulations. It would be of interest in future studies to identify the amebic antigens that stimulate protective T-cell clones in order to circumvent the immunosuppressive activity of the parasite.

4.2.2. *Macrophage Functions in Amebiasis*

Macrophages play an important effector role in cell-mediated immunity against *E. histolytica.* Murine bone marrow-derived macrophage precursors have strong amebicidal activity which is lost during *in vitro* maturation to mature macrophages.[73] Human monocyte-derived macrophages activated with crude lymphokines or IFN-γ kill *E. histolytica* trophozoites.[74] From lymphocytes isolated from patients treated for amebic liver abscess, amebic proteins or Con A can elicit lymphokines for the activation of naive monocytes to kill amoebae.[68] With murine macrophages, the acquisition of macrophage amebicidal activity occurs after treatment with IFN-γ alone or synergistically with LPS, rTNF-α, or CSF-1.[73] The mechanism by which cytokine-activated macrophages kill *E. histolytica* trophozoites is contact-dependent and involves both oxygen- and non-oxygen-dependent pathways. *E. histolytica* is susceptible to H_2O_2 and hyperchlorite when exogenously added to culture media. Studies with the glucose–glucose oxidase and xanthine–xanthine oxidase systems generating O_2, H_2O_2, and OH enzymatically have shown that amebic killing is inhibited by catalase, but not by superoxide dismutase or scavengers of the hydroxyl radical and singlet oxygen.[75] In IFN-γ- and LPS-activated macrophages, the addition of catalase inhibits amebic killing by 45 and 61%, respectively, with the protease inhibitor, tosyl-lysyl chloromethyl ketone.[73] We have recently demonstrated, however, that nitric oxide (NO, measured as NO_2^-) produced by activated macrophages is the major cytotoxic molecule for amoebae and that H_2O_2 and O_2^- may be cofactors for the NO effector molecule.[76] Evidence for this was suggested by the fact that NO-dependent amebicidal activity is abrogated by the addition of arginase (to deplete arginine) or N^G-monomethyl L-arginine (an L-arginine analogue) to culture medium, without affecting H_2O_2 production. The relative contribution of H_2O_2 and NO produced by activated macrophages for amebicidal activity is shown in Fig. 2. Amebicidal activity was only optimal under conditions of high NO_2^- production. In control or LPS-stimulated macrophages, even though H_2O_2 production was high, NO_2^- production remained low and trophozoite viability was not affected, indicating a major role of NO in killing amoebae. Interestingly, catalase, an H_2O_2 scavenger, markedly inhibits both H_2O_2 and NO production with a corresponding decrease in amebicidal activity, whereas superoxide dismutase, a scavenger of O_2^-, partially inhibits amebicidal activity without influencing NO production. More recently, we have determined that TNF-α produced by activated macrophages is involved in the induction of the inducible *macNOS* gene for NO-dependent cytotoxicity against *E. histolytica* which may be associated with the accumulation of TNF-α mRNA.[77] The evidence for this is as follows: (1) IFN-γ synergizes with TNF-α for NO-dependent amebicidal activity, (2) TNF-α plays an autocrine role for NO-dependent amebicidal activity by IFN-γ- plus LPS-activated macrophages, and (3) IFN-γ in combination with exogenous or endogenous TNF-α acts synergistically to increase the expression of both *macNOS* and TNF-α genes, with a concomitant increase in NO production. The kinetics of *macNOS* and

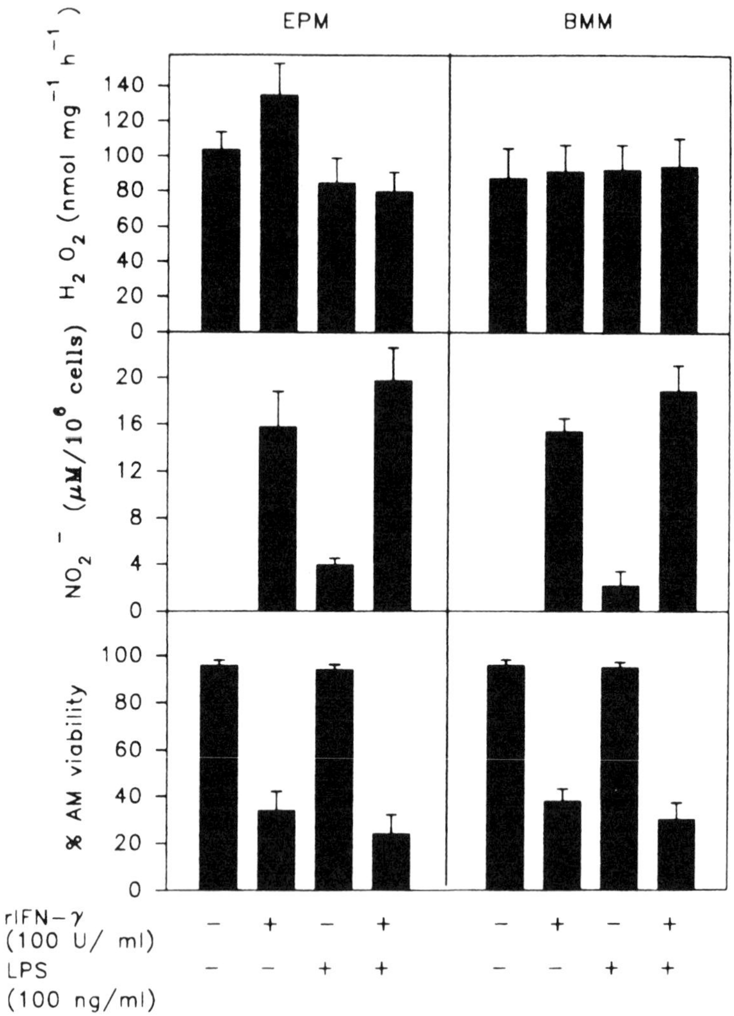

FIGURE 2. Relationship between amebicidal activity and NO_2^- and H_2O_2 production by activated macrophages. Murine macrophages (EPM, proteose peptone-elicited; BMM, bone marrow-derived) were exposed to IFN-γ alone or in combination with LPS for 24 hr before determining PMA-stimulated H_2O_2 release, NO_2^- production, and amebicidal activity. Reprinted with permission from Lin and Chadee.[76]

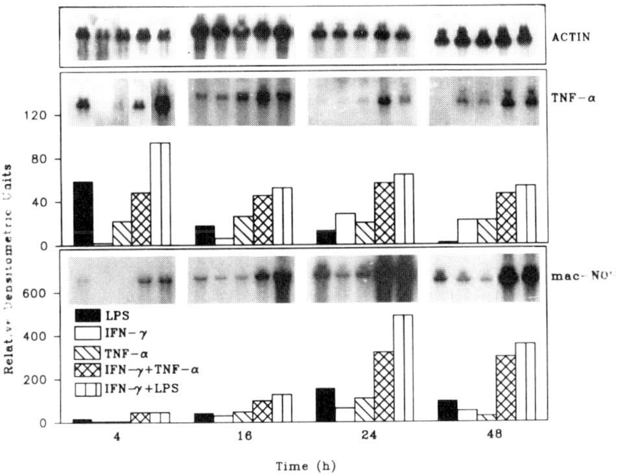

FIGURE 3. Kinetics of induction of both *macNOS* and TNF-α gene expression by IFN-γ plus TNF-α/LPS stimulation. Murine bone marrow macrophages were stimulated with IFN-γ (100 U/ml) plus TNF-α (100 U/ml)/LPS (100 ng/ml) for 4, 16, 24, and 48 hr. Levels of *macNOS* and TNF-α mRNAs were then determined by Northern blot analysis. Results of scanning densitometric analysis of the Northern blot are presented as a histogram. The relative levels of *macNOS* and TNF-α mRNA were determined after normalization of the respective actin signal to account for variability in the amount of RNA. Lane designations are identical for the blot and the histogram. Reprinted with permission from Lin *et al.*[77]

TNF-α mRNA synthesis in macrophages stimulated with IFN-γ plus TNF-α/LPS is shown in Fig. 3. IFN-γ plus TNF-α/LPS induces *macNOS* mRNA synthesis after 4 hr of stimulation. The response peaks at 24 hr and decreases slightly after 48 hr. High levels of TNF-α mRNA were observed after 4 hr of stimulation with IFN-γ plus LPS, and decreased slowly after 48 hr. However, the high TNF-α mRNA levels induced by IFN-γ plus TNF-α did not differ during 4 to 48 hr of stimulation. High TNF-α mRNA expression was observed after 4 hr of stimulation with LPS which returned to basal levels after 48 hr, whereas TNF-α mRNA levels remained low in response to IFN-γ or TNF-α stimulation. Taken together, these results suggest that TNF-α augments NO-dependent macrophage cytotoxicity against *E. histolytica* through elevated levels of *macNOS* and TNF-α gene expression.

4.3. Immunomodulation by *E. histolytica*

Invasive amebiasis is characterized by a transient state of immunosuppression during the early stages of the disease. However, it is unclear what cells are affected or the identity of the amebic molecules involved in immunomodulation. Early studies[78,79] have shown that *E. histolytica* components are mitogenic for murine splenic lymphocytes. Specifically, aqueous, delipidated amebic extracts induce the proliferation of a subset of non-T cells in the spleen of athymic nude mice and it was hypothesized that polyclonal activation of lymphocytes by amoebae might play a role in impairing cell-mediated immunity in amebiasis. Recently,[71] it was shown that injection of mice with virulent *E. histolytica*, but not the nonpathogenic *E. moshkovskii* (Laredo strain), proteins results in a decreased capacity of splenocytes to respond to mitogen-induced blastogenesis. A decrease in the ratio of $CD4^+/CD8^+$ T lymphocytes has been described in the peripheral blood of patients recently treated for amebic liver abscess that returned to normal 1 year following treatment.[68] Interestingly, factors are also present in the serum of patients with amebic liver abscess that inhibit *E. histolytica* antigen-induced lymphocyte proliferation and IFN-γ production.[72] Thus, suppressor T-cell activity as well as serum factors may contribute to the uncontrolled progression of disease observed in untreated amebic liver abscess patients.

Animal models of the disease and *in vitro* studies have unraveled the mechanism of immunomodulation by *E. histolytica* on macrophage effector and accessory cell functions. Macrophages isolated from the amebic liver granulomas were shown to be defective in their capacity to develop a respiratory burst, to secrete and express interleukin 1, to produce NO, and to kill amoebae as well as to respond to IFN-γ/LPS stimulation.[80,81] However, they produce low basal levels of TNF-α and high constitutive basal release of PGE_2 and LTC_4.[82,83] Interestingly, peritoneal and splenic macrophages isolated from infected animals are not defective and are in a heightened state of activation, suggesting that immunosuppression is a local event associated with macrophages in close proximity to the parasites.[81–83] A similar downregulation in macrophage functions is obtained when naive macrophages are pretreated with soluble amebic proteins *in vitro*. More specifically, amebic proteins enhance PGE_2 and LTC_4 production,[83] suppress LPS-induced TNF-α production,[82] decrease TNF-α and *macNOS* gene expression and NO production,[81] downregulate IFN-γ-induced macrophage Ia expression,[84] and inhibit the priming effect of LPS- or IFN-γ-induced macrophage respiratory burst potential.[85] At present, the mechanisms whereby *E. histolytica* modulates macrophage cytotoxic activity (effector functions) via suppression of TNF-α and NO production are not known. However, the fact that indomethacin, a cyclooxygenase inhibitor, can enhance TNF-α production by amebic granuloma macrophages and naive macrophages pretreated with amebic proteins [81,82] suggests that PGE_2 may be an important immunomodulator in amebiasis. It is apparent, therefore, that there are two pathways by which *E. histolytica* can inhibit macrophage effector and accessory functions, either through suppression of TNF-α, which is inhibited by PGE_2, or through NO by an unknown mechanism. It is also becoming clear that *E. histolytica* can stimulate the early expression of TNF-α, IL-1α/β, and *c-fos* gene expression via PKC signal transduction which rapidly degrades and, thus, the cell may be refractory to further activating signals.[86] Therefore, in addition to stimulating inflammatory mediators (e.g., PGE_2), the parasite may specifically modulate the expression and translation of monokines that are involved in macrophage activation (e.g., TNF-α, iNOS).

5. SUMMARY

E. histolytica trophozoites colonize the gut by adhering to the Gal and GalNAc residues of mucus glycoproteins via the Gal lectin on the surface of the amoebae. Following depletion of the protective mucus barrier, the trophozoites adhere to and lyse colonic epithelial cells and disseminate to systemic sites. Neutrophils in contact with amoebae are lysed and the released proteolytic enzymes destroy tissues, establishing the initial amebic lesions in soft organs. A prompt antibody response primarily of the IgG class occurs, but it is not protective. Rather, cell-mediated immunity mediated by macrophages is central in host defense against *E. histolytica.* NO produced by IFN-γ plus LPS/TNF-α-activated macrophages is the major cytotoxic molecule against amoebae. Reactive oxygen intermediates (O_2^-, H_2O_2) and proteases may synergize or act as cofactors for the NO effector molecule. *E. histolytica* evades the host immune response by suppressing macrophage TNF-α and NO production by modulating monokine (TNF-α, IL-1α/β) mRNA expression as well as stimulating the production of PGE_2 which acts in an autocrine fashion to suppress the immune system.

ACKNOWLEDGMENTS. Research in the laboratory of K.C. is supported by grants from the Natural Sciences and Engineering Research Council of Canada and the Medical Research Council of Canada. Financial support for S.A.-N. was provided by CONACYT and the University of Guanajuato (Mexico).

REFERENCES

1. Walsh, J. A., 1986, Problems in recognition and diagnosis of amebiasis: Estimation of the global magnitude of morbidity and mortality, *Rev. Infect. Dis.* **8:**228–238.

2. Diamond, L. S., and Clark, C. G., 1993, A redescription of *Entamoeba histolytica* Schaudinn, 1903 (Emended Walker, 1911) separating it from *Entamoeba dispar* Brump, 1925, *J. Eukaryotic Microbiol.* **40:**340–344.
3. Martínez-Palomo, A., 1988, Biology of amebiasis: Progress and perspectives, in: *The Biology of Parasitism* (P. T. Englund and A. Sher, eds.), Liss, New York, pp. 61–76.
4. Bakker-Grunwald, T., and Wöstmann, C., 1993, *Entamoeba histolytica* as a model for the primitive eukaryotic cell, *Parasitol. Today* **9:**27–31.
5. Ravdin, J. I., 1988, Pathogenesis of amebiasis: An overview, in: *Amebiasis: Human Infection by Entamoeba histolytica* (J. I. Ravdin, ed.), Wiley, New York, pp. 166–176.
6. Prathap, K., and Gilman, R., 1970, The histopathology of acute intestinal amebiasis, *Am. J. Pathol.* **60:**229–245.
7. Chadee, K., and Meerovitch, E., 1985, *Entamoeba histolytica*: Early progressive pathology in the cecum of the gerbil (*Meriones unguiculatus*), *Am. J. Trop. Med. Hyg.* **34:**283–291.
8. Chadee, K., and Meerovitch, E., 1985, The pathology of experimentally-induced cecal amebiasis in gerbils, *Am. J. Pathol.* **119:**485–494.
9. Tannich, E., Ebert, F., and Horstmann, R. D., 1991, Primary structure of the 170 kDa surface lectin of pathogenic *Entamoeba histolytica*, *Proc. Natl. Acad. Sci. USA* **88:**1849–1853.
10. Mann, B. J., Torian, B. E., Vedvick, T. S., and Petri, W. A., 1991, Sequence of a cysteine-rich galactose-specific lectin of *Entamoeba histolytica*, *Proc. Natl. Acad. Sci. USA* **88:**3248–3252.
11. Chadee, K., Petri, W. A., Innes, D. J., and Ravdin, J. I., 1987, Rat and human colonic mucins bind to and inhibit adherence lectin of *Entamoeba histolytica*, *J. Clin. Invest.* **80:**1245–1254.
12. Meza, I., Cázares, F., Rosales-Encina, J. L., Talamás-Rohana, P., and Rojkind, M., 1987, Use of antibodies to characterize a 220-kilodalton surface protein from *Entamoeba histolytica*, *J. Infect. Dis.* **156:**798–805.
13. Arroyo, R., and Orozco, E., 1987, Localization and identification of an *Entamoeba histolytica* adhesin, *Mol. Biochem. Parasitol.* **23:**151–158.
14. Talamás-Rohana, P., and Meza, I., 1988, Interaction between pathogenic amebas and fibronectin: Substrate degradation and changes in cytoskeleton organization, *J. Cell Biol.* **106:**1787–1794.
15. Stanley, S. L., Becker, A., Kunz-Jenkins, C., Foster, L., and Li, E., 1990, Cloning and expression of a membrane antigen of *Entamoeba histolytica* possessing multiple tandem repeats, *Proc. Natl. Acad. Sci. USA* **87:**4976–4980.
16. Stanley, S. L., Huizenga, H., and Li, E., 1992, Isolation and partial characterization of a surface glycoconjugate of *Entamoeba histolytica*, *Mol. Biochem. Parasitol.* **50:**127–138.
17. Bhattacharya, A., Prasad, R., and Sacks, D. L., 1992, Identification and partial characterization of a lipophosphoglycan from a pathogenic strain of *Entamoeba histolytica*, *Mol. Biochem. Parasitol.* **56:**161–178.
18. Ravdin, J. I., and Guerrant, R. L., 1981, Role of adherence in cytopathogenic mechanisms of *Entamoeba histolytica*. Study with mammalian tissue culture cells and human erythrocytes, *J. Clin. Invest.* **68:**1305–1313.
19. Tse, S.-K., and Chadee, K., 1991, The interaction between intestinal mucus glycoproteins and enteric infections, *Parasitol. Today* **7:**163–172.
20. Petri, W. A., Chapman, M. D., Snodgrass, T., Mann, B. J., Broman, J., and Ravdin, J. I., 1989, Subunit structure of the galactose and N-acetyl-D-galactosamine-inhibitable adherence lectin of *Entamoeba histolytica*, *J. Biol. Chem.* **264:**3007–3012.
21. Mann, B. J., Chung, C. Y., Dodson, J. M., Ashley, L. S., Braga, L. L., and Snodgrass, T. L., 1993, Neutralizing monoclonal antibody epitopes of the *Entamoeba histolytica* galactose adhesion map to the cysteine-rich extracellular domain of the 170-kilodalton subunit, *Infect. Immun.* **61:**1772–1778.
22. McKerrow, J. H., Sun, E., Rosenthal, P. J., and Bouvier, J., 1993, The proteases and pathogenicity of parasitic protozoa, *Annu. Rev. Microbiol.* **47:**821–853.

23. Lynch, E. C., Rosenberg, I. M., and Gitler, C., 1982, An ion-channel forming protein produced by *Entamoeba histolytica*, *EMBO J.* **1**:801–804.
24. Young, J. D., Young, T. M., Lu, L. P., Unkeless, J. C., and Cohn, Z. A., 1982, Characterization of a membrane pore-forming protein from *Entamoeba histolytica*, *J. Exp. Med.* **156**:1677–1690.
25. McGowan, K., Kane, A., Asarkof, N., Wicks, J., Guerina, V., Kellum, J., Baron, S., Gintzler, A. R., and Donowitz, M., 1983, *Entamoeba histolytica* causes intestinal secretion: Role of serotonin, *Science* **221**:762–764.
26. McGowan, K., and Donowitz, M., 1988, Secretagogues of *Entamoeba histolytica*: Role of diarrhea, in: *Amebiasis: Human Infection by Entamoeba histolytica* (J. I. Ravdin, ed.), Wiley, New York, pp. 283–296.
27. Saíd-Fernández, S., and López-Revilla, R., 1988, Free fatty acids released from phospholipids are the major heat-stable hemolytic factor of *E. histolytica* trophozoites, *Infect. Immun.* **56**:874–879.
28. Belley, A., and Chadee, K., 1995, Eicosanoid production by parasites: From pathogenesis to immunomodulation, *Parasitol. Today* **11**:327–334.
29. Chadee, K., Keller, K., Forstner, J., Innes, D. J., and Ravdin, J. I., 1991, Mucin and non-mucin secretagogue activity of *Entamoeba histolytica* and cholera toxin in rat colon, *Gastroenterology* **100**:986–997.
30. Petri, W. A., and Ravdin, J. I., 1988, *In vitro* models of amebic pathogenesis, in: *Amebiasis: Human Infection by Entamoeba histolytica* (J. I. Ravdin, ed.), Wiley, New York, pp. 191–204.
31. Weikel, C. S., Murphy, C. F., Orozco, E., and Ravdin, J. I., 1988, Phorbol esters specifically enhance cytolytic activity of *Entamoeba histolytica*, *Infect. Immun.* **56**:1485–1491.
32. Ravdin, J. I., Schlesinger, P. H., Murphy, C. F., Gluzman, I. Y., and Krogstad, D. J., 1986, Acid intracellular vesicles and the cytolysis of mammalian target cells by *Entamoeba histolytica* trophozoites, *J. Protozool.* **33**:478–486.
33. Ravdin, J. I., Murphy, C. F., and Schlesinger, P. H., 1988, The effect of increased vesicle pH on the cytopathogenicity of *Entamoeba histolytica*, *J. Protozool.* **35**:159–163.
34. Rosenberg, I., Bach, D., Loew, L. M., and Gitler, C., 1989, Isolation, characterization and partial purification of a transferable membrane channel (amebapore) produced by *Entamoeba histolytica*, *Mol. Biochem. Parasitol.* **33**:237–248.
35. Leippe, M., Ebel, S., Schoenberger, O. L., Horstmann, R. D., and Müller-Eberhard, H. J., 1991, Pore-forming peptide of pathogenic *Entamoeba histolytica*, *Proc. Natl. Acad. Sci. USA* **88**:7659–7663.
36. Leippe, M., Tannich, E., Nickel, R., van der Goot, G., Pattus, F., Horstmann, R. D., and Müller-Eberhard, H. J., 1992, Primary and secondary structure of the pore-forming peptide of pathogenic *Entamoeba histolytica*, *EMBO J.* **11**:3501–3506.
37. Long-Krug, S. A., Fischer, K. J., Hysmith, R. M., and Ravdin, J. I., 1985, Phospholipase A enzymes of *E. histolytica*: Description and subcellular localization, *J. Infect. Dis.* **152**:536–541.
38. Ravdin, J. I., Murphy, C. F., Guerrant, R. L., and Long-Krug, S. A., 1985, Effect of antagonists of calcium and phospholipase A on the cytopathogenicity of *E. histolytica*, *J. Infect. Dis.* **152**: 542–549.
39. Keene, W. E., Petit, M. G., Allen, S., and McKerrow, J. H., 1986, The major neutral proteinase of *Entamoeba histolytica*, *J. Exp. Med.* **163**:536–549.
40. McLaughlin, J., and Faubert, G., 1976, Partial purification and some properties of a neutral sulfhydryl and an acid proteinase from *Entamoeba histolytica*, *Can. J. Microbiol* **23**:420–425.
41. Muñoz, M. L., Calderón, J., and Rojkind, M., 1982, The collagenase of *Entamoeba histolytica*, *J. Exp. Med.* **155**:42–51.
42. Scholze, H., and Werries, E., 1984, A weakly acid protease has a powerful proteolytic activity in *Entamoeba histolytica*, *Mol. Biochem. Parasitol.* **11**:293–300.
43. Avila, E. E., and Calderón, J., 1993, *Entamoeba histolytica*: A surface-associated cysteine protease, *Exp. Parasitol* **76**:232–241.

44. Keene, W. E., Hidalgo, M. E., Orozco, E., and McKerrow, J. H., 1990, Correlation of the cytopathic effect of virulent trophozoites with secretion of a cysteine proteinase, *Exp. Parasitol.* **71:**199–206.
45. Chadee, K., and Meerovitch, E., 1984, The pathogenesis of experimentally-induced amebic liver abscess in the gerbil (*Meriones unguiculatus*), *Am. J. Pathol.* **117:**71–80.
46. Tsutsumi, V., Mena-López, R., Anaya-Velázquez, F., and Martínez-Palomo, A., 1984, Cellular bases of experimental liver abscess formation, *Am. J. Pathol.* **117:**81–91.
47. Chadee, K., Moreau, F., and Meerovitch, E., 1987, *Entamoeba histolytica*: Chemoattractant activity for gerbil neutrophils *in vivo* and *in vitro*, *Exp. Parasitol.* **64:**12–23.
48. Salata, R. A., and Ravdin, J. I., 1986, The interaction of human neutrophils and *Entamoeba histolytica* increases cytopathogenicity for liver cell monolayers, *J. Infect. Dis.* **154:**19–26.
49. Meerovitch, E., and Chadee, K., 1988, *In vivo* models of immunity of amebiasis, in: *Amebiasis: Human Infection by Entamoeba histolytica* (J. I. Ravdin, ed.), Wiley, New York, pp. 425–437.
50. Trissl, D., 1982, Immunology of *Entamoeba histolytica* in humans and animal hosts, *Rev. Infect. Dis.* **4:**1154–1184.
51. Joyce, M. P., and Ravdin, J. I., 1988, Antigens of *Entamoeba histolytica* recognized by immune sera from liver abscess patients, *Am. J. Trop. Med. Hyg.* **38:**74–80.
52. Schulz, T. F., Kollaritsch, H., Hengster, P., Stemberger, H., Scheiner, O., Wiedemann, G., and Dierich, M. P., 1987, Molecular weight analysis of *Entamoeba histolytica* antigens recognized by IgG and IgM antibodies in the sera of patients with amebiasis, *Trop. Med. Parasitol.* **38:**149–152.
53. Shaalan, M., and Baker, R. P., 1970, Detection of coproantibodies in amebiasis of the colon: A preliminary report, *Am. J. Clin. Pathol.* **54:**615–617.
54. Mahajan, R. C., Agarwal, S. C., Chhuttani, P. N., and Chitkara, N. L., 1972, Coproantibodies in intestinal amebiasis, *Indian J. Med. Res.* **60:**547–550.
55. Petri, W. A., Jr., Joyce, M. P., Broman, J., Smith, R. D., Murphy, C. F., and Ravdin, J. I., 1987, Recognition of the galactose- or N-acetylgalactosamine-binding lectin of *Entamoeba histolytica* by human immune sera, *Infect. Immun.* **55:**2327–2331.
56. Abd-Alla, M. D., El-Hawey, A. M., and Ravdin, J. I., 1992, Use of an enzyme-linked immunosorbent assay to detect anti-adherence protein antibodies in sera of patients with invasive amebiasis in Cairo, Egypt, *Am. J. Trop. Med. Hyg.* **47:**800–804.
57. Carrero, J. C., Díaz, M. Y., Viveros, M., Espinoza, B., Acosta, E., and Ortiz-Ortiz, L., 1994, Human secretory immunoglobulin A anti-*Entamoeba histolytica* antibodies inhibit adherence of amebae to MDCK cells, *Infect. Immun.* **62:**764–767.
58. Petri, W. A., and Ravdin, J. I., 1991, Protection of gerbils from amebic liver abscess by immunization with the galactose-specific adherence lectin of *Entamoeba histolytica*, *Infect. Immun.* **59:**97–101.
59. Zhang, T., and Stanley, S. L., 1994, Protection of gerbils from amebic liver abscess by immunization with a recombinant protein derived from the 170-kilodalton surface adhesin of *Entamoeba histolytica*, *Infect. Immun.* **62:**2605–2608.
60. Zhang, T., Cieslak, P. R., and Stanley, S. L., 1994, Protection of gerbils from amebic liver abscess by immunization with a recombinant *Entamoeba histolytica* antigen, *Infect. Immun.* **62:**1166–1170.
61. Zhang, T., Cieslak, P. R., Foster, L., Kunz-Jenkins, C., and Stanley, S. L., 1994, Antibodies to the serine rich *Entamoeba histolytica* protein (SREHP) prevent amoebic liver abscess formation in severe combined immunodeficient (SCID) mice, *Parasite Immunol.* **16:**225–230.
62. Calderón, J., Muñoz, M. L., and Acosta, H. M., 1980, Surface redistribution and release of antibody induced caps in *Entamoeba*, *J. Exp. Med.* **151:**184–193.
63. Reed, S. L., Sargeaunt, P. G., and Braude, A. I., 1983, Resistance to lysis by human serum of pathogenic *Entamoeba histolytica*, *Trans. R. Soc. Trop. Med. Hyg.* **77:**248–253.

64. Calderón, J., and Schreiber, R. D., 1985, Activation of the alternative and classical complement pathways by *Entamoeba histolytica, Infect. Immun.* **50:**560–565.
65. Braga, L. L., Ninomiya, H., McCoy, J. J., Eacker, S., Wiedmer, T., Pham, C., Wood, S., Sims, P. J., and Petri, W. A., 1992, Inhibition of the complement membrane attack complex by the galactose-specific adhesin of *Entamoeba histolytica, J. Clin. Invest.* **90:**1131–1137.
66. Denis, M., and Chadee, K., 1989, Human neutrophils activated by interferon-γ and tumor necrosis factor-α kill *Entamoeba histolytica* trophozoites *in vitro, J. Leuk. Biol.* **46:**270–274.
67. Salata, R. A., Cox, J. G., and Ravdin, J. I., 1987, The interaction of human T-lymphocytes and *Entamoeba histolytica*: Killing of virulent amoebae by lectin-dependent lymphocytes, *Parasite Immunol.* **9:**249–261.
68. Salata, R. A., Martínez-Palomo, A., Murray, H. W., Canales, L., Treviño, N., Segovia, E., Murphy, C. F., and Ravdin, J. I., 1986, Patients treated for amebic liver abscess develop cell-mediated immune response effective *in vitro* against *Entamoeba histolytica, J. Immunol.* **136:**2633–2639.
69. Schain, D. C., Salata, R. A., and Ravdin, J. I., 1992, Human T-lymphocyte proliferation, lymphokine production and amebicidal activity elicited by the galactose-inhibitable adherence protein of *Entamoeba histolytica, Infect. Immun.* **60:**2143–2146.
70. Chadee, K., Meerovitch, E., and Moreau, F., 1985, *In vitro* and *in vivo* interaction between trophozoites of *Entamoeba histolytica* and gerbil lymphoid cells, *Infect. Immun.* **49:** 828–832.
71. Chadee, K., Denis, M., and Keller, K., 1991, Down-regulation of murine lymphocytes responsiveness to mitogens after treatments with antigens of *Entamoeba histolytica, Parasitol. Res.* **77:**572–576.
72. Salata, R. A., Martínez-Palomo, A., Canales, L., Murray, H. W., Treviño, N., and Ravdin, J. I., 1990, Suppression of T-lymphocyte responses to *Entamoeba histolytica* antigen by immune sera, *Infect. Immun.* **58:**3941–3946.
73. Denis, M., and Chadee, K., 1989, Cytokine activation of murine macrophages for *in vitro* killing of *Entamoeba histolytica* trophozoites, *Infect. Immun.* **57:**1750–1756.
74. Salata, R. A., Murray, H. W., Rubin, B. Y., and Ravdin, J. I., 1987, The role of gamma interferon in the generation of human macrophages cytotoxic for *Entamoeba histolytica* trophozoites, *Am. J. Trop. Med. Hyg.* **37:**72–78.
75. Murray, H. W., Aley, S. B., and Scott, W. A., 1981, Susceptibility of *Entamoeba histolytica* to oxygen intermediates, *Mol. Biochem. Parasitol.* **3:**381–391.
76. Lin, J. L., and Chadee, K., 1992, Macrophage cytotoxicity against *Entamoeba histolytica* trophozoites is mediated by nitric oxide from L-arginine, *J. Immunol.* **148:**3999–4005.
77. Lin, J.-Y., Seguin, R., Keller, K., and Chadee, K., 1994, Tumor necrosis factor alpha augments nitric oxide-dependent macrophage cytotoxicity against *Entamoeba histolytica* by enhanced expression of the nitric oxide synthase gene, *Infect. Immun.* **62:**1534–1541.
78. Diamanstein, T., Klos, M., Gold, D., and Hahn, H., 1981, Interaction between *Entamoeba histolytica* and the immune system. I. Mitogenicity of *Entamoeba histolytica* extracts for human peripheral T lymphocytes, *J. Immunol.* **126:**2084–2086.
79. Diamanstein, T., Trissl, D., Klos, M., Gold, D., and Hahn, H., 1980, Mitogenicity of *Entamoeba histolytica* extracts for murine lymphocytes, *Immunology* **41:**347–352.
80. Denis, M., and Chadee, K., 1988, *In vitro* and *in vivo* studies of macrophage functions in amebiasis *Infect. Immun.* **56:**3126–3131.
81. Wang, W., Keller, K., and Chadee, K., 1994, *Entamoeba histolytica* modulates the nitric oxide synthase gene and nitric oxide production by macrophages for cytotoxicity against amoebae and tumor cells, *Immunology* **83:**601–610.
82. Wang, W., Keller, K., and Chadee, K., 1992, Modulation of tumor necrosis factor production by macrophages in *Entamoeba histolytica* infection, *Infect. Immun.* **60:**3169–3174.

83. Wang, W., and Chadee, K., 1992, *Entamoeba histolytica* alters arachidonic acid metabolism in macrophages *in vitro* and *in vivo*, *Immunology* **76:**242–250.
84. Wang, W., and Chadee, K., 1995, *Entamoeba histolytica* suppresses gamma interferon-induced macrophage class II major histocompatibility complex Ia molecule and I-Aβ mRNA expression by a prostaglandin E_2-dependent mechanism, *Infect. Immun.* **63:**1089–1094.
85. Lin, J.-Y., Keller, K., and Chadee, K., 1993, *Entamoeba histolytica* proteins modulate the respiratory burst potential by murine macrophages, *Immunology* **78:**291–297.
86. Seguin, R., and Chadee, K., 1992, *Entamoeba histolytica* modulates TNF-α, IL-1α/β and *c-fos* gene expression in macrophages, *Arch. Med. Res.* **23:**147–149.

12

Giardia Infections

MARTIN F. HEYWORTH

1. *GIARDIA* ORGANISMS AND SPECIES

Protozoan parasites of the genus *Giardia* colonize the intestinal lumen of vertebrate hosts. *Giardia* species are flagellates that appear to be genetically primitive, despite the fact that they are morphologically specialized. This assertion is based on the fact that genes encoding small subunit ribosomal RNA (SSU-rRNA) of *Giardia* organisms have considerable sequence homology with corresponding genes of bacteria.(1,2) In fact, among known eukaryotic SSU-rRNA gene sequences, those of *Giardia* organisms have the greatest homology with bacterial genes.(1) This observation suggests that *Giardia* species are evolutionarily ancient, and raises intriguing unanswered questions about their morphology and "life-style" during the period before vertebrate hosts had evolved. Available data suggest that the haploid genome of *Giardia intestinalis* is approximately 12 million base pairs in length.(3,4) As judged by pulsed-field gel electrophoresis, different isolates of *G. intestinalis* have four to six chromosomes.(4–6)

Giardia species have a bimorphic life cycle, in which motile trophozoites (Fig. 1) alternate with thick-walled immobile cysts. Previously uninfected hosts become infected by ingesting cysts. *Giardia* trophozoites each have two nuclei, and lack mitochondria.(7,8) These organisms generate energy by anaerobic metabolism.(8)

At present, five well-characterized *Giardia* species are recognized (Table I).(8) Differences in trophozoite morphology distinguish these five species of *Giardia* from each other.(8–11) *G. intestinalis*, which causes human giardiasis, is

---

MARTIN F. HEYWORTH • Department of Veterans Affairs Medical and Regional Office Center, Fargo, North Dakota 58102; and Department of Internal Medicine, University of North Dakota, Grand Forks, North Dakota 58203.

Enteric Infections and Immunity, edited by Lois J. Paradise *et al.* Plenum Press, New York, 1996.

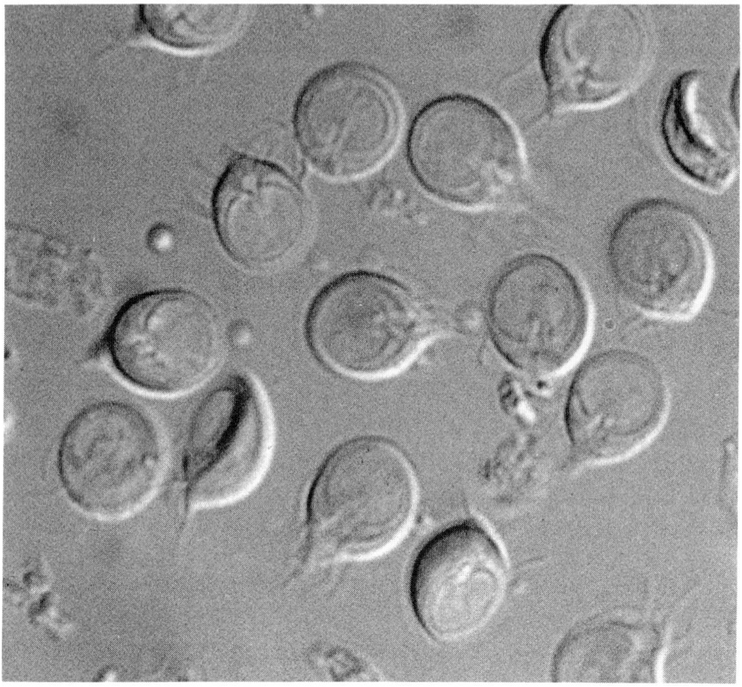

FIGURE 1. Light photomicrograph of *Giardia muris* trophozoites (Nomarski optics).

commonly known by the synonyms *G. lamblia* and *G. duodenalis.* There is substantial confusion about the correct nomenclature of this parasite, which was originally named *Cercomonas intestinalis* by Lambl in 1859, and designated *Hexamites duodenalis* by Davaine in 1875.[12] The designations *G. intestinalis* and *G. lamblia* date from 1914 and 1915, respectively.[12] There is agreement that *Giardia* is the correct generic name, and it is evident that *intestinalis* has

TABLE I
Recognized Species in the Genus *Giardia*

Species	Host(s)
Giardia intestinalis	Humans and other mammals
Giardia muris	Rodents
Giardia psittaci	Budgerigar
Giardia ardeae	Great blue heron
Giardia agilis	Amphibians

chronological priority over *duodenalis* and *lamblia*. Accordingly, *G. intestinalis* appears to be the correct name for the species of *Giardia* that parasitizes humans.

Trophozoites of *G. intestinalis* are 12–15 μm in length,(8) are flattened dorsoventrally, and have four pairs of flagella. *Giardia* trophozoites have a concave disk on their ventral surface; this disk enables them to adhere to the brush border (luminal surface) of intestinal epithelial cells. *Giardia* cyst formation occurs in the lumen of the small intestine, and cysts are subsequently excreted in the feces. *In vitro* studies indicate that emergence of trophozoites from *G. intestinalis* cysts (excystation) is triggered by acid pH.(13) This finding suggests that gastric acid is one of the main factors responsible for excystation of *Giardia* cysts, following their ingestion by a potential host.

Recently, biochemical and molecular biological techniques have been used to explore the degree of heterogeneity within the species *G. intestinalis*. These techniques include electrophoresis of trophozoite isoenzymes, and isoelectric focusing of total trophozoite proteins.(6,14,15) In addition, there is an extensive literature on the use of *G. intestinalis* chromosome electrophoresis (karyotyping), DNA restriction endonuclease analysis, and DNA sequencing to explore the heterogeneity of this parasite.(6,16–18) Finally, random amplified polymorphic DNA (RAPD) analysis has recently emerged as a powerful technique for examining the genetic diversity of *G. intestinalis*.(19,20) This method involves the use of single oligonucleotide primers to amplify random stretches of DNA, followed by gel electrophoresis of the amplification products.(21)

The techniques summarized above have shown that different isolates of *G. intestinalis* are biochemically and genetically diverse. This diversity, however, is not chaotic. Instead, there appear to be identifiable subgroups within the species *G. intestinalis*.(22,23) To what extent, if at all, these subgroups reflect geographical origin, preferred species of mammalian host, or pathogenic potential of different *G. intestinalis* isolates is not clear.

2. CLINICAL FEATURES AND PATHOPHYSIOLOGY OF GIARDIASIS

Clinically, human giardiasis ranges from an asymptomatic intestinal infection to a severely symptomatic enteritis, characterized by diarrhea, abdominal cramps, malabsorption, nausea, and occasionally vomiting.(4,24–27) Severe prolonged giardiasis is characteristically seen in individuals with antibody deficiency (common variable immunoglobulin deficiency and X-linked immunoglobulin deficiency).(28,29) Although the mechanisms responsible for diarrhea and malabsorption in giardiasis are incompletely understood, recent work has shed some light on these processes. There is evidence that human *G. intestinalis* infection can lead to shortening of villi in the small intestine.(30,31) In *Giardia*-infected rodents, shortening of microvilli on the luminal surface of intestinal epithelial cells has been seen.(32,33)

Giardia infections in humans and rodents lead to diminution in the activity of disaccharidase enzymes in the luminal membrane of intestinal epithelial cells.(32,34,35) Furthermore, *in vitro* work has shown that *G. intestinalis* trophozoites take up bile salts.(36) If this effect occurs *in vivo*, it might impair micellar solubilization of fat in the intestinal lumen and contribute to the fat malabsorption that has been documented in some individuals with giardiasis. Recent studies have shown that *G. intestinalis* trophozoites express one or more cysteine-rich surface proteins that are able to bind zinc.(37,38) Conceivably, binding of zinc by trophozoites in the intestinal lumen of *Giardia*-infected hosts might compromise the activity of zinc-requiring enzymes in intestinal epithelial cells.(37)

Human giardiasis can lead to impaired absorption of iron, folate, and vitamin B_{12}, with consequent anemia.(39–42) In addition, giardiasis occasionally causes hypoalbuminemia, with resulting edema and/or ascites. The hypoalbuminemia is a manifestation of protein-losing enteropathy caused by *G. intestinalis* infection. That giardiasis can cause protein-losing enteropathy has been confirmed by measurement of fecal α_1-antitrypsin levels, and by demonstrating excessive leakage of ^{51}Cr-labeled albumin from the plasma into the lumen of the gastrointestinal tract during *G. intestinalis* infection.(43,44)

The clinical literature on giardiasis includes case reports of patients with nonintestinal features, including skin lesions (urticaria and erythema nodosum), joint pains, liver granulomas, ocular inflammation, and psychiatric symptoms.(42,45–49) The mechanisms responsible for these nonintestinal features of giardiasis are largely unknown. However, at least one patient with psychiatric features (hallucinations and depression) was found to have vitamin B_{12} and folate deficiency associated with giardiasis, and recovered clinically following vitamin replacement and treatment of giardiasis with metronidazole.(42)

3. EPIDEMIOLOGY OF GIARDIASIS

Drinking of cyst-contaminated water is the usual mechanism by which human subjects become infected with *G. intestinalis*, although food-borne transmission of giardiasis has been described.(50–52) In addition, direct fecal–

oral transmission of *G. intestinalis* cysts can occur, as in day-care centers for infants and small children.[53]

Sources of drinking water contaminated with *G. intestinalis* cysts include springs, wells, and reservoirs.[51,54,55] *Giardia* cysts can be removed from water by filtration through sand or diatomaceous earth, or by using a membrane filter with a pore size $< 5\ \mu m$.[56,57] Cysts in water are killed by boiling. Beavers, muskrats, dogs, and cats can harbor *Giardia* infections, but the question of whether these animals transmit the infection to humans is largely unanswered.[58–60] However, in one recent study, a *Giardia*-infected beaver was implicated as the source of an outbreak of human giardiasis (in individuals drinking water contaminated by *Giardia* cysts), on the basis of karyotype identity between *G. intestinalis* isolated from the beaver and from infected human subjects in the outbreak.[6]

As noted above, antibody deficiency syndromes predispose to severe and prolonged giardiasis in humans. In some patients with common variable immunoglobulin deficiency and giardiasis, there is abnormal enlargement of intestinal lymphoid follicles [nodular lymphoid hyperplasia (NLH)], associated with absence of IgA-producing plasma cells in the intestinal mucosa.[28,61,62] The abnormally enlarged lymphoid follicles in NLH consist largely of IgM-expressing immature B lymphocytes, which fail to mature into IgA-expressing B cells. Under normal circumstances (but not in patients with NLH), intestinal lymphoid follicles contain IgA-expressing B lymphocytes, which are precursors of IgA-containing plasma cells that normally populate the intestinal lamina propria.[63] Persistence of giardiasis in patients with NLH almost certainly reflects inability to mount an intestinal IgA response to *G. intestinalis* trophozoites.

4. IMMUNOLOGICAL RESPONSES TO *GIARDIA* TROPHOZOITES

Most of the information about immunological responses to *Giardia* trophozoites has come from studies with *G. muris*-infected mice. Immunocompetent mice are able to clear *G. muris* infection; clearance of this infection is associated with the production of intestinal IgA antibody directed against *G. muris* trophozoites.[64,65] This antibody binds to trophozoite surfaces (Fig. 2).[65] Direct evidence that antitrophozoite IgA contributes to clearance of giardiasis, and to protection against this infection, has been obtained from studies with rodents. In this work, it has been shown that infusion of bile containing antitrophozoite IgA, into the duodenal lumen of *Giardia*-infected rats, diminishes the intensity of the infection.[66] Similarly, *in vitro* exposure of *G. muris* trophozoites to mouse intesti-

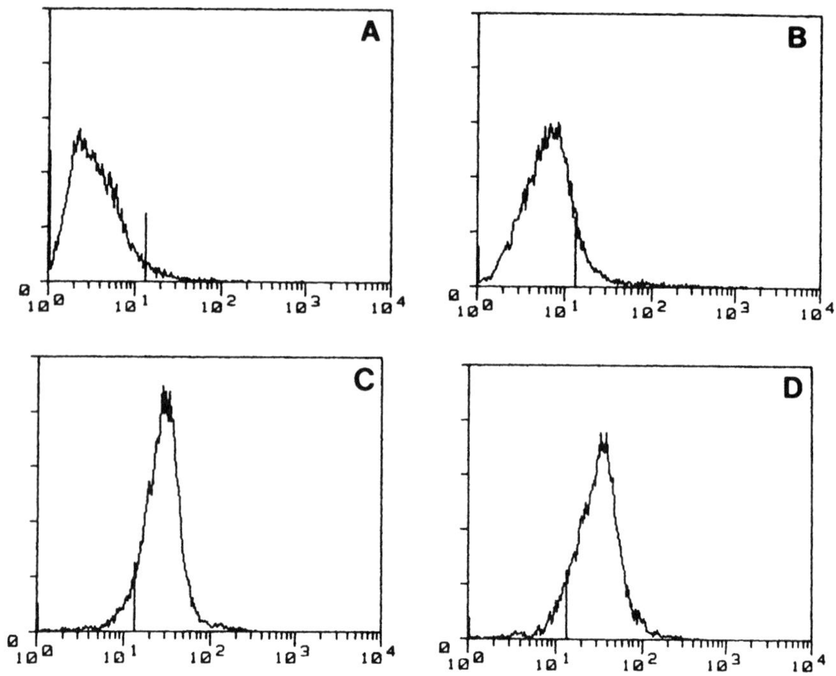

FIGURE 2. Flow cytometry of *Giardia muris* trophozoites incubated without (A) or with (B–D) mouse intestinal secretions *in vitro*, followed in each case by fluorescein-conjugated anti-mouse IgA. Intestinal secretions were obtained either from *G. muris*-infected mice (C, D) or from uninfected control mice (B). Vertical axes show relative numbers of trophozoites. Horizontal axes show fluorescence intensity (logarithmic scale). Shift of peaks in C and D to right compared with B indicates binding of IgA antibody to trophozoite populations in C and D, and confirms that such antibody is present in intestinal secretions from *G. muris*-infected mice. From Ref. 65 with permission.

nal secretions that contain antitrophozoite IgA reduces the infectivity of these organisms for mice.(67)

Human subjects with giardiasis develop serum IgM, IgG, and IgA antibody responses to *G. intestinalis* trophozoites.(27,68) It is, however, unclear whether these serum antibodies play any part in the clearance of *G. intestinalis* infections. There is evidence that immunocompetent individuals with giardiasis develop an intestinal IgA response to *G. intestinalis* trophozoites.(27) No other class of anti-*Giardia* antibody (besides IgA) has been reproducibly demonstrated in the intestinal secretions of any mammalian species.

Although *Giardia* trophozoites are killed by IgG and IgM antibodies *in vitro*, in the presence of complement, IgA antibodies have no discernible cytotoxic effect on these organisms.[69] Consequently, it is very unlikely that antibody-dependent cytotoxicity plays any part in the clearance of *Giardia* infections. Instead, it is likely that immunological clearance of these infections is mediated by peristaltic expulsion of IgA-coated trophozoites from the intestine of infected hosts. Experimental studies indicate that coating of *Giardia* trophozoites with antibody inhibits their ability to adhere to rodent intestinal epithelium.[70,71]

IgA antibody in the intestinal secretions of *G. muris*-infected immunocompetent mice binds to a trophozoite protein with an approximate molecular mass of 80 kDa.[65] It is intriguing to speculate that this protein may be a homologue of the cysteine-rich surface proteins of *G. intestinalis* trophozoites (see above),[37,72] which have molecular masses similar to that of the 80-kDa protein. Further characterization of trophozoite immunodominant proteins (including epitopes recognized by protective IgA antibody) might conceivably lead to the development of a vaccine against giardiasis.

5. DIAGNOSIS OF GIARDIASIS

Traditionally, giardiasis has been diagnosed by microscopic examination of fecal specimens for *G. intestinalis* cysts. Because this is a labor-intensive, subjective approach, alternative diagnostic techniques have been developed in recent years. These methods include immunofluorescence microscopy of fecal specimens incubated with fluorochrome-conjugated monoclonal antibodies that bind to cyst walls.[73,74] Immunofluorescence microscopy can also be used to examine environmental water samples for *G. intestinalis* cysts.[75] Although immunofluorescence microscopy is more sensitive than microscopic examination of unstained fecal specimens, it is also a labor-intensive and subjective technique.[73,74] Other diagnostic methods that rely on microscopy include examination of duodenal fluid or duodenal endoscopic biopsy specimens for *G. intestinalis* trophozoites.[76–78]

Enzyme-linked immunosorbent assay (ELISA) of fecal specimens for *G. intestinalis* antigen(s) has emerged as a sensitive and objective method for diagnosing giardiasis.[79,80] This approach is gradually replacing fecal microscopy as the standard technique for diagnosing giardiasis; several ELISA kits for detecting *G. intestinalis* antigen(s) are now available commercially.[74] Pilot studies suggest that polymerase chain reaction (PCR) amplification of *G. intestinalis* DNA is a sensitive method for detecting *Giardia* cysts in water samples.[81] It is also conceivable that PCR amplification of such DNA in fecal specimens might be applicable to the diagnosis of human giardiasis.[82]

As noted above, giardiasis leads to the development of serum antibody responses against *G. intestinalis* trophozoites. The presence or absence of detectable antitrophozoite antibodies, in human sera, can shed light on the prevalence of giardiasis in a population,(83,84) but has limited utility as a diagnostic test for giardiasis in an individual. Available information suggests that the presence of antitrophozoite antibodies in human serum is not necessarily an indication of current giardiasis.(74)

6. TREATMENT OF GIARDIASIS

Drugs for treating giardiasis include quinacrine (6 mg/kg body wt per day, for 5 days) or metronidazole (15 mg/kg per day, for 5 days). Alternatively, the infection can be treated with a single dose of tinidazole (50 mg/kg), or with furazolidone (6 mg/kg/day, for 7–10 days).(4,85)

The question of whether to treat asymptomatic *G. intestinalis* infection (documented by finding cysts in fecal specimens) has generated some controversy. However, it seems reasonable to treat asymptomatic cyst excretors with an anti-*G. intestinalis* drug (at least in temperate, industrialized countries such as the United States), as this approach reduces the likelihood of transmission of the infection to other persons.

ACKNOWLEDGMENTS. Research grant support from the Department of Veterans Affairs is gratefully acknowledged.

REFERENCES

1. Sogin, M. L., Gunderson, J. H., Elwood, H. J., Alonso, R. A., and Peattie, D. A., 1989, Phylogenetic meaning of the kingdom concept: An unusual ribosomal RNA from *Giardia lamblia*. *Science* **243:**75–77.
2. van Keulen, H., Gutell, R. R., Gates, M. A., Campbell, S. R., Erlandsen, S. L., Jarroll, E. L., Kulda, J., and Meyer, E. A., 1993, Unique phylogenetic position of Diplomonadida based on the complete small subunit ribosomal RNA sequence of *Giardia ardeae, G. muris, G. duodenalis* and *Hexamita* sp., *FASEB J.* **7:**223–231.
3. Fan, J.-B., Korman, S. H., Cantor, C. R., and Smith, C. L., 1991, *Giardia lamblia*: Haploid genome size determined by pulsed field gel electrophoresis is less than 12 Mb, *Nucleic Acids Res.* **19:**1905–1908.
4. Adam, R. D., 1991, The biology of *Giardia* spp., *Microbiol. Rev.* **55:**706–732.
5. Adam, R. D., Nash, T. E., and Wellems, T. E., 1988, The *Giardia lamblia* trophozoite contains sets of closely related chromosomes, *Nucleic Acids Res.* **16:**4555–4567.
6. Isaac-Renton, J. L., Cordeiro, C., Sarafis, K., and Shahriari, H., 1993, Characterization of *Giardia duodenalis* isolates from a waterborne outbreak, *J. Infect. Dis.* **167:**431–440.
7. Kabnick, K. S., and Peattie, D. A., 1990, *In situ* analyses reveal that the two nuclei of *Giardia lamblia* are equivalent, *J. Cell Sci.* **95:**353–360.
8. Thompson, R. C. A., Reynoldson, J. A., and Mendis, A. H. W., 1993, *Giardia* and giardiasis, *Adv. Parasitol.* **32:**71–160.

9. Feely, D. E., and Erlandsen, S. L., 1985, Morphology of *Giardia agilis*: Observation by scanning electron microscopy and interference reflexion microscopy, *J. Protozool.* **32:** 691–693.
10. Erlandsen, S. L., and Bemrick, W. J., 1987, SEM evidence for a new species, *Giardia psittaci*, *J. Parasitol.* **73:**623–629.
11. Erlandsen, S. L., Bemrick, W. J., Wells, C. L., Feely, D. E., Knudson, L., Campbell, S. R., van Keulen, H., and Jarroll, E. L., 1990, Axenic culture and characterization of *Giardia ardeae* from the great blue heron (*Ardea herodias*), *J. Parasitol.* **76:**717–724.
12. Ansari, M. A. R., 1952, Contribution à l'étude du genre *Giardia* Kunstler, 1882 (*Mastigophora: Octomitidae*): I. - Tableau synoptique des espèces connues et de quatre espèces nouvelles, *Ann. Parasitol.* **27:**421–479.
13. Bingham, A. K., and Meyer, E. A., 1979, *Giardia* excystation can be induced *in vitro* in acidic solutions, *Nature* **277:**301–302.
14. Korman, S. H., Le Blancq, S. M., Spira, D. T., El On, J., Reifen, R. M., and Deckelbaum, R. J., 1986, *Giardia lamblia*: Identification of different strains from man, *Z. Parasitenkd.* **72:**173–180.
15. Isaac-Renton, J. L., Byrne, S. K., and Prameya, R., 1988, Isoelectric focusing of ten strains of *Giardia duodenalis*, *J. Parasitol.* **74:**1054–1056.
16. Sarafis, K., and Isaac-Renton, J., 1993, Pulsed-field gel electrophoresis as a method of biotyping of *Giardia duodenalis*, *Am. J. Trop. Med. Hyg.* **48:**134–144.
17. Ey, P. L., Darby, J. M., Andrews, R. H., and Mayrhofer, G., 1993, *Giardia intestinalis*: Detection of major genotypes by restriction analysis of gene amplification products, *Int. J. Parasitol.* **23:**591–600.
18. Weiss, J. B., van Keulen, H., and Nash, T. E., 1992, Classification of subgroups of *Giardia lamblia* based upon ribosomal RNA gene sequence using the polymerase chain reaction, *Mol. Biochem. Parasitol.* **54:**73–86.
19. Morgan, U. M., Constantine, C. C., Greene, W. K., and Thompson, R. C. A., 1993, RAPD (random amplified polymorphic DNA) analysis of *Giardia* DNA and correlation with isoenzyme data, *Trans. R. Soc. Trop. Med. Hyg.* **87:**702–705.
20. van Belkum, A., Homan, W., Limper, L., and Quint, W. G. V., 1993, Genotyping isolates and clones of *Giardia duodenalis* by polymerase chain reaction: Implications for the detection of genetic variation among protozoan parasite species, *Mol. Biochem. Parasitol.* **61:**69–78.
21. Williams, J. G. K., Kubelik, A. R., Livak, K. J., Rafalski, J. A., and Tingey, S. V., 1990, DNA polymorphisms amplified by arbitrary primers are useful as genetic markers, *Nucleic Acids Res.* **18:**6531–6535.
22. Ey, P. L., Khanna, K., Andrews, R. H., Manning, P. A., and Mayrhofer, G., 1992, Distinct genetic groups of *Giardia intestinalis* distinguished by restriction fragment length polymorphisms, *J. Gen. Microbiol.* **138:**2629–2637.
23. Ey, P. L., Andrews, R. H., and Mayrhofer, G., 1993, Differentiation of major genotypes of *Giardia intestinalis* by polymerase chain reaction analysis of a gene encoding a trophozoite surface antigen, *Parasitology* **106:**347–356.
24. Ish-Horowicz, M., Korman, S. H., Shapiro, M., Har-Even, U., Tamir, I., Strauss, N., and Deckelbaum, R. J., 1989, Asymptomatic giardiasis in children, *Pediatr. Infect. Dis. J.* **8:**773–779.
25. Hoskins, L. C., Winawer, S. J., Broitman, S. A., Gottlieb, L. S., and Zamcheck, N., 1967, Chronic giardiasis and intestinal malabsorption, *Gastroenterology* **53:**265–279.
26. Wright, S. G., Tomkins, A. M., and Ridley, D. S., 1977, Giardiasis: Clinical and therapeutic aspects, *Gut* **18:**343–350.
27. Nash, T. E., Herrington, D. A., Losonsky, G. A., and Levine, M. M., 1987, Experimental human infections with *Giardia lamblia*, *J. Infect. Dis.* **156:**974–984.
28. Hermans, P. E., Diaz-Buxo, J. A., and Stobo, J. D., 1976, Idiopathic late-onset immunoglobulin deficiency: Clinical observations in 50 patients, *Am. J. Med.* **61:**221–237.

29. LoGalbo, P. R., Sampson, H. A., and Buckley, R. H., 1982, Symptomatic giardiasis in three patients with X-linked agammaglobulinemia, *J. Pediatr.* **101:**78–80.
30. Levinson, J. D., and Nastro, L. J., 1978, Giardiasis with total villous atrophy, *Gastroenterology* **74:**271–275.
31. Rosekrans, P. C. M., Lindeman, J., and Meijer, C. J. L. M., 1981, Quantitative histological and immunohistochemical findings in jejunal biopsy specimens in giardiasis, *Virchows Arch. A* **393:**145–151.
32. Buret, A., Gall, D. G., and Olson, M. E., 1990, Effects of murine giardiasis on growth, intestinal morphology, and disaccharidase activity, *J. Parasitol.* **76:**403–409.
33. Buret, A., Hardin, J. A., Olson, M. E., and Gall, D. G., 1992, Pathophysiology of small intestinal malabsorption in gerbils infected with *Giardia lamblia, Gastroenterology* **103:**506–513.
34. Belosevic, M., Faubert, G. M., and MacLean, J. D., 1989, Disaccharidase activity in the small intestine of gerbils (*Meriones unguiculatus*) during primary and challenge infections with *Giardia lamblia, Gut* **30:**1213–1219.
35. Welsh, J. D., Poley, J. R., Hensley, J., and Bhatia, M., 1984, Intestinal disaccharidase and alkaline phosphatase activity in giardiasis, *J. Pediatr. Gastroenterol. Nutr.* **3:**37–40.
36. Halliday, C. E. W., Clark, C., and Farthing, M. J. G., 1988, *Giardia*–bile salt interactions *in vitro* and *in vivo, Trans. R. Soc. Trop. Med. Hyg.* **82:**428–432.
37. Nash, T. E., and Mowatt, M. R., 1993, Variant-specific surface proteins of *Giardia lamblia* are zinc-binding proteins, *Proc. Natl. Acad. Sci. USA* **90:**5489–5493.
38. Zhang, Y.-Y., Aley, S. B., Stanley, S. L., and Gillin, F. D., 1993, Cysteine-dependent zinc binding by membrane proteins of *Giardia lamblia, Infect. Immun.* **61:**520–524.
39. De Vizia, B., Poggi, V., Vajro, P., Cucchiara, S., and Acampora, A., 1985, Iron malabsorption in giardiasis, *J. Pediatr.* **107:**75–78.
40. Ambrose, N. S., Hutchison, S., and Tejan, J., 1989, Folate deficiency due to giardiasis, *J. R. Soc. Med.* **82:**48–49.
41. Cordingley, F. T., and Crawford, G. P. M., 1986, *Giardia* infection causes vitamin B12 deficiency, *Aust. N. Z. J. Med.* **16:**78–79.
42. Heap, B., and Mumford, J. D., 1989, Chronic giardiasis with vitamin B_{12} and folate deficiency presenting with psychiatric symptoms, *J. R. Army Med. Corps* **135:**25–26.
43. Sutton, D. L., and Kamath, K. R., 1985, Giardiasis with protein-losing enteropathy, *J. Pediatr. Gastroenterol. Nutr.* **4:**56–59.
44. Korman, S. H., Bar-Oz, B., Mandelberg, A., and Matoh, I., 1990, Giardiasis with protein-losing enteropathy: Diagnosis by fecal α_1-antitrypsin determination, *J. Pediatr. Gastroenterol. Nutr.* **10:**249–252.
45. Clyne, C. A., and Eliopoulos, G. M., 1989, Fever and urticaria in acute giardiasis, *Arch. Intern. Med.* **149:**939–940.
46. Giordano, N., Fioravanti, A., Mariani, A., and Marcolongo, R., 1985, Erythema nodosum and giardia intestinalis, *Clin. Rheumatol.* **4:**481–483.
47. Goobar, J. P., 1977, Joint symptoms in giardiasis (Letter), *Lancet* **1:**1010–1011.
48. Roberts-Thomson, I. C., Anders, R. F., and Bhathal, P. S., 1982, Granulomatous hepatitis and cholangitis associated with giardiasis, *Gastroenterology* **83:**480–483.
49. Anderson, M. L., and Griffith, D. G., 1985, Intestinal giardiasis associated with ocular inflammation, *J. Clin. Gastroenterol.* **7:**169–172.
50. Jephcott, A. E., Begg, N. T., and Baker, I. A., 1986, Outbreak of giardiasis associated with mains water in the United Kingdom, *Lancet* **1:**730–732.
51. Kent, G. P., Greenspan, J. R., Herndon, J. L., Mofenson, L. M., Harris, J.-A. S., Eng, T. R., and Waskin, H. A., 1988, Epidemic giardiasis caused by a contaminated public water supply, *Am. J. Public Health* **78:**139–143.
52. Petersen, L. R., Cartter, M. L., and Hadler, J. L., 1988, A food-borne outbreak of *Giardia lamblia, J. Infect. Dis.* **157:**846–848.

53. Polis, M. A., Tuazon, C. U., Alling, D. W., and Talmanis, E., 1986, Transmission of *Giardia lamblia* from a day care center to the community, *Am. J. Public Health* **76:**1142–1144.
54. Craun, G. F., 1986, Waterborne giardiasis in the United States 1965-84 (Letter), *Lancet* **2:** 513–514.
55. Birkhead, G., and Vogt, R. L., 1989, Epidemiologic surveillance for endemic *Giardia lamblia* infection in Vermont: The roles of waterborne and person-to-person transmission, *Am. J. Epidemiol.* **129:**762–768.
56. Logsdon, G. S., 1988, Comparison of some filtration processes appropriate for *Giardia* cyst removal, in: *Advances in Giardia Research* (P. M. Wallis and B. R. Hammond, eds.), University of Calgary Press, Calgary, pp. 95–102.
57. Rose, J. B., Kayed, D., Madore, M. S., Gerba, C. P., Arrowood, M. J., Sterling, C. R., and Riggs, J. L., 1988, Methods for the recovery of *Giardia* and *Cryptosporidium* from environmental waters and their comparative occurrence, in: *Advances in Giardia Research* (P. M. Wallis and B. R. Hammond, eds.), University of Calgary Press, Calgary, pp. 205–209.
58. Monzingo, D. L., and Hibler, C. P., 1987, Prevalence of *Giardia* sp. in a beaver colony and the resulting environmental contamination, *J. Wildl. Dis.* **23:**576–585.
59. Kirkpatrick, C. E., and Benson, C. E., 1987, Presence of *Giardia* spp. and absence of *Salmonella* spp. in New Jersey muskrats (*Ondatra zibethicus*), *Appl. Environ. Microbiol.* **53:**1790–1792.
60. Winsland, J. K. D., Nimmo, S., Butcher, P. D., and Farthing, M. J. G., 1989, Prevalence of *Giardia* in dogs and cats in the United Kingdom: Survey of an Essex veterinary clinic, *Trans. R. Soc. Trop. Med. Hyg.* **83:**791–792.
61. van den Brande, P., Geboes, K., Vantrappen, G., van den Eeckhout, A., Vertessen, S., Stevens, E. A. M., and Ceuppens, J. L., 1988, Intestinal nodular lymphoid hyperplasia in patients with common variable immunodeficiency: Local accumulation of B and CD8(+) lymphocytes, *J. Clin. Immunol.* **8:**296–306.
62. Laszewski, M. J., Kemp, J. D., Goeken, J. A., Mitros, F. A., Platz, C. E., and Dick, F. R., 1990, Clonal immunoglobulin gene rearrangement in nodular lymphoid hyperplasia of the gastrointestinal tract associated with common variable immunodeficiency, *Am. J. Clin. Pathol.* **94:**338–343.
63. Crago, S. S., and Tomasi, T. B., 1988, Immunoglobulin circulation and secretion, in: *Immunology of the Gastrointestinal Tract and Liver* (M. F. Heyworth and A. L. Jones, eds.), Raven Press, New York, pp. 105–124.
64. Snider, D. P., Gordon, J., McDermott, M. R., and Underdown, B. J., 1985, Chronic *Giardia muris* infection in anti-IgM-treated mice. I. Analysis of immunoglobulin and parasite-specific antibody in normal and immunoglobulin-deficient animals, *J. Immunol.* **134:**4153–4162.
65. Heyworth, M. F., and Vergara, J. A., 1994, *Giardia muris* trophozoite antigenic targets for mouse intestinal IgA antibody, *J. Infect. Dis.* **169:**395–398.
66. Mayrhofer, G., and Sharma, A. W., 1988, The secretory immune response in rats infected with rodent *Giardia duodenalis* isolates and evidence for passive protection with immune bile, in: *Advances in Giardia Research* (P. M. Wallis and B. R. Hammond, eds.), University of Calgary Press, Calgary, pp. 49–54.
67. Underdown, B. J., Skea, D. L., Loney, G. M., and Snider, D. P., 1988, Murine giardiasis and mucosal immunity: A model for the study of immunity to intestinal protozoan parasites, *Monogr. Allergy* **24:**287–296.
68. Janoff, E. N., Taylor, D. N., Echeverria, P., Glode, M. P., and Blaser, M. J., 1990, Serum antibodies to *Giardia lamblia* by age in populations in Colorado and Thailand, *West. J. Med.* **152:**253–256.
69. Heyworth, M. F., 1992, Relative susceptibility of *Giardia muris* trophozoites to killing by mouse antibodies of different isotypes, *J. Parasitol.* **78:**73–76.

70. Inge, P. M. G., Edson, C. M., and Farthing, M. J. G., 1988, Attachment of *Giardia lamblia* to rat intestinal epithelial cells, *Gut* **29:**795–801.
71. Kaplan, B. S., and Altmanshofer, D., 1985, *Giardia muris* adherence to intestinal epithelium—The role of specific anti-*Giardia* antibodies, *Microecol. Ther.* **15:**133–140.
72. Gillin, F. D., Hagblom, P., Harwood, J., Aley, S. B., Reiner, D. S., McCaffery, M., So, M., and Guiney, D. G., 1990, Isolation and expression of the gene for a major surface protein of *Giardia lamblia, Proc. Natl. Acad. Sci. USA* **87:**4463–4467.
73. Tee, G. H., Moody, A. H., Hunt Cooke, A., and Chiodini, P. L., 1993, Comparison of techniques for detecting antigens of *Giardia lamblia* and *Cryptosporidium parvum* in faeces, *J. Clin. Pathol.* **46:**555–558.
74. Isaac-Renton, J. L., 1993, Giardiasis: Epidemiology and diagnosis, *Semin. Gastrointest. Dis.* **4:**206–213.
75. Sauch, J. F., 1985, Use of immunofluorescence and phase-contrast microscopy for detection and identification of *Giardia* cysts in water samples, *Appl. Environ. Microbiol.* **50:**1434–1438.
76. Jones, J. E., 1986, String test for diagnosing giardiasis, *Am. Fam. Physician* **34:**123–126.
77. Goka, A. K. J., Rolston, D. D. K., Mathan, V. I., and Farthing, M. J. G., 1990, The relative merits of faecal and duodenal juice microscopy in the diagnosis of giardiasis, *Trans. R. Soc. Trop. Med. Hyg.* **84:**66–67.
78. Oberhuber, G., and Stolte, M., 1990, Giardiasis: Analysis of histological changes in biopsy specimens of 80 patients, *J. Clin. Pathol.* **43:**641–643.
79. Green, E., Warhurst, D., Williams, J., Dickens, T., and Miles, M., 1990, Application of a capture enzyme immunoassay in an outbreak of waterborne giardiasis in the United Kingdom, *Eur. J. Clin. Microbiol. Infect. Dis.* **9:**424–428.
80. Addiss, D. G., Mathews, H. M., Stewart, J. M., Wahlquist, S. P., Williams, R. M., Finton, R. J., Spencer, H. C., and Juranek, D. D., 1991, Evaluation of a commercially available enzyme-linked immunosorbent assay for *Giardia lamblia* antigen in stool, *J. Clin. Microbiol.* **29:**1137–1142.
81. Mahbubani, M. H., Bej, A. K., Perlin, M. H., Schaefer, F. W., Jakubowski, W., and Atlas, R. M., 1992, Differentiation of *Giardia duodenalis* from other *Giardia* spp. by using polymerase chain reaction and gene probes, *J. Clin. Microbiol.* **30:**74–78.
82. Weiss, J. B., 1993, PCR detection of *Giardia lamblia,* in: *Diagnostic Molecular Microbiology: Principles and Applications* (D. H. Persing, T. F. Smith, F. C. Tenover, and T. J. White, eds.), American Society for Microbiology, Washington, D.C., pp. 480–485.
83. Gilman, R. H., Brown, K. H., Visvesvara, G. S., Mondal, G., Greenberg, B., Sack, R. B., Brandt, F., and Khan, M. U., 1985, Epidemiology and serology of *Giardia lamblia* in a developing country: Bangladesh, *Trans. R. Soc. Trop. Med. Hyg.* **79:**469–473.
84. Al-Tukhi, M. H., Ackers, J. P., Al-Ahdal, M. N., and Peters, W., 1993, Enzyme-linked immunosorbent assay for the detection of anti-*Giardia* specific immunoglobulin G in filter paper blood samples, *Trans. R. Soc. Trop. Med. Hyg.* **87:**36–38.
85. Drugs for parasitic infections, 1993, *Med. Lett. Drugs Ther.* **35**(Issue 911):111–122.

13

The Immune Regulation of Intestinal Helminthiases

ROBERT M. GENTA

1. INTRODUCTION

Intestinal helminthiases represent collectively the most prevalent parasitic infection of humankind. In developing countries, where nearly 75% of the global population lives, between one-third and one-half of the residents carry one or more intestinal parasitic worms; children born in endemic areas are likely to acquire these infections early and remain infected for the rest of their lives. The impact of these parasites on the health of their hosts covers the entire spectrum from negligible to catastrophic; therefore, there is a dearth of reliable information on the actual medical and financial burden caused by these infections. There is, however, little doubt that the combined morbidity resulting from intestinal worms represents a major obstacle to the full social and economic development of many tropical nations, and that the elimination of intestinal parasitoses would be a desirable goal.

This goal, however, like many ambitious attempts to control other parasitic infections, remains no more than a remote aspiration. Several different approaches can be taken in the fight against intestinal helminthiases. Every infected host could be treated with effective, broad-spectrum antihelminthic drugs. Chemotherapy may be expensive, requires an efficient continuous distribution network, and, most importantly, demands strict patient compliance. Even in the few instances when this approach has

---

ROBERT M. GENTA • Departments of Pathology, Medicine, and Microbiology and Immunology, Baylor College of Medicine, and Center for Infectious Diseases, University of Texas School of Public Health, Houston, Texas 77030.

Enteric Infections and Immunity, edited by Lois J. Paradise *et al.* Plenum Press, New York, 1996.

been close to success, as in the case of malaria, it has eventually been foiled by the emergence of resistant strains. Furthermore, the morbidity of intestinal helminthiasis is not immediately apparent to the patients and their relatives, and these infections are not regarded as life-threatening or even dangerous; thus, satisfactory compliance cannot be expected. A second approach, by far the most effective, is sanitation. Intestinal helminth nematodes are disseminated by contamination of the environment with the stools of infected subjects; these stools may contain eggs which will transmit the infection when ingested (*Ascaris lumbricoides, Trichuris trichiura, Enterobius vermicularis*), eggs which will liberate skin-penetrating larvae (hookworms), or skin-penetrating larvae (*Strongyloides stercoralis*). Cestodes (tapeworms) are acquired through the ingestion of infected meat. Effective disposal of human feces in the industrialized world has resulted in the virtual disappearance of all of these helminthiases without any other intervention. At the end of the 20th century, many tropical countries are torn by wars, civil unrest, and mass migrations of refugees. For the time being, the evolution of such countries into economies that will allow the development of adequate public health services seems like an unrealistic expectation. If parasites are to be eradicated, or at least considerably curtailed from such populations, we must develop a strategy that would circumvent the need for mass drug therapy and for a structured sanitation system. Such a strategy exists, and has been proven successful in the eradication or drastic reduction of many of the major infections of humankind, from smallpox to poliomyelitis to measles: protective immunization. The development of vaccines effective against helminths, however, is a much more complicated endeavor than the induction of protective immunity directed at viral or bacterial targets. Helminthic parasites have evolved a unique evolutionary relationship with their mammalian hosts. This has led to the establishment of a host–parasite balance that includes complex transmission routes, elaborate reproductive strategies, and immunologic arrangements that allow parasites to survive unharmed in their hosts despite ample evidence of immune recognition and attack.[1]

Understanding the wide array of mechanisms used by parasites to evade and modulate the host's defenses is pivotal to the development of both effective vaccines and new, improved drugs. Furthermore, it will contribute to expand our knowledge of the normal immune system, its evolution, and its disorders.

2. BIOLOGICAL FEATURES OF INTESTINAL HELMINTHS

The major human intestinal parasitic helminths discussed in this chapter belong to two phyla, nematodes and platyhelminths. Medically important nematodes include *Ascaris lumbricoides*, *Trichuris trichiura*, *Enterobius vermicularis*, *Ancylostoma duodenale*, *Necator americanus*, and *Strongyloides stercoralis*. Of the platyhelminths, only two members of the class cestodes, *Taenia saginata* and *T. solium*, will be briefly mentioned. Both these and other cestodes (*Echinococcus granulosus*, *Hymenolepis nana*, and *H. diminuta*) are highly prevalent parasites, but the host's immune responses against their intestinal phase in humans has received little or no attention. All of these worms share a common habitat, the human gut. Nevertheless, they have important biological differences. The first five of the nematodes reside in the intestinal lumen and lay eggs which need a maturation period in the external environment before they can infect another host. Consequently, they do not reproduce within their host.(1) In contrast, *S. stercoralis* parasitic females dwell within the intestinal mucosa where they lay eggs. After the eggs hatch, rhabditiform larvae emerge and migrate into the intestinal lumen, eventually passing with the feces into the external environment. Here, depending on poorly understood conditions of temperature and humidity, the rhabditiform larvae either molt directly into infective parasitic filariform larvae, capable of repenetrating the skin of a suitable host (direct cycle), or pass through four ecdyses (molts) to become adult male and female worms which mate and produce offspring that develop into filariform larvae and reenter parasitic life (indirect or heterogonical cycle). A small portion of the rhabditiform larvae that hatch from the eggs deposited in the intestine are believed to molt within the intestine into the filariform stage. These larvae penetrate the colonic wall or the perianal skin, complete the internal cycle, and establish themselves as mature adult females in the small intestine. This process, known as autoinfection, represents the mechanism by which *S. stercoralis* persists virtually indefinitely in its hosts.(2)

Similarly to *S. stercoralis*, but only in the external environment, the eggs of hookworms (*A. duodenale* and *N. americanus*) liberate larvae which will molt into skin-penetrating filariform larvae. Thus, for both *S. stercoralis* and hookworms the first immunologically active barrier they encounter in their host is the skin. In contrast, hosts are infected by ingesting the embryonated eggs of *A. lumbricoides*, *T. trichiura*, and *E. vermicularis*, and the cysticerci of tapeworms. The larvae of these parasites emerge in the human gastrointestinal tract, like the skin one of the body's most formidable immune compartments. Yet, apparently unencumbered by these potentially exacting first encounters, infective larvae proceed to complete their complex development and migratory pathways through the host internal organs and eventually settle in their final destination, the human gut.

Adult tapeworms are strictly intraluminal in humans, where they reside mostly in the small intestine.(1) These long worms consist of a head portion (scolex) which generates series of segments (proglottides), which mature as they grow farther from the scolex. Mature proglottides containing large numbers of eggs are passed with the human feces. When ingested by swine or cattle, the eggs liberate a larva, which penetrates the intestinal wall, enters the circulation, and eventually encysts in the animal's muscles, where it is called cysticercus. When a cysticercus reaches the human stomach, it emerges from its cystic life and matures into a scolex, the progenitor of a tapeworm.

3. SYSTEMIC IMMUNE RESPONSES TO INTESTINAL HELMINTHS

Virtually all human intestinal parasitic infections studied have been shown to induce both humoral and cellular immune responses in their hosts.(3) The most distinctive of these responses are those showing the features of immediate-type hypersensitivity reactions. These are characterized by IgE antibody production, tissue and peripheral blood eosinophilia, and the participation of inflammatory mediator-rich basophils and mast cells.(4,5) In atopic subjects these immediate hypersensitivity responses are directly involved in the pathogenesis of allergic diseases. In contrast, in parasitic helminth infections, not only are pathologic reactions apparently rare, but these responses even seem to confer some degree of immunologic protection from allergic manifestations.(6)

Our understanding of the regulation of these types of responses has been considerably enhanced by the identification of two functionally distinct subsets of differentiated $CD4^+$ T cells (helper cells) designated Th1 and Th2, characterized by the contrasting profiles of cytokines they secrete.(7–9) Th1 cells produce inflammatory mediators (including IL-2, IFN-γ, and TNF-β) and can selectively activate macrophages in response to antigen or mitogen stimulation. In contrast, Th2 cells produce cytokines (such as IL-4, IL-5, IL-6, and IL-10) capable of stimulating B cells, antibody production, and eosinophil production. These two populations of T cells originate from a common uncommitted precursor cell ($CD45RA^+$) which develops into Th0 cells with multipotent or intermediate secretory profiles and further differentiate into Th1 or Th2 cells following repeated antigenic stimulation. Mature Th1 and Th2 subsets are reciprocally cross-inhibitory, so that one cell type gains predominance while inhibiting the development of the other. This may explain the unbalanced Th1/Th2 ratio found in many helminthic infections in association with elevated eosinophilia and the enhanced IgE and IgG4 expression.(7)

3.1. IgE Responses

An almost universal feature of infections with tissue-invasive nematodes is the elevation of serum IgE levels.(3,5) Most of the IgE produced is not antigen-specific; rather, it represents the nonspecific potentiation or deregulation of a normally well-controlled immune response.[10,11] This potentiating effect seems to be selective for the IgE (and perhaps the IgG4) isotype.[10–13] Although the factors responsible for IgE elevation are not well defined, the increase most likely reflects a clonal expansion of IgE-producing B cells related to an imbalance of Th1/Th2 cells.[7,8,14] However, there seem to be not only quantitative differences but also qualitative ones. In contrast to atopic individuals, whose IgE tend to be directed toward only a few antigens, patients with invasive helminth infection produce IgE antibodies with specificity against a very broad range of parasite antigens. These responses, however, are very heterogeneous, and only rarely do patients recognize the same antigens. It has been proposed that such augmented polyclonal responses lead to the production of irrelevant IgE that saturate the mast cells' high-affinity Fcε receptor, rendering it refractory to triggering by parasite antigen. This interesting hypothesis has been extended to speculate that atopic subjects are protected against parasites. Studies conducted in Papua New Guinea, and Zambia, where asthmatic subjects were found to have a much lower prevalence of hookworm infection than nonasthmatic controls, and other more limited evidence from patients with strongyloidiasis provide some support to this theory.[6,15–17]

Of the intestinal helminthiases, ascariasis causes the most dramatic elevations of IgE serum levels; hookworms and *S. stercoralis* and *T. trichiura* infections are associated with lesser elevations, while little or none is found in patients with pinworms. Human intestinal cestode infections are not associated with increases of the total IgE component. Specific IgE antibodies, however, develop against *T. solium* antigens in patients with cysticercosis.[3]

The IgE responses in patients infected with *S. stercoralis* have been investigated in detail.[18,19] All patients studied had specific IgE antibodies directed against both *S. stercoralis* and *S. ratti* filariform larval somatic antigens. These patients also had specific IgE antibodies against a yet uncharacterized excretory/secretory (E/S) product of *S. stercoralis* filariform larvae, but not against the E/S product of *S. ratti.* When the histamine-release assay was performed in the presence of the patients' own serum, the percentage of release was considerably lower than the release obtained when the assay was conducted in the presence of the serum of a noninfected donor. One possible interpretation of this finding is that infected patients develop "blocking antibodies" that protect them from a continuous stimulation of IgE-bearing cells (mast cells and basophils) by the circulating parasite antigens. Such blocking factors have been demonstrated in patients with chronic schistosomiasis and bancroftian filariasis, and consist of IgG4 antibodies whose presumed function is to prevent a chronic anaphylactoid state.[20] Subsequent studies in patients with strongyloidiasis confirmed that antibodies of the IgG4 subclass indeed represented the predominant humoral parasite-specific response.[21] Because of its technical complexity, the histamine-release assay does not lend itself to the investigation of large numbers of patients. To examine the specific IgE responses in a larger population of infected people, we have developed a solid-phase radioallergosorbant assay.[19] Between 80 and 85% of patients with current *S. stercoralis* infection were found to have serum IgE antibodies that reacted with parasite antigens. Interestingly, even patients with normal total IgE serum levels had high parasite-specific IgE antibodies.

3.2. Other Humoral Responses

Other classes of immunoglobulins also participate in the specific humoral responses, and in some cases these responses have been exploited to develop serodiagnostic tests. For example, the development of enzyme-linked immunosorbant assays (ELISA) using soluble *S. stercoralis* filariform larval antigens,[22,23] and of an indirect immunofluorescence test using fresh or formalin-fixed whole filariform larvae,[24] resulted in the discovery that between 85 and 95% of the patients with chronic, uncomplicated strongyloidiasis had specific IgG antibodies against filariform larval antigens. The ELISA was later adapted for the detection of circulating parasite-specific IgA antibodies, and it was found that approximately 80% of all infected patients had serum IgA against *S. stercoralis* antigens.[25] When the recognition patterns of these antibodies were analyzed by immunoblotting techniques, it was found that part of the antigens reacting with IgA were different from those recognized by IgG. Analysis of subclasses of IgG involved in the responses to *S. stercoralis* revealed that although the total component of each subclass was within normal limits in virtually all patients, antibodies of the IgG4 subclass represented the predominant parasite-specific response. When the reactivity of each IgG subclass was analyzed by immunoblotting on SDS–PAGE-separated filariform larva antigens, IgG4 recognized a greater number of antigens than any of the other subclasses.[21,25] However, when attempts were made to correlate the magnitude of these humoral responses with the behavior of the infection, no correlation could be found.[26,27] Similarly, the congenital absence of IgA antibodies in dogs was not found to make these animals more susceptible to more severe infections.[28]

While several methods have been developed to measure specific antibodies against helminthic antigens for serodiagnostic purposes, recently there has been a new interest in exploring possible correlations between antibody levels, antigen recognition patterns, and protection. In a cohort of children with *T. trichiura* infection, Needham *et al.* demonstrated that changes in infection intensity over a period of 5 years correlated significantly with changes in all isotype levels, with the exception of IgA.[29] In other studies, the same investigators showed that both IgG1 and IgG2 levels related to the current intensity of infection as assessed by fecal egg counts.[30] Recently, a correlation between the levels of antibodies directed against excretory/secretory products of *A. lumbricoides* and the length of exposure to the parasite was observed in an endemic area,[31] and negative associations were demonstrated between the strength of recognition of certain *Ascaris* antigens and the intensity of infection.[32] Similar findings have been reported for patients with hookworm infection.[33] Such studies have rekindled interest in the search for the significant antigens, those capable of inducing protective responses that could be exploited for the design of vaccines.

3.3. Cellular Responses

The most apparent of the cellular responses is the diversion of the host's granulocytic resources to the production of eosinophils. Elevated peripheral eosinophilia, both absolute and relative, occurs in all infections caused by helminths that have a tissue phase as part of their life cycle. Thus, all human intestinal helminths except for *E. vermicularis* and the tapeworms are characterized by elevated peripheral eosinophilia.[4,34] These responses are probably mediated by increased production of IL-5, possibly in synergy with IL-3 and granulocyte–macrophage colony-stimulating factor (GM-CSF).[35] Eosinophil- associated tissue pathology has been found in association with the extreme hypereosinophilia that accompanies certain helminthic infections such as onchocerciasis, loaiasis in expatriates, and tropical (filarial) pulmonary eosinophilia syndrome. Eosinophil cationic proteins toxic to various normal tissues and cells both *in vitro* and *in vivo*, and the release of proinflammatory eicosanoids, most notably LTC4 and PAF-acether, have been implicated in the pathogenesis of this localized tissue damage. Interestingly, although the major source of the tissue destruction is the eosinophil and its toxic molecules, these same molecules can mediate larval killing of many helminths and thus, perhaps, exert also a protective effect.[36] There are no reports of eosinophil-associated tissue damage in humans infected with intestinal helminths.

Mast cells can also contribute to inflammatory reactions directed against invasive helminth parasites.[37] These cells express high-affinity Fcε receptors that are sensitized with specific antiparasite IgE and that can be triggered by parasite antigen. The presence of mast cells may thus represent an evolutionary adaptation to infection with parasitic helminths in that they confer a biologic advantage to the human host. In fact, it has been postulated that mediators specifically released by sensitized mast cells contribute to (1) the recruitment and activation of effector eosinophils, (2) increased local concentrations of antibody and complement, and (3) mucus hypersecretion and increased peristalsis in the gastrointestinal tract that can interfere with the attachment necessary for chronic infection of certain parasites.[38] Despite such apparent benefit, these same effector molecules may also mediate gastrointestinal pathology associated with intestinal parasites.

The *in vitro* proliferative responses of peripheral lymphocytes to parasite antigens were extensively studied in the 1970s and early 1980s. These studies consistently showed that the lymphocytes of most infected subjects were sensitized to parasite antigens and responded to them *in vitro* with proliferative responses of varying magnitude. These responses were not found to have protective function or to correlate with different manifestation of the infections, and this line of investigation has become obsolete.

4. INTESTINAL RESPONSES TO HELMINTHIC PARASITES

4.1. Experimental Models

As the first line of contact with myriad antigens equipped with a large lymphoid component, namely, the gut-associated lymphoid tissue (GALT), the gut is probably the body's most active immune compartment.[39] The GALT is composed of three distinct elements: the lymphoid aggregates known as Peyer's patches, the diffuse lymphoid cells of the lamina propria, and the intraepithelial lymphocytes. In concert with the draining mesenteric lymph nodes, these three components are involved in the process of antigen recognition, processing, and presentation, and with the subsequent effector mechanisms including the initiation of inflammatory reactions.[40] Responses to intestinal helminths involve the secretion of immunoglobulins (predominantly of the IgA class), a complex interplay of regulatory cytokines and their targets, and inflammation. This latter phenomenon, consisting of vascular permeability changes and multifarious leukocyte events, is for the most part initiated and regulated by the immune response to the parasites.[40,41]

The intestinal defense mechanisms against helminth parasites have been extensively explored in many rodent models. Among the most used nematodes are *Nippostrongylus brasiliensis, Strongyloides ratti, Trichinella spiralis,* and among cestodes *Hymenolepis diminuta.* One of the most investigated aspects has been the ability of rats and mice to expel adult worms after a certain period of time, usually from a few days to a few weeks postinfection.[42–44] The intestinal expulsion is mediated by immune mechanisms: both active immunization and previous infection can considerably shorten the interval between patency and expulsion; rapid expulsion can be passively induced by immune serum and sensitized cell transfers; immunosuppression either limits or prevents expulsion. Several mechanisms are involved in the induction of the expulsions of worms from the intestine, including mucosal inflammation and edema, mucus changes and hypersecretion, epithelial shedding, and increased intestinal peristalsis. These events are regulated by a complex and still incompletely understood interplay of mediators, which include leukotrienes, prostaglandins, histamine and serotonin, mast cell proteases, and lysosomal enzymes.

4.2. Human Parasites

The pathologic lesions specifically associated with chronic intestinal helminthiasis infections in otherwise healthy hosts are little known because only rarely are mucosal biopsy specimens from such patients examined and even more rarely do such cases come to autopsy. However, scattered information is available from the literature regarding patients in whom intestinal nematodes were an incidental findings. The evidence from these reports indicates that in the great majority of patients, worms can exist in the intestinal tract without causing inflammatory responses or other tissue damage. There are, however, some notable exceptions.

Trichuriasis is one of the few intestinal helminthiases that have been the object of carefully designed field studies addressing the relationship between intensity of infection and morbidity. These studies, conducted mostly in Malaysia[45] and the Caribbean,[46]have conclusively demonstrated that heavy burdens of *T. trichiura* (defined as more than 200 adult worms recovered after anthelminthic expulsion) cause a clinical syndrome characterized by chronic dysentery, predisposition to rectal prolapse, anemia, and finger clubbing.[47] Infections with lighter worm burdens have been associated with more subtle but equally devastating manifestations, particularly growth retardation, in children.[45–47] In the absence of systematic biopsy studies, the pathology of the intestinal lesions caused by *T. trichiura* remains inadequately defined. It is clear, however, that the cephalic head of the worms buried in the colonic mucosa elicits an inflammatory response quite similar to that seen in idiopathic ulcerative colitis. The inflammation results in mucosal erosion with increased exudation, production of mucus rich in neutrophils (pus), and microhemorrhage. Combined with decreased water absorption, these lesions result in the passage of numerous soft or liquid stools containing blood and pus. Recently, Cooper and colleagues at the Imperial College in London detected a considerable increase in the number of mast cells and the presence of parasite-specific immediate hypersensitivity responses in the colonic mucosa of children with *T. trichiura* dysentery.[49] The authors speculate that these reaginic mechanisms, the protective significance of which remains unclear, may be involved in the pathogenesis of diarrhea associated with trichuriases.

A. lumbricoides is a well-known cause of intestinal obstruction and perforation in children in developing countries,[49] and its role in protein energy malnutrition and growth retardation is reasonably established.[50] Intestinal ascariasis, however, has not been reported to be a cause of intestinal disease and no information is available about mucosal changes that may take place in the intestinal mucosa. Thus, we do not know whether a local immune or inflammatory response exists. A similar situation exists in hookworm disease: neither chronic *Ancylostoma duodenale* nor *Necator americanus* infection has been shown to cause intestinal disease in clinical studies. Recent detailed reviews by Gilles[51] on the clinical aspects of uncinariasis and by Schad[52] on the biology of hookworms fail to even mention that hookworm disease may have intestinal manifestations. Severe diarrhea, usually limited to the first few weeks after infection, has been described frequently in volunteers exposed to infective larvae of *A. duodenale, A. ceylanicum,* and *N. americanus.*[53,54] Although the mechanisms of this transient diarrhea are unknown, its timing coincides with the initial establishment of the juvenile worms in the host's duodenum. Since no mucosal penetration is known to occur, and most experimental infections have been produced by few worms only, mechanical damage (erosion, ulceration) or inflammatory changes directed against the worms are unlikely to account for the intestinal manifestations of early hookworm infection. A plausible hypothesis would be that local allergic reactions take place in response to the diffusion of worm antigens in the intestinal lumen. The resulting release of vasoactive mediators capable of altering the permeability and motility of the gut could cause the abdominal pain and diarrhea that seem to characterize some of these experimental infections.

Chronic strongyloidiasis does not cause intestinal inflammation in either humans or primates,[55,56] and the nature of the local immune responses in the intestinal mucosa is unknown. In contrast, severe bloody diarrhea is a major manifestation of hyperinfective or disseminated strongyloidiasis, a condition in which millions of filariform larvae penetrate the wall of small and large intestine, causing mucosal erosion, ulceration, interstitial edema, and hemorrhage.[55–57] These lesions, however, are the result of mechanical damage and superimposed bacterial infection and probably have no relationship with specific protective immune responses.

As mucosal biopsies become increasingly more available in areas endemic for intestinal parasites, new studies utilizing histochemical staining methods and *in vitro* proliferation techniques are conducted on the tissues of patients infected with a variety of helminths.[58,59] Although the results of such studies are still at early stages, these techniques appear promising and may contribute to answer many questions that previously could only be addressed in animal models.

5. DO IMMUNE RESPONSES AGAINST PARASITES INDUCE RESISTANCE IN HUMANS?

With the possible exception of trichinosis, an infection characterized by the rapid disappearance of adult worms from the intestine, no other human intestinal helminth is expelled in a fashion similar to that taking place in experimental rodent systems.[42] Furthermore, intestinal helminths have been and continue to be very successful in their spread among the human population. Thus, one would be tempted to conclude that the immune responses humans mount against these parasites have neither protective nor defensive function.

Even if it is difficult or impossible to demonstrate that the parasite-specific immune responses developing in any given individual human host confer resistance to subsequent infection, three sets of epidemiological observations indicate that immune responses do induce resistance in humans. The first one is the observation that in endemic populations younger subjects are the most susceptible to helminthic infections, and resistance appears to increase with age. This pattern has been best demonstrated with schistosomiasis and filariasis,[60–62] but there are indications that it is also detectable in some intestinal helminthiases such as *T. trichiura* infection.[63] A second concept in support of immune protection is the demonstration of concomitant immunity, a phenomenon whereby hosts are protected from new infecting parasites while tolerating an existing parasite load. Concomitant immunity is believed to play an important role in maintaining the "tolerable" levels of parasitemia found in most infected individuals in areas where malaria is endemic. Concomitant immunity is well documented in experimental infections with schistosomes, cestodes, and nematodes in rodent models.[64,65] In human helminthic infections, however, concomitant immunity is difficult to prove. Numerous factors have a confounding impact: the natural rate of parasite turnover and the self-regulation of parasite populations within the host are important determinants of the worm load; the possibility of autoinfection of the same host (in the case of *S. stercoralis* and, with a different mechanism, *E. vermicularis*) affects the size of the parasite population; and the pathophysiological consequences of infection (inflammation, altered motility, etc.) may alter the niche available to later waves of parasites.[7] In the case of human helminthiases, concomitant immunity would imply protection against newly invading larvae in a host who tolerates a certain adult worm load. This would indicate that helminths express stage-specific antigens, as well as an ability of adult worms to participate in the subversion of those immune effector mechanisms that kill the infective forms. This concept, although still largely hypothetical, provides the theoretical framework for the development of vaccines targeting the infective stages of helminths.[7] The third piece of evidence for immune resistance rests mostly on data obtained from studies on intestinal helminths, particularly *A. lumbricoides, T. trichiura,* and hookworms and schistosomes. These studies, based on the analysis of pretherapy worm burdens and reinfection rates following chemotherapy, have consistently shown a highly aggregated distribution of parasite numbers per person. This means that in a given population, few people harbor most of the worms, and that after everyone is treated, the same people who had the largest pretreatment parasite loads will again be infected with the largest numbers of worms.[66,67] Again, the influence of immunity is difficult to separate from that of genetic, behavioral, environmental and nutritional factors; however, recent theoretical work on the dynamics of the Th1–Th2 interaction in helminth infections and on the effects of parasite-mediated modulation of effector mechanisms suggest that early exposure in infancy may determine predisposition to heavy or light infection where high exposure generates anergy with elevated Th2 responses and permits high worm loads. Theory suggests that the quantitative details of the host's ability to recognize and respond to particular parasite antigens are of great importance in determining whether the host is persistently susceptible or develops protective immunity. Different responses to individual antigens among subjects with similar exposure experience are likely to be influenced by genetic, nutritional, and behavioral components. Recent data demonstrate individual variability, in antibody recognition of particular parasite antigens and in T-cell responses depending on the host major histocompatibility complex (MHC) genotype.[68] Similar work in animal models suggests that both MHC and non-MHC genes can be critical to the course of infection. In human populations, the combination of these variable factors is undoubtedly a major determinant of parasite aggregation and predisposition. Detailed longitudinal studies of parasitological and immunological changes in individuals within the community are necessary to understand the factors that control the observed patterns of infection. Recent field studies of this type are beginning to identify patterns that provide important clues to what, among many measurable responses to parasite antigens, may determine the buildup of acquired immunity which acts to reduce the intensity of infection in older individuals. For example, in studies of the gut-dwelling worm *T. trichiura,* comparison of age-dependent isotype responses and the age profile of infection intensity in two endemic communities with very different levels of transmission, suggests that parasite-specific secretory IgA responses best reflect the accumulated past experience of infection, whereas serum IgG isotypes simply correlate positively with current worm burden.[7,29,30,46,63]

6. EVASION OF THE HOST IMMUNE RESPONSES

Central to the ability of an infectious agent to succeed is the ability to evade, at least in part, the host immune system. Thus, it is hardly surprising that helminths have evolved an array of adaptations to coexist with the constant immune attacks mounted by their hosts. These parasite defense strategies have been reviewed recently by Maizels *et al.*,[7] who divided them into three broad categories: (1) avoiding the initial induction of damaging immune responses, (2) compromising selected branches of the immune system, and (3) disabling the short-range offensives mounted by various effector mechanisms. These mechanisms have been most extensively studied in animal models and, in humans, in schistosomiasis and filariasis. Although there is no doubt that such defense strategies must also be available to the highly successful intestinal nematodes and cestodes, little is known on how intestinal worms negotiate their survival in the host digestive tract.

Helminths reside in an extracellular niche. Therefore, the immune system depends on the exogenous pathway for antigen presentation. Helminths produce great quantities of polysaccharides, glycoconjugates, and phosphorylcholine; it has been suggested that these products may interfere with antigen processing by inhibiting activation and interfering with various mediators.[70] Both neutrophil chemotaxis and T-cell proliferation and production of IL-2 are inhibited by the serine protease inhibitor taeniaestatin, a product of *Taenia taeniaeformis* larvae capable of acting in the extracellular milieu.[71] These interferences with the early steps of antigen presentation may have a dual effect: they inhibit the responses and they may also induce tolerance, a state of specific immunological permissiveness, which allows parasites to be assimilated into their host environment.[7,72]

Another strategy used by parasites to evade the host responses is to selectively neutralize certain components of the immune system.[7,69] For example, some parasites can hinder the effects of antibodies by secreting products capable of degrading host immunoglobulins; other parasites, such as tapeworms, can compromise complement function by activating complement in the fluid phase with highly sulfated proteoglycans[73]; excretory/secretory products and live filariform larvae of *S. stercoralis* activate C3 and

C5 *in vitro*, an activity presumably aimed at neutralizing effective complement function during their penetration and migration in their host.(74) Helminths may also interfere with host inflammatory reactions.(7,69) For example, *Echinococcus granulosus* secretes an elastase inhibitor which blocks neutrophil attraction by C5a, and some filariae and taeniae release prostanoids (prostacyclin and PGE), which exert anti-inflammatory properties and may also inhibit T-cell proliferation.(75)

The third strategy involves the disabling of short-range offensives mounted by various effector mechanisms of the host defense system. These mechanisms are mostly physical, and include such stratagems as the thickening of the schistosome tegument during maturation, the rapid turnover of the nematode cuticle, and the inaccessibility of the nematode hypodermal plasma membrane beneath a thick extracellular cuticle.(69,76,77)

7. CONCLUSIONS

Much progress has been made in the past two decades toward understanding the mechanisms that regulate the balance between intestinal parasites and their hosts. Broad questions such as "How can parasites survive the host's immune attacks?" have been dissected into smaller components and many answers have been found. Most data, however, are generated from studying small-animal models, few of which replicate the conditions found in humans. Thus, it remains unclear how much of the information gathered from these studies can be usefully exploited to improve the human condition. While sanitation and mass chemotherapy remain unattainable short-term goals in many countries, and vaccines against metazoa continue to elude the intensive quest of immunologists, helminthic parasites, and particularly intestinal worms, will likely continue to lead parallel lives with their human hosts for many years to come. Animal and experimental studies represent indispensable tools for the development of drugs, techniques, and for the elucidation of mechanisms. However, field studies conducted on the very populations we hope to help will ultimately offer the best prospects for the development of successful preventive strategies.

REFERENCES

1. Noble, E. R., Noble, G. A., Schad, G. A., and MacGuiness, A. J., 1989, *Parasitology: The Biology of Animal Parasites*, Lea & Febiger, Philadelphia.
2. Genta, R. M., 1992, The dysregulation of strongyloidiasis: A new hypothesis, *Clin. Microbiol. Rev.* **5**:345–355.

3. Warren, K. S. (ed.), 1993, *Immunobiology and Molecular Biology of Parasitic Infections*, 3rd ed., Blackwell, Oxford.
4. Gleich, G. J., Ottesen, E. A., Leiferman, K. M., and Ackerman, S. J., 1989, Eosinophils and human disease, *Int. Arch. Allergy Appl. Immunol.* **88:**59–62.
5. Jarrett, E. E. E., and Miller, H. R. P., 1982, Production and activities of IgE in helminthic infections, *Prog. Allergy* **31:**178–233.
6. Lynch, N. R., Isturiz, G., Sanchez, Y., Perez, M., Martinez, A., and Castes, M., 1992, Bronchial challenge of tropical asthmatics with *Ascaris lumbricoides, J. Invest. Allergol. Clin. Immunol.* **2:** 97–105.
7. Maizels, R. M., Bundy, D. A. P., Selkirk, M. E., Smith, D. F., and Anderson, R. M., 1993, Immunological modulation and evasion by helminth parasites in human populations, *Nature* **365:**797–805.
8. Romagnani, S., 1992, Human Thl and Th2 subsets: Regulation of differentiation and role in protection and immunopathology, *Int. Arch. Allergy Appl. Immunol.* **98:**279–284.
9. Hermanek, J., Goyal, P. K., and Wakelin, D., 1994, Lymphocyte, antibody and cytokine responses during concurrent infections between helminths that selectively promote T-helper-l or T-helper-2 activity, *Parasite Immunol.* **16:**111–117.
10. Turner, K., Fedde, A. L., and Quinn, E. H., 1979, Non-specific potentiation of IgE by parasite infections in man, *Int. Arch. Allergy Appl. Immunol.* **58:**232–238.
11. Fraser, E. M., Christie, J. F., and Kennedy, M. W., 1993, Heterogeneity amongst infected children in IgE antibody repertoire to the antigens of the parasitic nematode *Ascaris, Int. Arch. Allergy Immunol.* **100:**283–286.
12. Egwang, T. G., Nguiri, C., Kombila, M., Duong, T. H., and Richard-Lenoble, D., 1993, Elevated antifilarial IgG4 antibody levels in microfilaremic and microfilarermic Gabonese adults and children, *Am. J. Trop. Med. Hyg.* **49:**135–142.
13. Kurniawan, A., Yazdanbakhsh, M., van Ree, R., Aalberse, R., Selkirk, M. E., Partono, F., and Maizels, R. M., 1993, Differential expression of IgE and IgG4 specific antibody responses in asymptomatic and chronic human filariasis, *J. Immunol.* **150:**3941–3950.
14. Urban, J. F., Jr., Madden, K. B., Svetic, A., Cheever, A., Trotta, P. P., Gause, W. C., Katona, I. M., and Finkelman, F. D., 1992, The importance of Th2 cytokines in protective immunity to nematodes, *Immunol. Rev.* **127:**205–220.
15. Capron, A., Dessaint, J. P., and Capron, M., 1992, Allergy and immune defense: Common IgE mechanisms or divergent pathways, in: *Allergy and Immunity to Helminths: Common Mechanisms or Divergent Pathways?* (R. Moqbel, ed.), Taylor & Francis, London, pp. 1–16.
16. Pritchard, D. I., 1992, Parasites and allergic diseases: A review of the field and experimental evidence for a cause-and-effect relationship, in: *Allergy and Immunity to Helminths: Common Mechanisms or Divergent Pathways?* (R. Moqbel, ed.), Taylor & Francis, London, pp. 38–50.
17. Wehner, J. H., Kirsch, C. M., Kagawa, F. T., Jensen, W. A., Campagna, A. C., and Wilson, M., 1994, The prevalence and response to therapy of *Strongyloides stercoralis* in patients with asthma from endemic areas, *Chest* **106:**762–766.
18. Genta, R. M., Ottesen, E. A., Poindexter, R., Gam, A. A., Neva, F. A., Tanowitz, H. B., and Wittner, M., 1983, Specific allergic sensitization to *Strongyloides* antigens in human strongyloidiasis, *Lab. Invest.* **48:**633–638.
19. McRury, J., de Messias, I. T., Walzer, P. D., Huitger, T., and Genta, R. M., 1986, Specific IgE responses in human strongyloidiasis, *Clin. Exp. Immunol.* **65:**631–638.
20. Hussain, R., Poindexter, R. W., and Ottesen, E. A., 1992, Control of allergic reactivity in human filariasis. Predominant localization of blocking antibody to the IgG4 subclass, *J. Immunol.* **148:**2731–2737.
21. Genta, R. M., and Lillibridge, J. P., 1989, Prominence of IgG4 antibodies in the human responses to Strongyloides stercoralis infection, *J. Infect. Dis.* **160:**692–699.

22. Genta, R. M., 1988, Predictive value of an enzyme-linked immunosorbent assay (ELISA) for the serodiagnosis of strongyloidiasis, *Am. J. Clin. Pathol.* **89:**391–394.
23. Conway, D. J., Lindo, J. F., Robinson, R. D., Bundy, D. A., and Bianco, A. E., 1994, *Strongyloides stercoralis*: Characterization of immunodiagnostic larval antigens, *Exp. Parasitol.* **79:**99–105.
24. Genta, R. M., and Weil, G. J., 1982, Antibodies to *Strongyloides stercoralis* larval surface antigens in chronic strongyloidiasis, *Lab. Invest.* **47:**87–90.
25. Genta, R. M., Frei, D. F., and Linke, M. J., 1987, Demonstration and partial characterization of parasite-specific immunoglobulin A responses in human strongyloidiasis, *J. Clin. Microbiol.* **25:**1505–1510.
26. Badarò, R., Carvalho, E. M., Santos, R. B., Gam, A. A., and Genta, R. M., 1987, Parasite-specific humoral responses in different clinical forms of strongyloidiasis, *Trans. R. Soc. Trop. Med. Hyg.* **81:**149–150.
27. Genta, R. M., Gatti, S., Linke, M. J., Cevini, C., and Scaglia, M., 1988, Endemic strongyloidiasis in northern Italy: Clinical and immunological aspects, *Q. J. Med.* **68:**679–690.
28. Mansfield, L. S., and Schad, G. A., 1992, *Strongyloides stercoralis* infection in IgA-deficient dogs, *Am. J. Trop. Med. Hyg.* **47:**830–836.
29. Needham, C. S., Lillywhite, J. E., Didier, J. M., Bianco, A. E., and Bundy, D. A., 1994, Temporal changes in *Trichuris trichiura* infection intensity and serum isotype responses in children, *Parasitology* **109:**197–200
30. Needham, C. S., Bundy, D. A., Lillywhite, J. E., Didier, J. M., Simmons, I., and Bianco, A. E., 1992, The relationship between *Trichuris trichiura* transmission intensity and the age-profiles of parasite-specific antibody isotopes in two endemic communities, *Parasitology* **105:**273–283.
31. Haswell-Elkins, M. R., Leonard, H., Kennedy, M. W., Alkenes, D. B., and Maizels, R. M., 1992, Immunoepidemiology of *Ascaris lumbricoides*: Relationships between antibody specificities, exposure and infection in a human community, *Parasitology* **104:**153–159.
32. Haswell-Elkins, M. R., Kennedy, M. W., Maizels, R. M., Elkins, D. B., and Anderson, R. M., 1989, The antibody recognition profiles of humans naturally infected with *Ascaris lumbricoides*, *Parasite Immunol.* **11:**615–627.
33. Pritchard, D. I., Walsh, E. A., Quinell, R. J., Raiko, A., Edmonds, P., and Keymer, A. E., 1992, Isotypic variation in antibody responses in a community in Papua New Guinea to larval and adult antigens during infection, and following reinfection, with the hookworm Necator americanus, *Parasite Immunol.* **14:**617–631.
34. Capron, M., 1992, Dual function of eosinophils in pathogenesis and protective immunity against parasites, *Mem. Inst. Oswaldo Cruz* **87**(Suppl 5):83:–89.
35. Cromwell, O., Wardlaw, A. J., Champion, A., Moqbel, R., Osei, D., and Kay, A. B., 1990, IgG-dependent generation of platelet-activating factor by normal and low-density human eosinophils, *J. Immunol.* **145:**3862–3868.
36. Butterworth, A. E., 1984, Cell-mediated damage to helminths, *Adv. Parasitol.* **23:**143–235.
37. Miller, H. R. P., 1992, Mast cells: Their function and heterogeneity, in: *Allergy and Immunity to Helminths: Common Mechanisms or Divergent Pathways?* (R. Moqbel, ed.), Taylor & Francis, London, pp. 228–248.
38. Madden, K. B., Urban, J. F., Jr., Ziltener, H. J., Schrader, J. W., Finkelman, F. D., and Katona, I. M., 1991, Antibodies to IL-3 and IL-4 suppress helminth-induced intestinal mastocytosis, *J. Immunol.* **15:**1387–1391.
39. Marsh, M. N. (ed.), 1987, *Immunopathology of the Small Intestine*, Wiley, New York.
40. Moqbel, R., and MacDonald, A., 1990, Immunological and inflammatory responses in the small intestine associated with helminthic infections, in: *Parasites: Immunity and Pathology* (J. M. Behnke, ed.), Taylor & Francis, London, pp. 249–282.

41. Brandtzaeg, O., Valnes, K., Scott, H., Rognum, T. O., Bjerke, K., and Baklien, K., 1985, The human gastrointestinal secretory immune system in health and disease, *Scand. J. Gastroenterol.* **20**(Suppl. 114):17–38.
42. Castro, G. A., 1990, Intestinal pathology, in: *Parasites: Immunity and Pathology* (J. M. Behnke, ed.), Taylor & Francis, London, pp. 282–310.
43. Abe, T., and Nawa, Y., 1988, Worm expulsion and mucosa mast cells response induced by repetitive IL-3 administration in *Strongyloides ratti*-infected nude mice, *Immunology* **63**:181–185.
44. Kassai, T., Redl, P., Jecsai, G., Balla, E., and Harangozo, E., 1989, Studies on the involvement of prostaglandins and their precursors in the rejection of *Nippostrongylus brasiliensis* from the rat, *Int. J. Parasitol.* **10**:115–120.
45. Gilman, R. H., Chong, Y. H., Davis, C., Greenberg, B., Virik, H. K., and Dixon, H. B., 1983, The adverse consequences of heavy *Trichuris* infection, *Trans. R. Soc. Trop. Med. Hyg.* **77**: 432–438.
46. Bundy, D. A. P., 1986, Epidemiological aspects of *Trichuris* and trichuriasis in Caribbean communities, *Trans. R. Soc. Trop. Med. Hyg.* **80**:706–718.
47. Cooper, E. S., and Bundy, D. A. P., 1987, Trichuriasis, in: *Bailliere's Clinical Tropical Medicine and Communicable Diseases, Intestinal Helminthic Infections* (Z. S. Pawlowski, ed.), Bailliere Tindall, London, pp. 629–643.
48. Cooper, E. S., Spencer, J., Whyte-Alleng, C. A. M., Cromwell, O., Whitney, P., Venugopal, S., Bundy, D. A., Haynes, B., and MacDonald, T. T., 1992, Immediate hypersensitivity in colon of children with chronic *Trichuris trichiura* dysentery, *Lancet* **338**:1104–1107.
49. Pawlowski, Z. B., 1982, Ascariasis: Host–pathogen biology, *Rev. Inf. Dis.* **4**:806–814.
50. Schulz, M. G., 1982, Ascariasis: Nutritional implications, *Rev. Inf. Dis.* **4**:815–819.
51. Gilles, H. M., 1990, Naturally acquired infections: What's needed? in: *Hookworm Disease, Current Status and New Directions* (G. A. Schad and K. S. Warren, eds.), Taylor & Francis, London, pp. 221–230.
52. Schad, G. A., 1991, The parasite, in: *Hookworm Infections* (H. M. Gilles and P. A. J. Ball, eds.), Elsevier, Amsterdam, pp. 15–50.
53. Miller, T. A., 1979, Hookworm infection in man, *Adv. Parasitol.* **17**:315–384.
54. Behnke, J. M., 1991, Pathology, in: *Hookworm Infections* (H. M. Gilles and P. A. J. Ball, eds.), Elsevier, Amsterdam, pp. 51–92.
55. Genta, R. M., and Caymmi-Gomes, M., 1989, Pathology, in: *Strongyloidiasis, a Major Roundworm Infection of Man* (D. A. Grove, ed.), Taylor & Francis, London, pp. 105–132.
56. Harper, J. S., Genta, R. M., Gam, A. A., London, W. J., and Neva, F. A., 1984, Experimental disseminated strongyloidiasis in *Erythrocebus patas*. Part I. Pathology, *Am. J. Trop. Med. Hyg.* **33**:431–443.
57. Longworth, D. L., and Weller, P. F., 1986, Hyperinfection syndrome with strongyloidiasis, in: *Current Clinical Topics in Infectious Diseases* (J. S. Remington and M. N. Swartz, eds.) McGraw–Hill, New York, pp. 1–26.
58. MacDonald, T. T., 1992, Epithelial proliferation in response to gastrointestinal inflammation, *Ann. N.Y. Acad. Sci.* **664**:202–209.
59. MacDonald, T. T., Choy, M. Y., Spencer, J., Richman, P. I., Diss, T., Hanchard, B., Venugopal, S., Bundy, D. A., and Cooper, E. S., 1991, Histopathology and immunohistochemistry of the caecum in children with the *Trichuris* dysentery syndrome, *J. Clin. Pathol.* **44**:194–199.
60. Butterworth, A. E., Sturrock, R. F., and Ouma, J. H., 1992, On the use of age-intensity data to detect immunity to parasitic infections, with special reference to *Schistosoma mansoni* in Kenya, *Parasitology* **105**(Pt. 2):219–227.
61. Day, K. P., Gregory, W. F., and Maizels, R. M., 1991, Age-specific acquisition of immunity to infective larvae in a bancroftian filariasis endemic area of Papua New Guinea, *Parasite Immunol.* **13**:277–290.

62. Capron, A., and Dessaint, J. P., 1992, Immunologic aspects of schistosomiasis, *Annu. Rev. Med.* **43**:209–218.
63. Bundy, D. A., Lillywhite, J. E., Didier, J. M., Simmons, I., and Bianco, A. E., 1991, Age-dependency of infection status and serum antibody levels in human whipworm (*Trichuris trichiura*) infection, *Parasite Immunol* **13**:629–638.
64. Smithers, S. R., and Terry, R. J., 1969, Immunity in schistosomiasis, *Ann. N.Y. Acad. Sci.* **160**(2):826–840.
65. Mitchell, G. F., 1990, A note on concomitant immunity in host–parasite relationships: A successfully transplanted concept from tumor immunology, *Adv. Cancer Res.* **54:** 319–332.
66. Schad, G. A., and Anderson, R. M., 1985, Predisposition to hookworm infection in humans, *Science* **228**:1537–1541.
67. Chan, L., Bundy, D. A., and Kan, S. P., 1994, Aggregation and predisposition to *Ascaris lumbricoides* and *Trichuris trichiura* at the familial level, *Trans. R. Soc. Trop. Med. Hyg.* **88:** 46–48.
68. Bennett, K., Levine, T., Ellis, J. S., Peanasky, R. J., Samloff, I. M., Kay, J., and Chain, B. M., 1992, Antigen processing for presentation by class II major histocompatibility complex requires cleavage by cathepsin E, *Eur. J. Immunol.* **22**:1519–1524.
69. Leid, R. W., Suquet, C. M., and Tanigoshi, L., 1987, Parasite defense mechanisms for evasion of host attack; a review, *Vet. Parasitol.* **25**:147–162.
70. Else, K. J., Finkelman, F. D., Maliszewski, C. R., and Grencis, R. K., 1994, Cytokine-mediated regulation of chronic intestinal helminth infection, *J. Exp. Med.* **179**:347–351.
71. Leid, R. W., Suquet, C. M., Bouwer, H. G., and Hinrichs, D. J., 1986, Interleukin inhibition by a parasite proteinase inhibitor, taeniaestatin, *J. Immunol.* **137**:2700–2702.
72. Fowlkes, B. J., and Ramsdell, F., 1993, T-cell tolerance, *Curr. Opin. Immunol.* **5**:873–879.
73. Fishelson, Z., 1991, Complement evasion by parasites: Search for "Achilles' heel," *Clin. Exp. Immunol.* **86**(Suppl. 1):47–52.
74. De Messias, I. J. T., Genta, R. M., and Mohren, W. D., 1994, Adherence of monocytes and polymorphonuclear cells to infective larvae of *Strongyloides stercoralis* after complement activation, *J. Parasitol.* **80**:267–274.
75. Liu, L. X., Serhan, C. N., and Weller, P. F., 1990, Intravascular filarial parasites elaborate cyclooxygenase-derived eicosanoids, *J. Exp. Med.* **172**:993–996.
76. Maizels, R. M., Blaxter, M. L., and Selkirk, M. E., 1993, Forms and functions of nematode surfaces, *Exp. Parasitol.* **77**:380–384.
77. Page, A. P., Rudin, W., Fluri, E., Blaxter, M. L., and Maizels, R. M., 1992, Toxocara canis: A labile antigenic surface coat overlying the infective larvae, *Exp. Parasitol.* **75**:72–86.

Index

Brian Henderson

7 50 —

638285